CW00506606

Ground and Soil Improvement

Edited by

C. A. Raison

THE INSTITUTION OF CIVIL ENGINEERS

Published by Thomas Telford Publishing, Thomas Telford Ltd, 1 Heron Quay, London E14 4JD.
URL: http://www.thomastelford.com

Distributors for Thomas Telford books are
USA: ASCE Press, 1801 Alexander Bell Drive, Reston, VA 20191-4400
Japan: Maruzen Co. Ltd, Book Department, 310 Nihonbashi 2-chome, Chuo-ku, Tokyo 103
Australia: DA Books and Journals, 648 Whitehorse Road, Mitcham 3132, Victoria

First published 2004

A catalogue record for this book is available from the British Library

ISBN: 0 7277 3170 X

Typeset by Keytec Typesetting Ltd, Bridport, Dorset
Printed and bound in Great Britain by MPG Books, Bodmin, Cornwall

Preface

C. A. RAISON

The first *Géotechnique* Symposium in Print was published in March 1975. Since then the Symposium in Print has become a regular feature with topics chosen by the *Géotechnique* Advisory Panel to reflect current issues and research interests covering a wide range of subjects, (see **Table 1**).

The first Symposium was published on the subject of Ground Treatment by Deep Compaction. The topic of soil improvement was addressed again in March 1981 when the third *Géotechnique* Symposium in Print was published on Vertical Drains. Both issues contained a significant number of papers dealing with case histories and trials, reflecting the empirical nature of the design process and the lack of a rigorous theoretical framework.

Methods for improving ground and soil have undergone significant developments since the first Symposium particularly in terms of application and usage, and many innovative techniques have been introduced. However, it is of significance that in many areas the design process still lacks a theoretical framework. It is also clear that ground and soil improvement has received little input from the research community in the last two decades despite the immense practical importance of the subject.

Against this background, the *Géotechnique* Advisory Panel decided to call for a Symposium in Print focussed again on Ground and Soil Improvement. The theme for the Symposium was to be restricted to the mechanical improvement of ground using Vibro Compaction, Vibro Replacement, Dynamic Compaction, Soil Mixing, Compaction Grouting, Surcharging, Deep Drainage and other improvement techniques used to consolidate or reinforce the soil. One of the aims of the Symposium was to collect together data and experience in the hope of stimulating more interest from the research community.

Since 1975, substantial research efforts have been made in understanding fundamental soil behaviour, in developing powerful numerical analysis methods and improving testing and modelling techniques, particularly in the centrifuge. These developments are beginning to be applied to ground and soil improvement techniques and this Symposium includes at least two examples. However, the influence and importance of the construction methods, many of them proprietary, and the linkage with final performance still needs to be addressed. Because of these difficulties, much development is still based on observations and site specific trials.

The *Géotechnique* Advisory Panel set up an organising sub-committee comprising three members of the panel and three external members. The sub-committee was responsible for reviewing and assessing papers with the full support of the Advisory Panel, the Institution of Civil Engineers Secretariat and external referees.

The response to the Symposium was substantial with over sixty five papers offered by authors throughout the world reflecting the interest and importance of this topic to many engineers. Unfortunately due to time restraints and the very tight deadline imposed by the Advisory Panel, many of these papers did not achieve fruition. The Panel was also acutely aware of the difficulties many of the authors had in meeting the high standards required by *Géotechnique* particularly in regard to completeness and extent of supporting data. Despite these difficulties, a total of fourteen papers were selected covering a range of topics from vacuum preloading, consolidation by vertical drains, dynamic compaction to vibro replacement. As reflects the practical nature of the subject matter, many of these papers are case histories. A significant proportion of the papers are from international contributors.

The eleventh *Géotechnique* Symposium in Print was published in December 2000, and was followed by a discussion meeting held at the Institution of Civil Engineers in London on 6th February 2001. This full day meeting was attended by about 140 delegates and followed the format of brief presentations of each paper by the authors followed by periods to allow questions and discussion from the audience. The meeting was divided into four sessions roughly following themes as follows:

Consolidation	Vacuum consolidation
	Vertical drains
	Preloading
Compaction	Explosive compaction
	Dynamic compaction
	Compaction grouting
Stone columns	Group effects
	Instrumented trial
	Drainage and reinforcement
Jet grouting	Single fluid grouting

In addition to the published papers, two key note presentations were made by Dr Andrew Charles of the Building Research Establishment and by Dr Alan Bell of Keller

Table 1. *Géotechnique* **Symposia in Print**

1st	March 1975	Ground treatment by deep compaction
2nd	March 1976	Piles in weak rock
3rd	March 1981	Vertical drains
4th	June 1983	The influence of vegetation on the swelling and shrinking of clays
5th	December 1984	Performance of propped and cantilevered rigid walls
6th	March 1987	The engineering application of direct and simple shear testing
7th	March 1992	The geotechnical aspects of contaminated land
8th	June 1992	Bothkennar soft clay test site: characterization and lessons learned
9th	December 1994	The observational method in geotechnical engineering
10th	August 1997	Pre-failure deformation behaviour of geomaterials
11th	December 2000	Ground and soil improvement

Ground Engineering. Dr Charles introduced session 1 with a paper dealing with the interaction between engineering based on science and theory and that based on experience and empiricism as related to ground improvement. Dr Bell introduced session 3 with a paper dealing with the importance of construction technique in successful deep vibratory ground improvement. Both papers are included in this volume.

This new publication has the aim of bringing together in one volume the fourteen papers originally published in *Géotechnique* in December 2000, the two keynote papers together with a full record of the informal discussion and questions. The publication also includes written discussion received and published by *Géotechnique* subsequent to the meeting. It is hoped that publication of this volume provides a valuable record of the current state of the art in ground and soil improvement, covers a typical cross section of problems faced by many practising engineers and lastly, gives advice and guidance on how these problems can be dealt with in a practical manner.

Chris Raison

Contents

Session 1

Charles, J. A. (2002). *Géotechnique* **52**, No. 7, 527–532

Ground improvement: the interaction of engineering science and experience-based technology*

J. A. CHARLES†

An increasing proportion of building development takes place on poor ground, which presents the geotechnical engineer with the challenge of providing satisfactory foundation performance at low cost. Ground behaviour can be modified by ground treatment so that ground properties are improved and heterogeneity is reduced. Ground treatment has developed largely as an experience-based technology, whereas the scientific aspects of ground engineering have been the principal concern of the *Géotechnique* journal. In a keynote lecture for a *Géotechnique* Symposium in Print on the subject of ground and soil improvement, it is pertinent to ask how well the engineering science of soil mechanics is serving the practice of ground improvement. The mystique that sometimes surrounds ground treatment methods can be dispelled only if ground treatment is implemented in a rational context in which the required ground behaviour for a particular use of the ground is defined, any likely deficiencies in ground behaviour are identified, and appropriate ground treatment is designed and implemented to remedy those deficiencies.

KEYWORDS: compaction; consolidation; ground improvement; soil stabilisation; vibration

INTRODUCTION

The value of an engineering science is determined by what it can accomplish as a tool in the hands of the practising engineer. (Terzaghi, 1939)

This statement, made by Karl Terzaghi when presenting the 45th James Forrest Lecture, 'Soil mechanics—a new chapter in engineering science', at the Institution of Civil Engineers on 2 May 1939, is apposite to this *Géotechnique* Symposium in Print. Ground improvement has developed largely in an empirical, contractor-led fashion, and, for the most part, it can be described as an experience-based technology. In contrast, *Géotechnique* is principally concerned with the scientific aspects of ground engineering. It is pertinent to ask how well the engineering science of soil mechanics has served, and is serving, the practice of ground improvement.

The subject of this symposium is termed 'ground treatment' or 'ground improvement', or sometimes 'ground modification':

(*a*) The process is ground treatment.

(*b*) The purpose of the process is ground improvement.

(*c*) The result of the process is ground modification; for better or for worse the treatment has modified the ground properties.

Mitchell & Jardine (2000) gave the following definition of ground treatment/improvement:

the controlled alteration of the state, nature or mass behaviour of ground materials in order to achieve an intended satisfactory response to existing or projected environmental and engineering actions.

By way of introduction to the subject, two axioms are advanced:

(*a*) Ground treatment is of major importance in geotechnical engineering.

(*b*) Ground treatment has not received sufficient attention in the past from geotechnical engineers.

Virtually all construction is done on, in or with soil, but the soil in its natural state is not always adequate for the required application. There are a number of aspects of ground performance that may be deficient, and this keynote lecture is restricted to ground treatment that aims to improve the load-carrying properties of the ground.

Where construction is to take place on ground with inadequate load-carrying properties, there are several alternatives:

(*a*) The site may be abandoned and the structure located elsewhere.

(*b*) Deep foundations may be used to carry the weight of the building down to a competent stratum.

(*c*) The structure may be redesigned to survive the ground movements.

(*d*) The properties of the ground may be improved prior to construction.

A consideration of these four alternatives shows the practical importance of alternative (*d*), ground treatment. As the supply of good building land decreases and land hitherto considered marginal is required for building development, it is surely self-evident that ground treatment is of major importance in geotechnical engineering.

It may be more debatable whether ground treatment has received sufficient attention in the past from geotechnical engineers. Ground treatment has a long history, and it has been suggested that soil improvement is probably the oldest of all common execution methods in civil engineering (Van Impe, 1989). In the introduction to the ASCE (1978) report on *Soil improvement: history, capabilities and outlook* it was asserted that:

the basic concepts of soil improvement—densification, cementation, reinforcement, drainage, drying, and heating—were developed hundreds or thousands of years ago and remain unchanged today.

Manuscript received 2 March 2001; revised manuscript accepted 27 March 2002.
* Keynote Lecture presented to *Géotechnique* Symposium in Print on Ground and Soil Improvement, 6 February 2001.
† Building Research Establishment, Watford.

There has been no lack of attention paid to ground improvement in engineering practice, but there has been a lack of attention from the perspective of theoretical soil mechanics. This can be demonstrated by the lack of space devoted to the subject in many early textbooks on soil mechanics and in journals, and by the content of the First *Géotechnique* Symposium in Print, *Ground treatment by deep compaction,* which was held at the Institution of Civil Engineers on 29 May 1975. The nine papers in the first symposium generally indicated a subject in a relatively primitive state, as viewed from a *Géotechnique* perspective, and some of the symposium papers lacked the erudition normally associated with papers published in *Géotechnique.* In the preface to the published version of the first symposium, the editors, Burland *et al.* (1976), commented:

> Nearly all the papers have been produced by contractors specialising in these techniques and, not unnaturally, they have concentrated on the successes obtained by their methods. The failures, or at least the lack of apparent successful applications, have remained unrecorded. The result of this has been the growth of a certain mystique surrounding the techniques, and claims have been made on their ability to 'strengthen' ground which cannot always be substantiated when subjected to a critical review.

GROUND TREATMENT IN CONTEXT

Burland *et al.* (1976) referred to a mystique surrounding ground treatment methods. It is essential that ground treatment is implemented in a rational context with, typically, the following stages:

(a)   Define the required ground behaviour for a particular use of the ground.
(b)   Identify any deficiencies in ground behaviour.
(c)   Design and implement appropriate ground treatment to remedy any deficiencies.

The scope of this lecture is restricted to situations where treatment is applied before construction commences on the site in order to improve the load-carrying properties of the ground. Treatment techniques, such as compensation grouting or underpinning, that are applied subsequent to construction are not considered.

Required ground behaviour

It is necessary to define what ground behaviour is required for a particular end-use. Inadequate ground performance is then related to that specific end-use. Both the imposed loads and the tolerance of the structure to differential movements need to be realistically assessed. If the magnitude of the loads is overestimated and the tolerance of the structure to differential movement is underestimated, a solution involving ground treatment may be rejected unnecessarily. Conversely, where the magnitude of the loads is underestimated and the tolerance of the structure to differential movement is overestimated, an inadequate ground treatment may be adopted.

The different situations in which ground treatment is implemented range from small housing developments to large civil engineering works. Significant differences in resources and expertise are available in these two situations. Performance criteria for buildings can be in terms of settlement, for example total settlement, differential settlement, tilt, or relative deflection (Wahls, 1994). Unfortunately, ground movements defined in terms of the more technically useful criteria, such as relative deflection, are the most difficult to predict.

Deficiencies in ground behaviour

Having defined what is required, any deficiencies in ground behaviour need to be identified and the nature of the problems diagnosed. This assumes that ground behaviour without treatment can be predicted. Two types of inadequate performance are associated with:

(a)   ground movements attributable to the weight of building or structure
(b)   ground movements not attributable to the weight of the building or structure.

The concept of bearing capacity relates exclusively to the former type of ground movements. It is based on the premise that, under a certain bearing pressure, ground movements will be tolerable, but at some point, as the pressure applied to the ground increases, the movements will become intolerable. Fig. 1 shows settlements monitored at a housing development on clay fill that had been treated by dynamic compaction. The houses have been subjected to 50 mm settlement over a period of 25 years. However, ground adjacent to the houses (described in the figure as 'unloaded', in the sense of not loaded) has settled by a comparable amount, indicating that the settlement of the houses is unrelated to the weight of the buildings and that the ground movements are attributable to other causes. Foundation design based solely on an allowable bearing pressure would not address the principal cause of ground movements at this site.

Risk mitigation

There are situations in which ground treatment is used not so much to remedy a clearly identified deficiency as to reduce the risks posed by heterogeneous ground conditions such as non-engineered fills:

> ... the most fundamental principle of humility in civil engineering design is as follows. As a first step, use every possible means to avoid having to deal with conditions determined by the statistics and probabilities of extreme values. (De Mello, 1977)

It is sensible to invest in risk mitigation measures rather than refine the calculation of risk. For example, for a heterogeneous granular fill it may be preferable to improve the ground by vibro compaction rather than carry out further site investigation in an attempt to prove that treatment is not necessary. In this situation, treatment may be considered to be a form of insurance. Some types of ground treatment act as a further site investigation. Nevertheless, effective treatment requires some knowledge of the likely hazards.

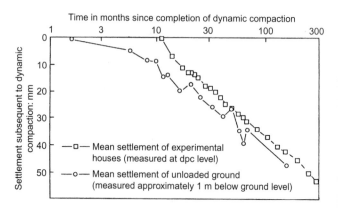

Fig. 1. Settlement of clay fill subsequent to dynamic compaction (after Charles *et al.,* 1981 (updated))

The perception that there is a problem may be incorrect and, in reality, the ground may be fit for purpose. This type of application distorts success/failure statistics. In many applications little may be required of the treatment, and this may hide the true risk in a situation where much is required.

Appropriate ground treatment

When the nature of the deficiency has been identified and the extent of the improvement that is required has been evaluated, appropriate ground treatment can be designed and implemented. Two major forms of ground treatment are examined in this lecture:

(*a*) densification
(*b*) stiffening columns.

DENSIFICATION

The utility of densification as a means of ground improvement rests primarily on three factors:

(*a*) An increase in the density of a soil generally results in improved ground behaviour, such as increased stiffness and strength.
(*b*) Many of the hazards for buildings on untreated ground are associated with volumetric compression of the ground: therefore appropriate ground treatment by densification can act as a type of inoculation by subjecting the ground to a form of the hazard before construction on the ground takes place.
(*c*) Soils are generally inelastic and strains are non-recoverable: therefore once the ground has been densified by ground treatment it will remain densified, and its subsequent vulnerability to volume compression will be greatly reduced.

There are two principal forms of densification: compaction and consolidation. From a technical standpoint compaction may be more effective in reducing vulnerability to liquefaction, whereas consolidation may be more effective in reducing compressibility, because in both these cases the remedy closely resembles the problem. Thus treatment by compaction is particularly effective in reducing subsequent susceptibility to self-compaction due to vibration, and treatment by consolidation in reducing settlement due to subsequent loading. However, the selection of a particular compaction or consolidation technique will depend on economic factors and practicalities, as well as on technical performance.

Compaction

The term 'compaction' is used to describe processes in which densification of the ground is achieved by some mechanical means such as rolling, vibration, ramming, or explosion. If the principal deficiency of the ground is related to its loose state, *in situ* compaction may be the most appropriate type of treatment. Loose sandy soils usually respond best to vibratory methods. Vibratory rollers will densify the soil to only a very limited depth: therefore this approach requires excavation of the deposit and compaction in thin layers. The compaction of sands can be monitored by *in situ* test measurements, which can be correlated with density index and hence used to characterise shear strength and compressibility.

The following methods are used for *in situ* depth compaction of granular soils:

(*a*) deep vibratory compaction
(*b*) impact compaction
(*c*) explosive compaction
(*d*) compaction grouting.

Deep vibratory compaction is most commonly effected using vibro compaction. This technique, utilising depth vibrators generating horizontal vibrations, was first used in Germany in the mid-1930s (Greenwood, 1970), and the successful application of vibro methods to an increasing number of poor ground conditions has been strongly dependent on the development of equipment and treatment procedures. Field investigations indicate that vibro compaction typically can reduce the compressibility of a granular fill to as little as 50% of its original value. Generally, it has been considered that vibro compaction is effective only where the fines content does not exceed about 15%, but Slocombe *et al.* (2000) have presented case histories to demonstrate that, with new depth vibrators, sands with significantly higher fines content can be compacted, illustrating the dominant role of equipment development. Renton-Rose *et al.* (2000) have described the treatment of a hydraulically placed calcareous sand and gravel.

Impact compaction of loose heterogeneous fills and sandy soils can be achieved using techniques such as dynamic compaction and rapid impact compaction. In dynamic compaction a heavy weight is dropped onto the ground surface, whereas in rapid impact compaction the steel compacting foot remains in contact with the ground and is subjected to rapidly applied impacts from a modified piling hammer. Dynamic compaction does not require the same type of specially developed, relatively complex equipment as does deep vibratory compaction, and therefore can be more readily investigated by model testing and mathematical analysis. Feng *et al.* (2000) have compared the relative effectiveness of a conical-based tamper and a flat-based tamper in compacting laboratory sand samples. Merrifield & Davies (2000) have studied the rapid impact compactor in the field and in centrifuge model tests, using dynamic stiffness measurements to assess the depth of effectiveness.

A major limitation of the method is the limited depth to which surface impact loading can densify the soil. The uncertainty of the depth to which treatment is effective is also a problem. In the first *Géotechnique* Symposium in Print, Menard & Broise (1975) stated that the depth that could be compacted, H (in metres), could be determined from the relationship $Mh > H^2$, where a weight of mass M tonnes is dropped from a height of h metres. This curious inequality merely informs us that $H < (Mh)^{0.5}$. Despite this strange origin, the relationship between H and $(Mh)^{0.5}$ has dominated analyses of dynamic compaction (Mayne *et al.*, 1984; Chow *et al.*, 2000).

Explosive compaction can be used to compact sandy soils at great depth. A theoretical basis for explosive compaction has been presented by Gohl *et al.* (2000), together with examples of the use of the method in different parts of the world. Doubtless, considerable practical experience is required for the successful application of this method, and Faraco (1981) suggested that lack of experience was the reason for poor results when the method was used on a granular hydraulic fill in Spain.

No classification scheme for ground treatment techniques is without problems, and, with the definitions adopted in this lecture, compaction grouting might be classified as a consolidation method or as stiffening columns. The design of compaction grouting has been based on empiricism and experience. In a low-permeability soil excess pore pressures may be generated, and Kovacevic *et al.* (2000) have analysed the effect of the development of undrained pore pressure on the efficiency of compaction grouting.

Consolidation

Consolidation can be defined as a process in which densification is achieved by an increase in effective stress applied by some form of static loading. One of the most fundamental methods of ground improvement is to consolidate the ground by temporary preloading prior to construction. Consolidation makes the ground stiffer under subsequent applied loads that are smaller than the maximum pressure applied by preloading. Soils are not linear elastic materials, and densification produced by consolidation is unlikely to be reversed. Some small amount of heave may follow removal of a temporary surcharge of fill, but usually this will occur immediately following removal of the surcharge and have little effect on subsequent construction.

Preloading can be usefully applied to a wide range of ground conditions, and there are a number of methods of preloading:

(*a*) applying a surcharge of fill without installing drains
(*b*) applying a surcharge of fill with the installation of drains
(*c*) lowering the groundwater level
(*d*) vacuum preloading.

The simplest approach is to consolidate using a temporary surcharge of fill. A relatively large area is needed for the process to be feasible, and the cost will be controlled largely by the haul distance, so a local supply of fill is required. Where the soil has high permeability, or is partially saturated, the response to loading should be rapid. BRE has carried out a number of field investigations to determine the effectiveness of the method when used on loose, partially saturated fills (Charles *et al.*, 1986; Charles, 1996). Where the method is used on relatively free-draining natural soils, the main concern is likely to be secondary compression. Alonso *et al.* (2000) have carried out a preload test to investigate the effectiveness of overconsolidation in reducing secondary settlement.

With saturated soft clay soils, the weight of the surcharge will generate excess pore pressures, which may take a long time to dissipate because of the low permeability of the clay. Consequently, it is often necessary to install vertical drains to increase the rate of consolidation. Almeida *et al.* (2000) have described the consolidation of a very soft organic clay, with vertical band drains on a 1·7 m grid, under a 3 m high embankment. Some 2 m of settlement has been monitored over a four-year period, and the results have been analysed using the work of Barron (1948), who utilised the Terzaghi theory of consolidation in developing a mathematical analysis of radial flow towards vertical drains.

Preloading with a surcharge of fill effectively converts a normally consolidated soil into an overconsolidated soil, with the improvement in soil properties that are consequent on this change. It might be claimed, therefore, that this form of treatment is one for which modification will inevitably mean improvement. However, an adverse consequence could occur if the placement of surcharge triggers instability in the treated soil. Methods that induce consolidation through a lowering of pore water pressure eliminate this hazard. With a high-permeability soil it may be feasible to increase the applied stresses by lowering the groundwater level. With low-permeability soil this will not usually be practical, but vacuum preloading may be used.

Vacuum preloading provides a good example of the interaction of theory and practice, because both a good understanding of soil behaviour and appropriate tools and techniques are required. The papers for the Symposium include two examples from China. At a 50 000 m² site in Tianjin, a 20 m depth of soft clay settled by 1 m (Chu *et al.*, 2000). At Yaoqiang Airport a 2–3 m thick silty sandy layer at the ground surface necessitated the installation of a slurry trench cut-off wall (Tang & Shang, 2000).

Preloading can greatly reduce compressibility. Provided that the structural load is significantly smaller than the preloading, the treated ground has been effectively overconsolidated, and the compressibility may be reduced to as little as 10–20% of its untreated value. The ratio of the ground heave that occurs when the preloading is removed to the settlement that had occurred when the ground was preloaded gives a crude indication of the likely improvement in constrained modulus. This is illustrated in Fig. 2, which shows the initial loading and unloading of three samples of sandstone rockfill (a, b and c) compacted to different initial densities in a large oedometer. The induced vertical strain (ε_v) is plotted against the applied vertical stress (σ_v), and the dry density (γ_d) and density index (I_D) of each sample are given. The compression on first loading is very dependent on the initial dry density, but the unloading is not. The reloading curve 'ar' of sample 'a' indicates a much stiffer response than the initial loading of any of the three samples and follows a similar line to the unloading curve. To illustrate this superior behaviour, the reloading curve 'ar' of sample 'a' has been replotted from the origin as though it were a new sample formed by preloading.

STIFFENING COLUMNS

Not all soils are easy to densify. The installation of stiffening elements into the ground effectively forms a new composite soil structure. The objective is that the composite soil structure will have superior load-carrying properties compared with the untreated ground. Installation of the columns may modify the properties of the surrounding ground.

The most common form of stiffening element is in the form of a column, and four types of stiffening column are considered:

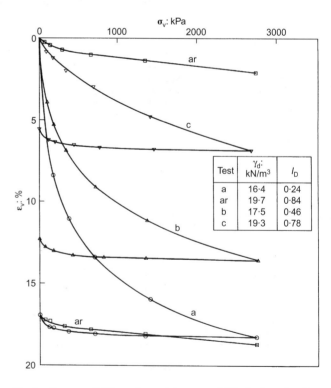

Fig. 2. **Compressibility of sandstone rockfill during loading, unloading and reloading (after Charles, 1991)**

(a) vibrated stone columns (sometimes termed vibro stone columns)

(b) soil-stabilised columns

(c) jet-grouted columns

(d) vibrated concrete columns (sometimes termed vibro concrete columns).

The stiffness of vibrated stone columns is dependent entirely on the support of the surrounding ground, which may or may not be compacted by the process of column installation. Watts et al. (2000) have described an instrumented trial in a heterogeneous fill. The group effects of stone columns in clay have been investigated by Muir Wood et al. (2000) using model tests and numerical analysis. Craig & Al-Khafaji (1997) and Al-Khafaji & Craig (2000) have described centrifuge model tests on the behaviour of a model tank foundation on sand columns in clay.

Unlike stone columns, soil-stabilised columns and jet-grouted columns do have some inherent strength and stiffness, in addition to that associated with the support of the surrounding soil. The effectiveness of both types of technique is very dependent on the details of the treatment process. Croce & Flora (2000) have analysed single-fluid jet grouting.

Vibrated concrete columns can be regarded as piles rather than as a type of ground treatment. Essentially, loads are transmitted by the columns through the poor ground to a stiffer underlying stratum.

Field and laboratory data for stone columns are presented in Fig. 3 as a plot of s_r against A_r, where the settlement ratio, s_r, is the ratio of the settlement with stone columns to the settlement that would have occurred if no stone columns had been installed, and the area ratio, A_r, is the proportion of the total area of the treated ground occupied by columns. By definition, when $A_r = 0$, $s_r = 1$. The field and laboratory data presented in Fig. 3 show that, typically, the installation of vibrated stone columns (with $A_r = 0.3$) can produce a composite soil structure with a compressibility about 50% of the untreated soil ($s_r = 0.5$). It might be assumed that the stiffer soil-stabilised columns would give a much greater improvement in performance for the same area ratio. However, many case histories do not show this (Charles & Watts, 2002). The performance of soil-stabilised columns and jet-grouted columns is critically dependent on the efficacy of the mixing process.

CONCLUSIONS

An increasing proportion of building development takes place on poor ground, presenting the geotechnical engineer with the challenge of providing satisfactory foundation performance at low cost. Ground treatment provides a means of modifying ground behaviour so that ground properties are improved and heterogeneity is reduced. Treatment of the ground prior to building can reduce uncertainty, and, consequently, ground treatment is of increasing importance within the practice of geotechnical engineering.

There are relatively few reports of ground treatment failing. However, an absence of reports is not the same as an absence of failures. Furthermore, in many applications not much is required of the treatment. When much is required from the treatment, the possibility of failure will be much higher. Where failures do occur, they may result from a failure to identify the principal deficiency in the ground rather than from a technical shortcoming of the treatment. There are also problems of procurement, which point to the need for better specifications.

The geotechnical engineer is not just an applied scientist. Whereas through observation and experiment the scientist can continue to gather and systematise knowledge, and has the intellectual satisfaction that goes with this activity, the engineer must make decisions despite uncertainty, using engineering judgement that brings together experience and scientific knowledge. The successful application of ground treatment requires an appropriate interaction between practical tools and techniques and an understanding of ground behaviour.

In the 26 years that have passed since the first Géotechnique Symposium in Print on ground treatment much has happened in the practice of ground treatment, and the papers published in the present symposium show a marked advance since the first symposium. However, there are still many areas where the the application of better engineering science could interact beneficially with experience-based technology. Desirable consequences of such interaction include:

(a) an improved understanding of soil behaviour and diagnosis of deficiencies

(b) an improved understanding of physical treatment processes (such as impact compaction) through numerical analysis and testing of physical models

(c) a clearer relationship between field performance of treated ground and improvement indicated by in situ testing before and after treatment

(d) a more realistic appreciation of what can be achieved by ground treatment through the study of well-documented case histories of long-term performance.

This lecture commenced with a quotation from Terzaghi's 45th James Forrest Lecture. It is appropriate to draw to a close with a quotation from the First James Forrest Lecture, presented by Dr William Anderson on 4 May 1893. The lecture was entitled 'The interdependence of abstract science and engineering'. Dr Anderson described the competent and successful engineer as 'one who is careful to make theory and practice walk side by side, the one ever aiding and guiding the other, neither asserting undue supremacy'.

Complete equality is always difficult to realise. Ground treatment is pre-eminently a practical subject that, for the most part, is commerce- and technology-led. It may be best

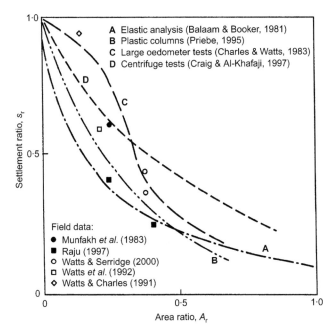

Fig. 3. Stiffening effect of stone columns (after Charles & Watts, 2002)

that the subject should be experience-based and that engineering science should be the junior partner. Nevertheless, there is room for improvement in the junior partner, and for closer integration within the partnership.

REFERENCES

Al-Khafaji, Z. A. & Craig, W. H. (2000). Drainage and reinforcement of soft clay tank foundation by sand columns. *Géotechnique* **50**, No. 6, 709–713.

Almeida, M. S. S., Santa-Maria, P. E. L., Martins, I. S. M., Spotti, A. P. & Coelho, L. B. M. (2000). Consolidation of a very soft clay with vertical drains. *Géotechnique* **50**, No. 6, 633–643.

Alonso, E. E., Gens, A. & Lloret, A. (2000). Precompression design for secondary settlement reduction. *Géotechnique* **50**, No. 6, 645–656.

American Society of Civil Engineers (1978). *Soil improvement: history, capabilities and outlook*, Report by the committee on placement and improvement of soils of the Geotechnical Engineering Division. New York: ASCE.

Anderson, W. (1893). The interdependence of abstract science and engineering. First James Forrest Lecture. *Minutes of Proceedings of Institution of Civil Engineers* **114** (1892–1893 session, part IV), May, 255–283.

Balaam, N. P. & Booker, J. R. (1981). Analysis of rigid rafts supported by granular piles. *Int. J. Numer. Anal. Methods Geomech.* **5**, No. 4, 379–403.

Barron, R. A. (1948). Consolidation of fine-grained soils by drain wells. *ASCE Trans.* **113**, 718–753.

Burland, J. B., McKenna, J. M. & Tomlinson, M. J. (1976). Preface: Ground treatment by deep compaction. *Géotechnique* **25**, No. 1, 1–2.

Charles, J. A. (1991). Laboratory compression tests and the deformation of rockfill structures. In *Advances in rockfill structures* (ed. E. Maranha das Neves), Ch. 5, pp. 73–96. Dordrecht: Kluwer.

Charles, J. A. (1996). The depth of influence of loaded areas. *Géotechnique* **46**, No. 1, 51–61.

Charles, J. A. & Watts, K. S. (1983). Compressibility of soft clay reinforced with granular columns. *Proc. 8th Eur. Conf. Soil Mech. Found. Engng, Helsinki* **1**, 347–352.

Charles, J. A. & Watts, K. S. (2002). *Treated ground: engineering properties and performance*, CIRIA Funders Report/C572. London: Construction Industry Research and Information Association.

Charles, J. A., Burford, D. & Watts, K. S. (1981). Field studies of the effectiveness of 'dynamic consolidation'. *Proc. 10th Int. Conf. Soil Mech. Found. Engng, Stockholm* **3**, 617–622.

Charles, J. A., Burford, D. & Watts, K. S. (1986). Improving the load carrying characteristics of uncompacted fills by preloading. *Munic. Engr* **3**, No. 1, 1–19.

Chow, Y. K., Yong, D. M., Yong, K. Y. & Lee, S. L. (2000). Improvement of granular soils by high-energy impact. *Ground Improvement* **4**, 31–35.

Chu, J., Yan, S. W. & Yang, H. (2000). Soil improvement by the vacuum preloading method for an oil storage station. *Géotechnique* **50**, No. 6, 625–632.

Craig, W. H. & Al-Khafaji, Z. A. (1997). Reduction of soft clay settlement by compacted sand columns. *Proc. 3rd Int. Conf. ground improvement geosystems, London*, 218–231.

Croce, P. & Flora, A. (2000). Analysis of single-fluid jet grouting. *Géotechnique* **50**, No. 6, 739–748.

De Mello, V. F. B. (1977). Reflections on design decisions of practical significance to embankment dams: 17th Rankine Lecture. *Géotechnique* **27**, No. 3, 279–355.

Faraco, C. (1981). Deep compaction field tests in Puerto de la Luz. *Proc. 10th Int. Conf. Soil Mech. Found. Engng, Stockholm* **3**, 659–662.

Feng, T.-W., Chen, K.-H., Su, Y.-T. & Shi, Y.-C. (2000). Laboratory investigation of efficiency of conical-based pounders for dynamic compaction. *Géotechnique* **50**, No. 6, 667–674.

Gohl, W. B., Jefferies, M. G., Howie, J. A. & Diggles, D. (2000). Explosive compaction: design, implementation and effectiveness. *Géotechnique* **50**, No. 6, 657–665.

Greenwood, D. A. (1970). Mechanical improvement of soils below ground surface. *Proceedings of conference on ground engineering*, Institution of Civil Engineers, London, pp. 11–22.

Institution of Civil Engineers (1976). *Ground treatment by deep compaction*. London: ICE. (The papers in the volume were first published as a Symposium in Print in *Géotechnique*, March 1975.)

Kovacevic, N., Potts, D. M. & Vaughan, P. R. (2000). The effect of the development of undrained pore pressure on the efficiency of compaction grouting. *Géotechnique* **50**, No. 6, 683–688.

Mayne, P. W., Jones, J. S. & Dumas, J. C. (1984). Ground response to dynamic compaction. *ASCE J. Geotech. Engng* **110**, No. 6, 757–774.

Menard, L. & Broise, Y. (1975). Theoretical and practical aspects of dynamic consolidation. *Géotechnique* **25**, No. 1, 3–18.

Merrifield, C. M. & Davies, M. C. R. (2000). A study of low-energy dynamic compaction: field trials and centrifuge modelling. *Géotechnique* **50**, No. 6, 675–681.

Mitchell, J. M. & Jardine, F. (2000). *A guide to ground improvement*, Report RP 373. London: Construction Industry Research and Information Association.

Muir Wood, D., Hu, W. & Nash, D. F. T. (2000). Group effects in stone column foundations: model tests. *Géotechnique* **50**, No. 6, 689–698.

Munfakh, G. A., Sarkar, S. K. & Castelli, R. J. (1983). Performance of a test embankment founded on stone columns. *Proc. Int. Conf. piling and ground treatment*, London, pp. 259–265.

Priebe, H. J. (1995). The design of vibro replacement. *Ground Engng* **28**, December, 31–37.

Raju, V. R. (1997). The behaviour of very soft cohesive soils improved by vibro replacement. *Proc. 3rd Int. Conf. Ground Improvement Geosystems, London*, 253–259.

Renton-Rose, D. G., Bunce, G. C. & Finlay, D. W. (2000). Vibro replacement for industrial plant on reclaimed land, Bahrain. *Géotechnique* **50**, No. 6, 727–737.

Slocombe, B. C., Bell, A. L. & Baez, J. I. (2000). The densification of granular soils using vibro methods. *Géotechnique* **50**, No. 6, 715–725.

Tang, M. & Shang, J. Q. (2000). Vacuum preloading consolidation of Yaoqiang Airport runway. *Géotechnique* **50**, No. 6, 613–623.

Terzaghi, K. (1939). Soil mechanics—a new chapter in engineering science: 45th James Forrest Lecture. *J. Instn Civ. Engrs* **12**, 106–141.

Van Impe, W. F. (1989). *Soil improvement techniques and their evolution*. Rotterdam: Balkema.

Wahls, H. E. (1994). Tolerable deformations. *Proc. ASCE Conf. Settlement '94, Texas* **2**, 1611–1628.

Watts, K. S. & Charles, J. A. (1991). The use, testing and performance of vibrated stone columns in the United Kingdom. In *Deep foundation improvement: design, construction and testing*, STP 1089, pp. 212–233. Philadelphia: ASTM.

Watts, K. S. & Serridge, C. J. (2000). A trial of vibro bottom-feed stone column treatment in soft clay soil. In *Grouting, soil improvement: geosystems including reinforcement* (ed H. Rathmayer), pp. 549–556. Helsinki: Building Information Ltd, Helsinki.

Watts, K. S., Saadi, A., Wood, L. A. & Johnson, D. (1992). Preliminary report on a field trial to assess the design and performance of vibro ground treatment with reinforced strip foundations. *Proc 2nd Int. Conf. Polluted and Marginal Land, London*, 215–221.

Watts, K. S., Johnson, D., Wood, L. A. & Saadi, A. (2000). An instrumented trial of vibro ground treatment supporting strip foundations in a variable fill. *Géotechnique* **50**, No. 6, pp. 698–708.

Tang, M. and Shang, J. Q. (2000). *Géotechnique* **50**, No. 6, 613–623

Vacuum preloading consolidation of Yaoqiang Airport runway

M. TANG* and J. Q. SHANG†

Yaoqiang Airport is an international airport serving the city of Jinan, China. The soil on the site consists of alternate layers of silty sand, silty clay, silt and clay. An under-consolidated soft clay layer approximately 4 m thick is located at between 7·5 m and 11·5 m depth above a silty clay layer. Soil improvement was proposed to consolidate the site prior to the construction of a runway to eliminate excessive settlement under static and dynamic loads on the runway. The results from field pilot tests for both surcharge preloading and vacuum preloading under the same consolidation pressure of 80 kPa are presented. Soil settlement in the range 20–30 cm was achieved in 80–90 days. For the vacuum treatment, in-situ deep mixing slurry cut-off walls were installed to control vacuum loss in a shallow silty sand layer. Data from the successful full-scale vacuum treatment of the site are also presented.

KEYWORDS: case history; consolidation; cut-off walls and barriers; full-scale tests; geosynthetics; ground improvement.

L'aéroport de Yaoqiang est un aéroport international desservant la ville chinoise de Jinan. Le sol de ce site est constitué par couches alternées de sable silteux, d'argile silteuse, de limon et d'argile. Une couche sous-consolidée d'argile tendre d'une épaisseur d'environ 4 m est située entre environ 7·5 et 11·5 m de profondeur au-dessus d'une couche d'argile silteuse. Il a été proposé d'améliorer le sol pour consolider le site avant la construction d'une piste d'atterrissage, afin d'empêcher les excès de tassement dus aux charges statiques et dynamiques sur la piste. Nous présentons les résultats des essais pilotes sur le terrain pour la précharge de surcharge et la précharge sous vide avec une même pression de consolidation de 80 kPa. On a obtenu un tassement du sol se situant dans une gamme de 20 à 30 cm, en 80–90 jours. Pour le traitement sous vide, des murs écrans ont été installés, en mélangeant en profondeur une boue argileuse et le sol en place, afin de contrôler la perte de vide dans une couche de sable silteux peu profonde. Nous présentons aussi les données obtenues une fois que le traitement sous vide de grande envergure a été mené à bien.

INTRODUCTION

Yaoqiang Airport is an international airport serving the city of Jinan, China. It is located 26 km north of Jinan (Fig. 1). The airport runway is laid out in a south–north direction, 2600 m long and 60 m wide, as shown in Fig. 2. The soil profile and properties on the site as measured from 54 subsurface survey boreholes along the runway are summarized in Table 1. The soil on the site consists of alternate layers of silty sand, silty clay, silt and clay. An under-consolidated soft clay layer approximately 4 m thick is present at depths between 7·5 m and 11·5 m, lying above a silty clay layer that extends to a depth of 20 m.

Kjellman (1952) of the Swedish Geotechnical Institute first introduced the principle of vacuum preloading consolidation of fine-grained soils in the early 1950s. The technique has been studied and applied quite extensively in Asia since the 1980s for consolidation of soft, highly compressible soils (Qian *et al.*, 1992; Harvey, 1997; Bergado *et al.*, 1998; Shang *et al.*, 1998; Shang & Zhang, 1999). When a vacuum is applied to a soil mass, it draws down the soil pore water pressure. When the total stress remains unchanged, the decrease in the pore pressure results in an increase in the effective stress in the soil and consolidation. A diagram of the vacuum preloading method is shown in Fig. 3 (Shang *et al.*, 1998). The working platform consists of a sand layer through which vertical drains are placed in the soil. The treatment area is sealed by a flexible membrane, which is keyed into an anchor trench surrounding the area. A perforated pipe system is placed beneath the liner to collect water. Vacuum pumps with sufficient capacity to generate a vacuum in the soil and capable of pumping water and air are connected to the collection system. It is essential that the site to be treated be securely sealed and isolated from any surrounding permeable soils in order to avoid leaks in the membrane and loss of the vacuum. As pinholes or cracks in the geomembrane are difficult to locate and repair, the geomembrane should be checked before it is placed. In order to obtain and sustain a high vacuum, the membrane is covered with water. This minimizes ageing of the membrane and reduces the risk of damage from pedestrians and wildlife. When the required preloading pressure is higher than the capacity of the vacuum pumps, a surcharge fill may be used in conjunction with the vacuum method, as shown in Fig. 3. The fill must be free from stones or sharp objects to minimize the risk of punch holes in the membrane.

The technique of vacuum preloading consolidation has been successfully applied in land reclamation (Shang *et al.*, 1998), consolidation of soda ash tailings (Shang & Zhang, 1999) and other projects in China since the late 1980s. Based on the condition of the Yaoqiang site, excessive settlement would be induced by static and dynamic loading of the runway. Therefore soil improvement was proposed to consolidate the site prior to construction. Before full-scale treatment, two field pilot tests were performed to assess the feasibility of vacuum and surcharge preloading techniques. The two methods were both found to be effective, and yielded similar results under the same

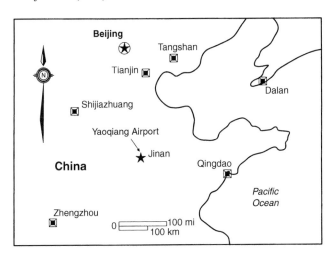

Fig. 1. Location of Yaoqiang Airport

Manuscript received 26 January 2000; revised manuscript accepted 25 April 2000.
* Tianjin Port Engineering Institute, China
† Department of Civil and Environmental Engineering, University of Western Ontario

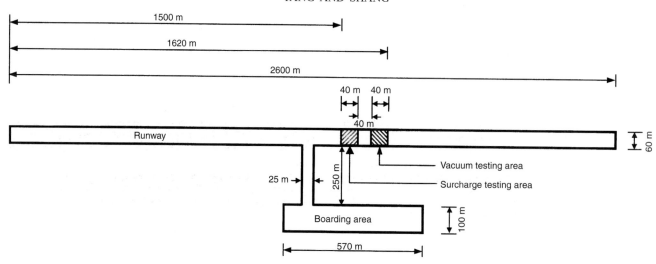

Fig. 2. Yaoqiang Airport runway layout

Table 1. Soil profile and properties[*]

Layer no.	Description†	Depth: m	w: %	γ: kN/m³	e	c_v§: cm²/s	k§: cm/s
1	Silty sand	0–3	12–18	15·6–19·0	0·83–0·98	–	$2·8 \times 10^{-2}$
2	Silty clay	3–5	26–31	19·0–19·5	0·85–0·88	0·042	$1·87 \times 10^{-5}$
3	Silt	5–7·5	28–32	19·0–19·2	0·73–0·81	0·035	$7·85 \times 10^{-5}$
4	Soft clay	7·5–11·5	32–40	17·2–19·2	1·1–1·6	0·001‡	$1·55 \times 10^{-6}$‡
5	Silty clay	11·5–20	22–28	19·3–20·3	0·60–0·70	0·021	2×10^{-5}

Elevation of ground surface   +21·3 to +22·2 m
Elevation of groundwater table   +19·1 m

* Averages from 54 boreholes along the runway.
† Based on the Chinese soil classification system:
Silty sand: 50% by weight exceeds a particle size of 0·0075 mm
Silty clay: the plasticity index is greater than 10 but not more than 17
Silt: the plasticity index is greater than 3 but not more than 10
Soft clay: the plasticity index is greater than 17, the natural water content is greater than the liquid limit and the void ratio is greater than 1·0.
‡ Estimated (Shelby tube sample was too soft to perform consolidation tests).
§ Measured from standard consolidation tests.

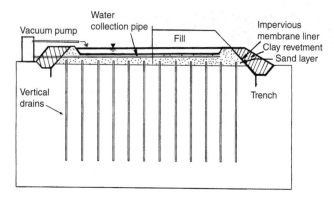

Fig. 3. Schematic of vacuum preloading consolidation (after Shang et al., 1998)

design preloading pressure of 80 kPa. The site owner decided to use the vacuum technique because there was no requirement for fill materials, the treatment time was shorter, and an experienced team was available for the design and implementation of the vacuum preloading.

The primary objectives of the soil improvement at the Yaoqiang site were to consolidate the soil and to reduce runway settlement under static and dynamic loads. It was specified by the owner that the average degree of consolidation of all clayey and silty soil layers (layers 2–5, Table 1) after treatment should be not less than 95% under the design preloading pressure of 80 kPa.

A major difficulty at this site was the existence at the surface

of a silty sand layer approximately 2–3 m thick. The silty sand layer was mostly unsaturated, as the groundwater table was located at about 2–2·5 m below the ground surface. Generally, the surface membrane is keyed in to a trench less than 1 m deep (see Fig. 3). A trench 1 m deep would not be sufficient to seal the silty sand layer, which extended down to a depth of approximately 3 m. This was overcome by the construction of an in-situ deep mixing slurry cut-off wall, which penetrated at least 1·5 m below the base of the silty sand layer to seal the vacuum treatment area. Also, the treatment area was reduced to 600 m² to minimize the pressure loss due to leakage.

The performance of the vacuum treatment was assessed by monitoring the vacuum pressure, pore pressures, soil settlement, and lateral displacements. In addition, the field vane shear strength, CPT resistance and physical properties of the soil at various depths were recorded before and after treatment. The results of the field pilot tests using surcharge and vacuum preloading methods are reported below, followed by a description of the implementation and operation of the vacuum preloading at the site. The analysis and assessment of soil improvement are reported in terms of the real-time responses of the vacuum pressure and pore pressure, soil settlement, and lateral displacement. In addition, the field vane shear strength, CPT resistance and physical properties of the soil at various depths are reported before and after treatment.

FIELD PILOT TEST
The field pilot tests were conducted using surcharge and vacuum preloading methods in two areas, each 60 m × 40 m, as shown in Fig. 2. The two testing areas were set 40 m apart to

avoid interference during treatment. The top silty sand layer, with a hydraulic conductivity of approximately $2 \cdot 8 \, \text{m} \times 10^{-2}$ cm/s, was used as the surface drain in both the surcharge and vacuum tests. The design preloading pressure was 80 kPa, based on the dead load of the runway (26·2 kPa) and the dynamic load of airplanes (51·8 kPa). This is within the capacity of the vacuum technique: hence the additional surcharge fill was not needed. Pumping was carried out continuously during treatment, leading to an immediate decrease in pore pressure and an increase in the effective stress without the danger of bearing capacity failure. In the case of surcharge preloading, the surcharge fill was applied in three stages—30 kPa, 55 kPa and 80 kPa—over three months to avoid bearing capacity failure caused by the excess (positive) pore pressure.

In both test areas, prefabricated vertical drains (PVDs) were installed to a depth of 12 m based on an analysis of the stress distribution of the structural and dynamic loads on the runway. The PVDs comprised a grooved polyethylene core and non-woven textile filter 100 mm wide and 4 mm thick. The coefficient of permeability of the filter was 5×10^{-4} cm/s. The PVDs were installed in a square grid 1·3 m apart. The PVDs penetrated through the soft clay layer and reached the underlying silty clay. Horizontal collectors comprising perforated PVC pipes, 75 mm in diameter and 5 mm in wall thickness, were placed in the silty sand in trenches 30 cm wide and 20 cm deep at a spacing of 3·9 m along the runway and 30 m across the runway.

It was found in the vacuum treatment trial run that the degree of vacuum did not reach the design specification (80 kPa). This was thought to be due to air leakage through the silty sand layer, which was unsaturated and located mostly above the groundwater table. To control this leakage a cut-off wall 120 cm wide and 4·5 m deep was constructed by deep mixing clay slurry around the vacuum trial area. The wall was constructed in two phases by (1) preparing the clay slurry and (2) deep soil mixing. A diagram of the operation is shown in Fig. 4. A mixer of 4 m³ capacity was used to prepare the slurry, which was then filtered to eliminate coarse grains and roots and subsequently stored in an 8 m³ storage tank. The deep mixing equipment has a hollow stem auger and two sets of mixing blades. During both penetration and withdrawal, the mixing blades mix the slurry

with the soil in situ to form overlapping sand–clay mixing columns. The criteria used for the construction and quality control of the cutoff wall are summarized as follows:

(a) clay soil used for making slurry: 30–40% clay, clear of roots and gravel
(b) clay–water ratio of the slurry: 1:2 mass/mass
(c) slurry mixing time: 15–20 min
(d) area of mixing column: 0·71 m² (0·7 m thickness, 1·2 m width)
(e) mixing blades penetration and withdrawal speed: 0·8–1·25 m/min
(f) slurry injection speed: 160–250 l/min
(g) volume of slurry injection: 0·57 m³ per metre of mixing column
(h) clay fraction of the cut-off wall: 15%
(i) coefficient of permeability of the wall: $< 1 \times 10^{-5}$ cm/s.

After completion of the cut-off wall, samples were taken along the wall at 50–60 m intervals and tested for grain size analysis and hydraulic conductivity. The clay content of the cut-off wall was found to be greater than 15%, and the hydraulic conductivity was in the range $1 \cdot 0 \times 10^{-5}$ to $1 \cdot 0 \times 10^{-7}$ cm/s in most samples tested. This was two to three orders of magnitude less permeable than the silty sand. Sealing membranes were placed on top of the drainage pipes and keyed into an anchor trench. The membrane was sealed off by a clay revetment, which was compacted on top of the cut-off wall, as shown in Fig. 4. To further minimize the effects of leakage, the coverage area of each vacuum pump (7·5 kW) was reduced to 600 m². This compares with the 1000–1500 m²/pump used in the Xingang Port land reclamation project (Shang et al., 1998).

The consolidation theory for vertical drains developed by Barron (1948) was used to estimate the time required to achieve the required average degree of consolidation and settlement. The design equations are:

within the treatment depth (< 12 m)

$$U_{\text{avg}} = 1 - \exp\left(-\frac{8T_{\text{h}}}{F(n)}\right) \quad (1)$$

below the treatment depth (12 m $<$ depth $<$ 20 m)

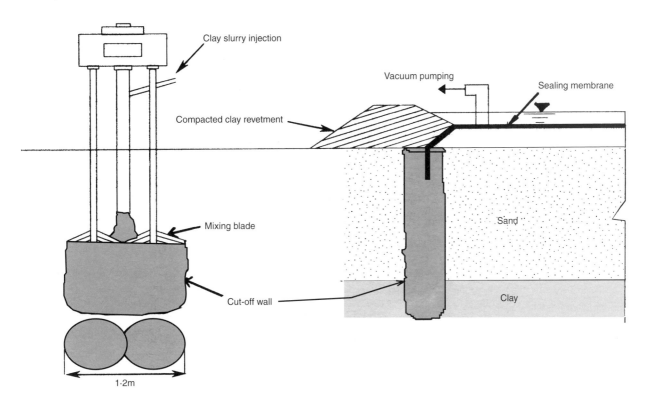

Fig. 4. Schematic of slurry cut-off wall

$$U_{avg} = 1 - \frac{8}{\pi^2} \exp\left(-\frac{\pi^2 T_h}{4}\right) \qquad (2)$$

where U_{avg} = average degree of consolidation throughout the soil mass at time t, T_h = time factor;

$$T_h = \frac{c_h t}{D_e^2} \qquad (3a)$$

$$F(n) = \frac{n^2}{n^2 - 1} \ln(n) - \frac{3n^2 - 1}{4n^2} \qquad (3b)$$

n = drain spacing ratio, $n = D_e/d_w$ $\qquad (3c)$

D_e = diameter of equivalent soil cylinder,

$D_w = S$ = spacing of PVDs‡ $\qquad (3d)$

d_w = equivalent diameter of PVDs, $d_w = (a + b)/2$ $\quad (3e)$

where a and b are dimensions of the PVD cross-section (a = 100 mm and b = 4 mm); c_h = horizontal coefficient of consolidation, assuming the soil is homogeneous, i.e. $c_v = c_h$.

The settlement of the silty sand layer was estimated based on the design load and the stress–strain relationship of the silty sand using the following equation:

$$s = \frac{e_0 - e_f}{1 + e_0} h \qquad (4)$$

where e_0 = initial void ratio, e_f = final void ratio under the design load, h = layer thickness.

Calculations using equations (1) and (2) and the soil parameters listed in Table 1 show that, for a period of 90 days, the average degree of consolidation would reach:

(a) in the silty clay and silt layers (layers 2 and 3): U_{avg} = 100% (eq. (1))
(b) in the soft clay layer (layer 4): U_{avg} = 77·4% (eq. (1))
(c) in the silty sand layer (layer 5): U_{avg} = 100% (using eq. (2))

The overall weighted average degree of consolidation over the depth between 3 m and 20 m would be 95%, satisfying the design requirement of U_{avg} = 95%. Based on the above assessment, the preloading time was set at about 3 months.

Monitoring during the preloading consolidation for both the surcharge and vacuum testing areas included surface settlement across the runway, settlement at various depths, lateral displacement, pore water pressure, and groundwater table. The soil property tests conducted before and after treatment included field vane tests, cone penetration tests (CPT), and soil physical property tests (water content, void ratio and bulk unit weight). Table 2 summarizes the instrumentation used giving type, location, elevation, depth and number of tests carried out during and after treatment.

For the vacuum treatment area, pumping began on 1 July

Table 2. Summary of instrumentation*

	Location	Elevation: m		Depth: m		No. of tests		Remarks
						Before	After	
Field vane	Centre			20		1	2	1 shear/m
Shelby tube sample	Centre			20		1	2	1 sample/m
CPT	Centre			20		1	2	
Inclinometer 1	See diagram			20				
Inclinometer 2	See diagram			20				
Surface settl. pins	See diagram			0				
		Vacuum	Surcharge	Vacuum	Surcharge			
Extensometer	Centre	+21·6	+22·1	0	0			
		+19·1	+20·1	2·5	2			
		+16·6	+16·1	5	6			
		+14·1	+14·6	7·5	7·5			
		+11·6	+11·6	10	10·5			
		+9·6	+10·1	12	12			
Piezometer	Centre	+19·6	+19·0	2	3			
		+17·6	+17·1	4	5			
		+15·6	+15·6	6	6·5			
		+13·1	+13·6	8·5	8·5			
		+11·6	+12·0	10	10			
		+7·6	+8·1	14	14			
Vacuum gauges	Along centre line every 10 m in vacuum testing area							

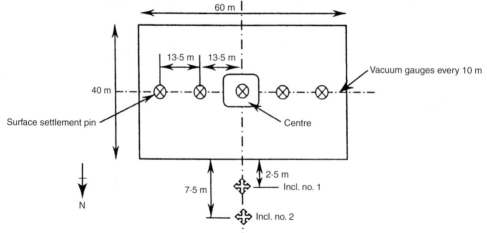

* For both surcharge and vacuum testing areas unless noted otherwise (not to scale)

‡ It is now agreed upon that $D_e = 1·13S$ (Rixner et al., 1986).

1989 and reached 550 mmHg under the membrane on the same day. The design vacuum pressure of 600 mmHg (80 kPa) was reached on 14 July and remained stable for the rest of the treatment period. The surface settlement rate had reduced to 0·17 mm/day on 20 September 1989 and remained constant for the next 5 days, which met the design requirement of settlement less than 1 mm/day. The vacuum was turned off on 27 September 1989, giving an overall time of vacuum application of 83 days (excluding time of power interruption).

For the surcharge area preloading started on 5 August 1989, with the load applied in three stages over a period of two months. The design load of 80 kPa was achieved on 16 September 1989. The registered surface settlement was 0·14 mm/day on 2 November 1989 and was maintained at the same level for the next 5 days. The overall treatment time was 89 days.

DEVELOPMENT OF SOIL PORE PRESSURE

The drawdown of soil pore pressure by vacuum application results in increases in the effective stress in the soil mass and consolidation, the fundamental principle of vacuum preloading consolidation. Therefore the development of pore pressure with time during a vacuum preloading process is of particular interest for engineers. Fig. 5 presents plots of continuous pore water pressure measurement against time at various depths for the vacuum preloading test. The data were recorded by piezometers installed at depths of 2 m, 8·5 m, 10 m and 14 m. Piezometers installed at depths of 4 m and 6 m were damaged in an early stage of testing. The groundwater table on the vacuum treatment area was 2·5 m below the ground surface. Hence the static pore pressures at depths 2 m, 8·5 m, 10 m and 14 m would be 0, 59 kPa, 75 kPa and 115 kPa respectively, which is in reasonable agreement with the initial readings recorded on the figure. The pore pressures were drawn down by the vacuum application, as seen on the figure, and reached equilibrium 5 days after the design vacuum pressure of 600 mmHg (80 kPa) had been reached on 14 July 1989. There was a power interruption on 8

August 1989 that lasted for three days. The piezometers responded to the pore pressure change due to the power interruption, indicating that they were in good working condition.

It was intended that the piezometers would be installed at the midpoint between PVDs at specific depths. However, it was difficult to install the piezometer precisely at a specific location, especially in deep boreholes. Hence certain piezometers might be installed fairly close to PVDs. The pore water pressure drawdown recorded by the piezometers appeared to be fairly rapid, and may not be representative for the analysis of the time rate of consolidation. On the other hand, the piezometer data at the end of treatment should be fairly representative of the pore pressure at equilibrium and are used in the following discussion.

The pore pressure drawdown is defined as the difference between the measured in-situ pore pressure at equilibrium and the hydrostatic pore pressure before vacuum application that is

$$\Delta u_v(z) = u_v(z) - u_0(z) \tag{5}$$

where $\Delta u_v(z)$ = pore pressure drawn down by vacuum pumps at depth z (negative); $u_v(z)$ = measured in-situ pore pressure at equilibrium at depth z; and $u_0(z)$ = initial pore pressure at depth z. The pore pressure drawdown distribution with depth is discussed further in a later section.

CONSOLIDATION SETTLEMENT

The results of vertical and lateral soil deformations during and after surcharge and vacuum preloading are presented in Figs 6, 7 and 8 respectively. Fig. 6 shows the surface settlement with the treatment time for (a) surcharge preloading and (b) vacuum preloading respectively. The surface settlement approached equilibrium after about 90 days from the time the fill was placed in the surcharge preloading test (Fig. 6(a)), and after about 50 days from the vacuum application in the vacuum preloading test (Fig. 6(b)). It may be seen from Fig. 6(b) that

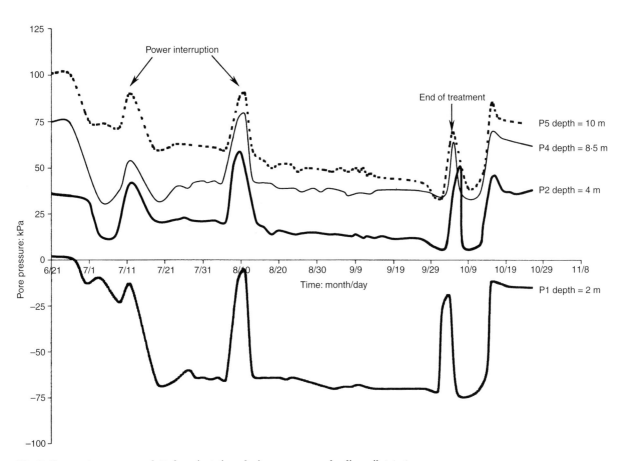

Fig. 5. Pore water pressure plotted against time during vacuum preloading pilot test

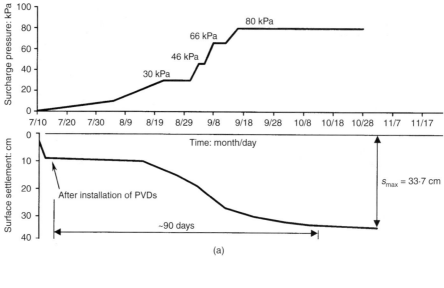

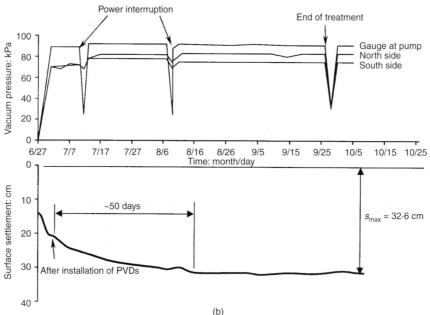

Fig. 6. Surface settlement during field pilot tests: (a) surcharge preloading test; (b) vacuum preloading test

the vacuum pressure was about 90 kPa at the vacuum pumps, and was maintained at between 70 and 80 kPa underneath the membrane. The measured maximum surface settlement was 33·7 cm and 32·6 cm for surcharge and vacuum preloading treatment respectively.

The results of soil settlement at various depths measured by extensometers after surcharge and vacuum tests are presented in Figs 7(a) and 7(b). It is shown that the vacuum preloading generated a settlement profile fairly similar to that generated by the surcharge fill. A major distinction between the surcharge preloading and vacuum consolidation methods is the direction of the lateral displacement. In the former, the lateral displacement is outward from the treatment area (soil is squeezed out), whereas in the latter, the lateral displacement is towards the treatment area (soil is sucked in). While the advantage of the vacuum treatment is that there is no danger of bearing capacity failure by applying the vacuum pressure instantaneously, one must be aware of the possible development of tensile cracks adjacent to the treatment area (Shang *et al.*, 1998). In this paper, the lateral displacement is denoted as positive for outward displacement away from the treatment area, as in the case

of surcharge preloading, and negative for inward displacement towards the treatment area, as in the case of vacuum preloading.

The results of inclinometer measurements at the end of treatment for surcharge and vacuum tests are presented in Fig. 8 (see the diagram in Table 2 for the locations of the inclinometers). Fig. 8(a) shows that the two inclinometers located at 2·5 m and 7·5 m from the north edge of the surcharge preloading area registered fairly consistent outward lateral displacements of 20–30 mm. For the vacuum treatment area, inward lateral displacement registered by inclinometer 1 reached 100 mm at the soil surface at a distance of 2·5 m from the treatment area, reducing to about 20 mm at 7·5 m from the treatment area, as recorded by inclinometer 2.

The measured settlements shown in Figs 6 and 7 represent the combined effect of one-dimensional consolidation settlement and soil lateral displacement. The actual one-dimensional settlement would be smaller for surcharge preloading because of soil being squeezed out, while it would be greater for vacuum preloading because of soil being sucked in. The following empirical equation was used to compute the one-dimensional

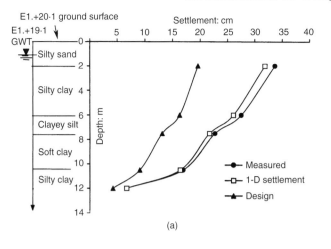

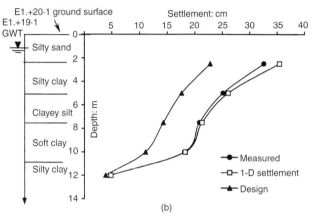

Fig. 7. Settlement profile at the end of field pilot tests: (a) surcharge preloading test; (b) vacuum preloading test

settlement with compensation for the lateral displacement, based on the mass equilibrium of the volume of the soil in the treatment zone:

$$s_{1D} = s - \frac{2(L + B)\sum_i h_i x_i}{LB} \qquad (6)$$

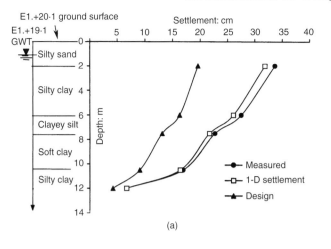

Fig. 8. Lateral displacement at the end of field pilot tests

where s_{1D} = one-dimensional settlement; s = measured settlement; L and B = length and width of the treatment area respectively; h_i = thickness of soil layer i; x_i = average lateral displacement of soil layer i: $x_i > 0$ for surcharge preloading and $x_i < 0$ for vacuum consolidation.

The results of one-dimensional settlement calculated from equation (6), along with the settlements measured in situ, and design settlements computed using equations (1), (2) and (3) for layers 2–5 and equation (4) for layer 1, are presented in Table 3 and Fig. 7, respectively. The overall settlements were 1·62 and 1·55 times greater than design specification in the surcharge and vacuum preloading tests respectively.

It is clearly shown from the measured settlement results that the design settlement based on the Barron (1948) analysis underestimated the average degree of consolidation and corresponding settlement. A back calculation based on Terzaghi's coefficient of consolidation, c_v, indicated that c_v values are one to two orders of magnitude higher than those listed in Table 1, if the hydraulic conductivities listed in Table 1 are used in the calculation. It should be noted that the hydraulic conductivity of the soil was measured indirectly from standard consolidation

Table 3. Settlement analysis
Surcharge preloading test

Layer no.	Description	Location of extensometers		Settlement: cm				
		Elevation: m	Depth: m	Measured s in equation (6)	Lateral dspl. 2nd term, equation (6)	1-D s_{1D} in equation (6)	Calculated from design load,* s_d	s_{1D}/s_d
1	Silty sand	+20·1	2	6·2	0·4	5·8	3·34 (equation (4))	1·71
2	Silty clay	+16·1	6	4·8	0·44	4·36	3·15	1·38
3	Silt	+14·6	7·5	5·8	0·48	5·32	4·05	1·31
4	Soft clay	+11·6	10·5	10·2	0·53	9·67	4·85	1·99
5	Silty clay	+10·1	12	6·7	0	6·7	4·22	1·59
			Total	33·7	1·85	31·85	19·61	1·62

Vacuum preloading test

Layer no.	Description	Location of extensometers		Settlement: cm				
		Elevation: m	Depth: m	Measured s in equation (6)	Lateral dspl. 2nd term, equation (6)	1-D s_{1D} in equation (6)	Calculated from design load,* s_d	s_{1D}/s_d
1	Silty sand	+19·1	2·5	7·5	−1·86	9·36	5·2 (equation (4))	1·8
2	Silty clay	+16·6	5	4·3	−0·48	4·78	3·32	1·44
3	Clayey silt	+14·1	7·5	2·7	−0·36	3·06	3·2	0·97
4	Soft clay	+11·6	10	13·3	−0·11	13·41	7·19	1·87
5	Silty clay	+9·6	12	4·8	0	4·8	3·88	1·24
			Total	32·6	−2·81	35·41	22·79	1·55

* Calculated from equations (1) and (2) at $U_{avg} = 95\%$ unless noted otherwise.

tests on Shelby tube samples. The values may considerably differ from those measured in-situ. This may contribute to the inconsistency of the predicted and measured settlements. Other possible reasons for the discrepancy include:

(a) the assumptions made in the derivation of the theory, such as the soil being saturated and homogeneous
(b) the hydraulic conductivity of soil being independent of location and soil void ratio ($k =$ constant, small strain theory)
(c) no excess pore pressure in PVDs (the hydraulic conductivity of PVD is much higher than the surrounding soil mass)
(d) no smear zone surrounding PVDs
(e) the zone of influence of each PVD being a cylinder
(f) all compressive strains within the soil mass occurring in a vertical direction.

The development of a numerical model is needed for better design of the vacuum preloading consolidation.

VANE SHEAR STRENGTH, CPT RESISTANCE AND PHYSICAL PROPERTIES

Figures 9 and 10 present the results of the field vane test and cone penetration tests (CPT) before and after treatment conducted on the surcharge and vacuum treatment areas respec-

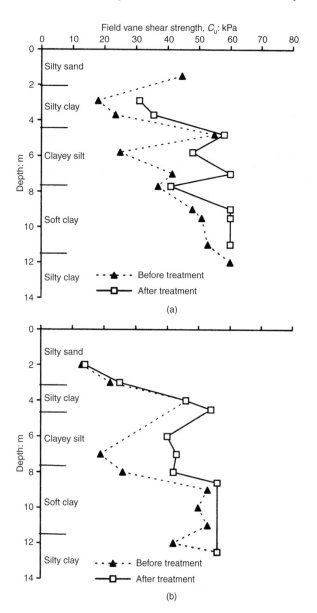

Fig. 9. Field vane shear strength profiles: (a) surcharge preloading test area; (b) vacuum preloading test area

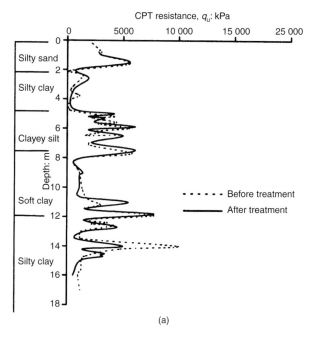

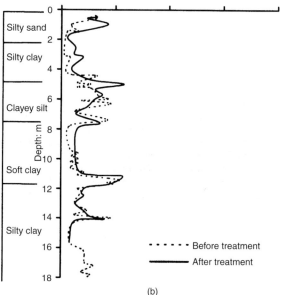

Fig. 10. CPT resistance profiles (a) surcharge preloading test area (b) vacuum preloading test area

tively. Table 4 presents the results of laboratory tests on Shelby tube samples recovered from five soil layers, including water content (w), bulk unit weight (γ), and void ratio (e). The data demonstrate clearly the overall improvement of the soil's geotechnical and physical properties. In particular, the most significant increase in undrained shear strength was detected in the soft clay layer (layer 4), as shown in Fig. 9.

As evidenced in the field pilot tests, both surcharge preloading and vacuum preloading treatments achieved the goal of soil improvement in terms of settlement under the design preloading pressure. The deep-mixing slurry cut-off wall was proven to be effective in isolating the top silty sand layer and maintaining the vacuum pressure. Further, for a coverage area of 600 m², 7·5 kW pump would be required for vacuum preloading consolidation of the site.

IMPLEMENTATION OF VACUUM TREATMENT ON YAOQIANG SITE

Based on the field pilot test results, the vacuum consolidation method was chosen to consolidate Yaoqiang Airport runway.

Table 4. Results of laboratory tests
Surcharge preloading test

Layer no.	Elevation: m	Description	Water content, w:%			Bulk unit wt, γ: kN/m³			Void ratio, e		
			BT	AT	Change	BT	AT	Change	BT	AT	Change
1*	+22·138 +20·138	Silty sand	12	23·1	92·5%	16·3	18·6	14·1%	0·83	0·77	−7·2%
2	+17·638	Silty clay	31·4	25	−20·4%	19	19·6	3·2%	0·85	0·72	−15·3%
3	+14·638	Clayey silt	31·7	27·2	−14·2%	19·2	19·6	2·1%	0·81	0·74	−8·6%
4	+10·638	Soft clay	39·6	36·6	−7·6%	18·3	18·8	2·7%	1·09	0·92	−15·6%
5	+2·138	Silty clay	24·4	23·9	−2·0%	20·3	20·3	0·0%	0·66	0·64	−3·0%

Vacuum preloading

Layer no.	Elevation: m	Description	Water content, w:%			Bulk unit wt, γ: kN/m³			Void ratio, e		
			BT	AT	Change	BT	AT	Change	BT	AT	Change
1*	+21·629 +18·629	Silty sand	20·8	27·2	30·8%	19·2	19·5	1·6%	−	0·73	
2	+16·629	Silty clay	29·0	26·9	−7·2%	19·5	19·7	1·0%	0·78	0·73	−6·4%
3	+14·129	Clayey silt	26·9	24·7	−8·2%	19·1	19·6	2·6%	0·73	0·69	−5·5%
4	+10·129	Soft clay	48·7	41·2	−15·4%	17·2	17·7	2·9%	1·40	1·30	−7·1%
5	+1·629	Silty clay	26·4	25·6	−3·0%	19·8	20·0	1·0%	0·70	0·69	−1·4%

BT: before treatment
AT: after treatment
* Saturated after treatment

The treatment was executed on eleven individually sealed areas, as shown in Fig. 11, using the same design parameters and configurations as described in the previous section for the trial. The project began on 9 June 1990 and was completed on 14 August 1991. The total treatment area was 145 000 m², including the entire runway and part of the connection road to the boarding area. A total of 86 970 PVDs were installed to a depth of 12 m in a 1·3 m square grid, with a total length of over 1000 km. The coverage area of each 7·5 kW vacuum pump was 600 m². Cut-off walls were constructed surrounding the treatment areas, with a width of 120 cm and depths varying from 3·5 m up to 7·5 m, depending on the thickness of the top sand layer. The total length of the cut-off walls was 6200 m. After treatment, the cut-off walls were excavated to 1·5 m deep, filled with silty sand, and compacted to a degree of compaction of 95%.

ASSESSMENT OF VACUUM PRELOADING TREATMENT OF YAOQIANG SITE

The vacuum pressure, surface settlement, settlement at depth, and lateral displacement were measured in all areas treated. In addition to these parameters, pore pressure and laboratory soil testing (water content, bulk unit weight, and void ratio) before and after treatment was performed on Areas 1, 3, 7 and 10. Table 5 presents a summary of the eleven treatment areas,

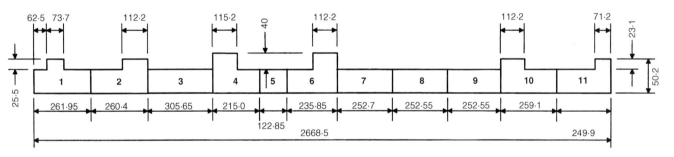

Fig. 11. Layout of vacuum treatment areas. Unit: m

Table 5. Summary of vacuum preloading consolidation on eleven areas

Area No	Area: m²	No. PVDs	Length of cut-off wall: m	Treatment time: days	Design settl. s_d: cm	Measured settl.,§ s_{1D}: cm	s_{1D}/s_d
1	14 655	8 844	629	81	18·8	20·9	1·11
2	15 257	9 092	617	90	16·9	22·0	1·30
3	14 977	8 968	662	84	17·9	23·2	1·30
4	15 752	9 426	555	85	19	22·7	1·19
5*	2 480	1 584	246	85	22·8†	12·9‡	
6	15 997	9 544	617	80	18·3	24·0	1·31
7	12 382	7 372	556	84	17·9	24·6	1·37
8	12 375	7 372	530	81	15·8	17·6	1·11
9	12 375	7 372	530	81	1°·⌐	24·0	1·31
10	15 193	9 110	615	81	15·6	19·8	1·27
11	13 801	8 286	649	81	18·1	20·9	1·15

* Including two field pilot test areas.
† Based on design value in field pilot tests.
‡ Part of the area has been treated in field pilot tests.
§ With lateral displacement adjustment using equation (6).

which includes the treatment area, number of PVDs installed, length of the cut-off wall, treatment time, design settlement, measured total settlement after correction for lateral displacement, and the maximum differential settlement across the runway. Fig. 12 presents the measured total settlements. Note that Area 5 was partially treated in two field pilot tests performed earlier that resulted in much less settlement. The ratios of the measured and design settlements are in the range 1·11 to 1·37, as shown in Table 5, which indicates that the design specification was exceeded in all areas treated.

The measured pore pressure drawdown versus depth for the field pilot test and for four of the treatment areas is presented in Fig. 13. From the figure, it may be seen that the distribution of pore pressure drawdown is fairly independent of depth, indicating that the PVDs functioned efficiently as a transfer medium for the vacuum pressure. Area 10 registered the highest magnitude of pore pressure drawdown at approximately −60 to −65 kPa, and this remained constant up to a depth of 12 m. In other areas, the pore pressure drawdown was within −40 to −50 kPa at a depth of 3 m and below. Higher-magnitude drawdowns (∼ 60 kPa) were recorded in the silty sand layer (depth ∼ 2 m). The stable pore pressure in the silty sand layer is a good indication that the cut-off walls were effective in preventing vacuum pressure loss in the sand layer. Table 6 summarizes the results of soil properties before and after treatment, taken as averages measured in Areas 1, 3, 7 and 10 for five soil layers. The results show consistent improvement of soil properties in terms of increases in the bulk unit weight decreases in soil water content and void ratio after vacuum preloading.

CONCLUSIONS

The application of vacuum preloading consolidation of Yaoqiang Airport runway has been discussed in this case study. Ground improvement was required on the site to eliminate excessive settlement under the static and dynamic loading on

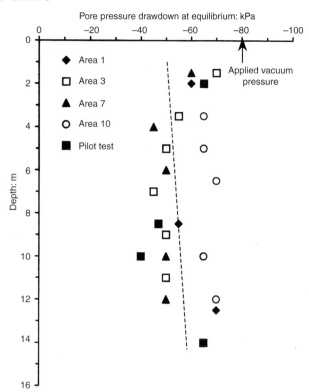

Fig. 13. Distribution of pore pressure drawdown at equilibrium

the runway. The challenge facing geotechnical engineers in this project was the existence of a silty sand layer, 2–3 m thick, at the ground surface, which imposed difficulties in sealing the treatment area and in reaching the design vacuum pressure. The installation of in-situ deep mixing slurry cut-off walls along with reduced coverage area of vacuum pumps enabled the soil improvement objectives to be achieved by vacuum preloading treatment. The project went through three stages: preliminary design, field pilot tests and full-scale treatment. The target degree of consolidation under the design preloading pressure was achieved in 80–90 days with soil settlements of the order of 20–30 cm.

It is shown from this case study that vacuum preloading consolidation can be used with confidence in soil improvement projects for highly compressive and fine-grained soils.

ACKNOWLEDGEMENTS

The authors wish to acknowledge the Tianjin Port Engineering Institute for giving its permission for publication of the results contained in this paper. The case study was funded by the Natural Sciences and Engineering Research Council, Canada under the research grant No. 203017-98 RGPIN.

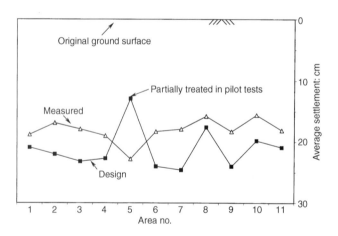

Fig. 12. Average settlement after vacuum preloading treatment along the runway

Table 6. Summary of laboratory tests*

Layer no.	Depth: m	Description	Water content, w:%			Bulk unit Wt, γ: kN/m³			Void ratio, e		
			BT	AT	Change	BT	AT	Change	BT	AT	Change
1	2·0–2·5	Silty sand†	13·0	24·0	84·6%	16·0	19·4	21·3%	0·884	0·744	−15·8%
2	4·0–5·0	Silty clay	27·9	27·2	−2·5%	19·3	19·5	1·0%	0·766	0·760	−0·8%
3	7·0–7·5	Silt	30·5	27·0	−11·5%	19·0	19·4	2·1%	0·889	0·763	−14·2%
4	11·0–12·5	Soft clay	33·8	28·7	−15·1%	18·8	19·5	3·7%	0·944	0·826	−12·5%
5	19·0–20·0	Silty clay	25·7	23·3	−9·3%	19·3	19·6	1·6%	0·700	0·661	−5·6%

* Averages from data of testing areas 1, 3, 7 and 10.
† Saturated after treatment.
BT: before treatment.
AT: after treatment.

REFERENCES

Barron, R. A. (1948). Consolidation of fine-grained soils by drain wells. *Trans. ASCE* **113**, 718–754.

Bergado, D. T., Chai, J. C., Miura, N. and Balasubramaniam, A. S. (1998). PVD improvement of soft Bangkok clay with combined vacuum and reduced sand embankment preloading. *Geotech. Engng. J.* **29**, No. 1, 95–122.

Harvey, J. A. F. (1997). Vacuum drainage to accelerate submarine consolidation at Chek Lap Kok, Hong Kong. *Ground Engng*, July, 34–36.

Kjellman, W. (1952). Consolidation of clayey soils by atmospheric pressure. *Proceedings of a conference on soil stabilization*, Massachusetts Institute of Technology, Boston, pp. 258–263.

Qian, J. H., Zhao, W. B., Cheung, W. B., Cheung, Y. K. and Lee, P. K. K. (1992). The theory and practice of vacuum preloading. *Comput. Geotech.* **13**, 103–118.

Rixner, J. J., Kraemer, S. R. and Smith, A. D. (1986). *Prefabricated vertical drains*, Vol. I, *Engineering guidelines* Report No. FHWA/RD-86/168. Washington, DC: Federal Highway Administration.

Shang, J. Q. and Zhang, J. (1999). Vacuum consolidation of soda-ash tailings. *Ground Improvement* **3**, No. 1, 169–177.

Shang, J. Q., Tang, M. and Miao, Z. (1998). Vacuum preloading consolidation of reclaimed land: a case study. *Can. Geotech. J.* **35**, No. 5. 740–749.

Chu, J., Yan, S. W. & Yang, H. (2000). *Géotechnique* **50**, No. 6, 625–632

Soil improvement by the vacuum preloading method for an oil storage station

J. CHU*, S. W. YAN† AND H. YANG†

This paper presents a case study on the application of the vacuum preloading method for a soil improvement project in Tianjin, China. The site area was 50 000 m², and the soft clay treated was about 20 m thick, including a very soft muddy clay layer 4–5 m thick, formed from dredged slurry. The water contents of the clays were higher than or as high as the liquid limits. A vacuum load of 80 kPa was applied for 4 months. The ground settled nearly 1 m, and the average degree of consolidation was greater than 80%. The undrained shear strength of the soil increased two- to three-fold after treatment. The procedures used for soil improvement, the field instrumentation programme, and the field monitoring data are described and discussed.

KEYWORDS: case history; consolidation; field instrumentation; ground improvement; soil stabilization.

Cet exposé présente une étude de cas sur l'application de la méthode de précharge sous vide dans un projet d'amélioration de sol à Tianjin en Chine. La superficie du site était de 50 000 m² et l'argile tendre traitée avait environ 20 m d'épaisseur, avec une argile boueuse très tendre sur une épaisseur de 4 à 5 m, argile formée à partir de produits dragués. Les teneurs en eau de ces argiles étaient plus élevées ou aussi élevées que les limites liquides. Une charge de vide de 80 kPa a été appliquée pendant 4 mois. Le sol s'est tassé de près d'un mètre et le degré moyen de consolidation était supérieur à 80%. On a constaté que la résistance au cisaillement non drainé du sol était multipliée par deux ou trois après le traitement. Nous décrivons et analysons les procédés utilisés pour améliorer le sol, le programme d'instrumentation sur le terrain et les données du suivi sur le terrain.

INTRODUCTION

An oil storage station was constructed in 1996 near the coast of Tainjin, China, on a site that had recently been reclaimed using clay slurry dredged from the seabed. The soil had a high water content, was very soft, and was still undergoing consolidation. The seabed soil on which the dredged slurry was deposited was also soft. The site soil conditions needed to be improved before any construction work could be carried out.

Several soil improvement schemes were considered. Preloading using a fill surcharge was not feasible as it is difficult to build a fill embankment several metres high on soft clay. Vacuum preloading was adopted as it was considered the most suitable and cost-effective method for this project.

The vacuum preloading method has been used widely in Tianjin for land reclamation and soil improvement work since about 1980. The technique has been well developed over the years as a result of intensive research and field trials (Chen & Bao, 1983; Ye *et al.*, 1983; Yan & Chen, 1986; Choa, 1990; TPEI, 1995). Sand drains and more recently prefabricated vertical drains (PVDs) are normally used together with vacuum loading to distribute the vacuum and dissipate the pore water pressure. Experience has shown that for an appropriately designed scheme, the vacuum can be maintained at 80 kPa or above for a long period. Compared with the fill surcharge method for the equivalent load, the vacuum preloading method has been shown to be cheaper and faster. According to a comparison made by TPEI (1995), the cost of soil improvement using vacuum preloading is only two-thirds of that by fill surcharge, based on the local prices of electricity and materials.

The principles and mechanism of vacuum preloading have been discussed in the literature (e.g. Kjellman, 1952; Holtz, 1975; Chen & Bao, 1983; Qian *et al.*, 1992). This paper presents the results of the application of the vacuum preloading method for an oil storage station project. The site conditions, the soil improvement scheme, and the field instrumentation used are described. The field monitoring data are presented, and the achieved degree of consolidation and the effect of the soil improvement are evaluated. Several issues concerning the vacuum preloading method are also discussed.

SITE CONDITIONS

The site for the oil storage station is shown in Fig. 1. It covered a total area of approximately 50 000 m². For the purposes of soil improvement, the site was divided into two sections: Section I of 30 000 m² and Section II of 20 000 m², as shown in Fig. 1.

The soil profile included two layers that required improvement. The first layer was soft clay consolidated from dredged slurry. It was about 4–5 m thick. The second layer below this was seabed marine clay. It was about 10–16 m thick, and was underlain by a stiff sandy silt layer. The marine clay layer was further divided into three sub-layers in accordance with the USCS classification system: a low-plasticity silty clay (ML) layer 2–4 m thick, a low-to-medium plasticity clay (CL) layer 7–8 m thick, and a low-plasticity silty clay (ML) layer, which was relatively stiff. The basic engineering properties of the soils are shown in Figs 2(a) and 2(b) for Sections I and II respectively. It can be seen from Figs 2(a) and 2(b) that, except at the bottom of the marine clay layer, the water content of the soils

Manuscript received 1 February 2000; revised manuscript accepted 25 April 2000.
* School of Civil and Structural Engineering, Nanyang Technological University, Singapore.
† Geotechnical Research Institute, Tianjin University, China.

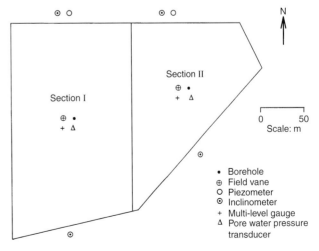

Fig. 1. Project site and plan view of instrumentation

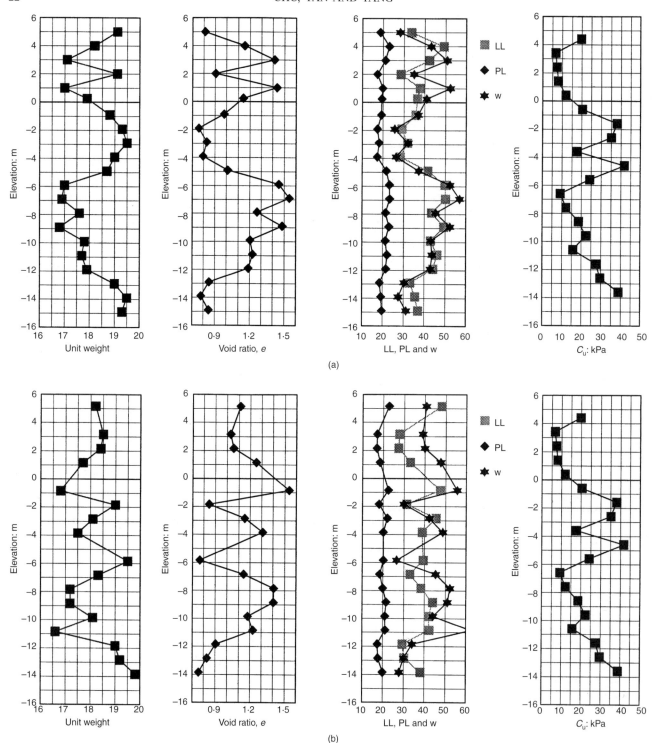

Fig. 2. Basic soil properties: (a) at Section I; (b) at Section II

was generally at or above the liquid limit, and the undrained shear strengths of the soils were generally low. The coefficient of consolidation in the horizontal direction was in the range $1 \cdot 1 – 4 \cdot 7 \times 10^{-3} \, cm^2/s$.

SOIL IMPROVEMENT WORK

Treatment of the soft soil was required to meet the following specifications:

(*a*) a minimum bearing capacity of 80 kPa
(*b*) an average degree of consolidation greater than 80% under a minimum surcharge of 100 kPa
(*c*) average settlement for a consecutive 10 day period less than 2 mm/day.

The soil improvement work was carried out as follows. The ground surface was very soft. To overcome this, a partially dried clayey fill 2 m thick with a 0·3 m sand blanket on top was first placed. PVDs were then installed on a square grid at a spacing of 1·0 m to a depth of 20 m. Corrugated flexible pipes 100 mm in diameter were laid horizontally in the sand blanket to link the PVDs to the main vacuum pressure line. The pipes were perforated and wrapped with a permeable fabric textile to act as a filter layer. Three layers of thin PVC membrane were laid to seal each section. Vacuum pressure was then applied using vacuum pumps.

The schematic arrangement of the vacuum preloading method used is shown in Fig. 3. The vacuum pressure was applied continuously for 4 months until the required degree of consolidation was achieved and the settlement rate for a consecutive

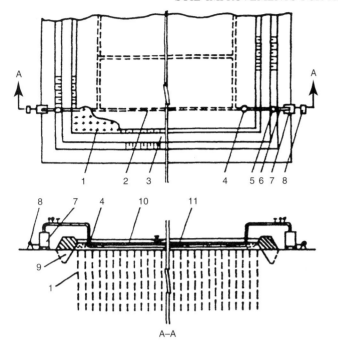

Fig. 3. Schematic arrangement of vacuum preloading method: 1, drains; 2, filter piping; 3, revetment; 4, water outlet; 5, valve; 6, vacuum gauge; 7, jet pump; 8, centrifugal gauge; 9, trench; 10, horizontal piping; 11, sealing membrane

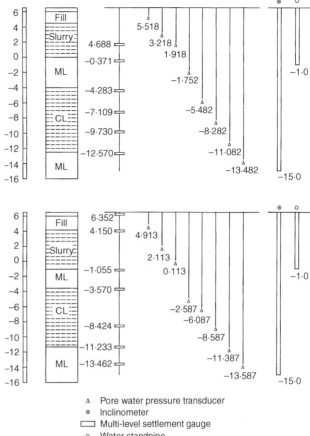

△ Pore water pressure transducer
⊚ Inclinometer
▭ Multi-level settlement gauge
○ Water standpipe

Fig. 4. Elevation view of instrumentation: (a) at Section I; (b) at Section II

10 day period was less than 2 mm/day. The average vacuum pressure was 80 kPa. The total surcharge applied was about 120 kPa including the 2 m of fill and 0·3 m sand blanket.

INSTRUMENTATION

After installation of the PVDs, instruments including pore water pressure gauges, surface settlement plates, multi-level settlement gauges, piezometers and inclinometers were installed in both sections to monitor the system performance. The instrumentation scheme is shown schematically in Fig. 1 (plan view) and Fig. 4 (elevation view). Undisturbed soil samples were taken, and field vane shear tests were conducted both before and after the soil improvement.

Monitoring data
Settlements and lateral displacement. It was observed that the ground settled 14·9 cm in Section I and 26·8 cm in Section II during installation of the PVDs. This was due to the further consolidation of the slurry clay under the influence of the surcharge effect of the 2 m of fill and 0·3 m sand blanket.

The settlement at different depths was monitored during vacuum preloading. Curves of the monitored settlement against time for the two sections are presented in Figs 5(a) and 5(b). The surface settlement at the end of preloading was 85·6 cm and 92·5 cm for Sections I and II respectively.

The vacuum load caused an inward lateral movement in the soil. The lateral displacements monitored at various depths and at different times are presented in Fig. 6. The data for the south boundary of Section I were not properly recorded as there were some malfunctions in the instruments.

Pore water pressures. Under vacuum load, the pore water pressure in the soil reduced. The reductions in the pore water pressure at different depths are plotted against time in Figs 7(a) and 7(b) for Sections I and II respectively. In general, the pore water pressures at various depths approached constant after 1·5–2·5 months of vacuum treatment. The initial and final pore water pressures together with hydrostatic pore water pressures are plotted in Figs 8(a) and 8(b). It can be seen that the final pore water pressures match the amount of suction applied throughout

the full depth, indicating that the vacuum preloading was very effective.

RESULTS
Degree of consolidation
The consolidation of soils under vacuum preloading was very effective in both Sections I and II. Nearly 1 m of settlement was induced, and this occurred throughout the thickness of the soft soils (see Figs 5(a) and 5(b)). The rate of settlement reduced towards the end of the vacuum preloading period, indicating that a substantial proportion of the ultimate consolidation should have taken place within the four-month loading period. The ultimate settlements for both Section I and Section II were estimated using the hyperbolic method, a procedure described by Sridharan & Rao (1981) and Tan (1993). In this method, the time–settlement relationship for the consolidating layer is expressed by a hyperbolic equation. The ultimate settlement is predicted by establishing the hyperbolic equation using field settlement monitoring data. Based on the monitored settlements and the ultimate settlements calculated, the average degree of consolidation for the whole layer and for each sub-layer was calculated, and the values are given in Tables 1 and 2 for Sections I and II respectively. It can be seen that the average degree of consolidation calculated was greater than 80% for all layers.

The suction line, which is the line given by the hydrostatic pore water pressure line minus the suction of 80 kPa, is also plotted in Figs 8(a) and 8(b). It can be seen that the final pore water pressures were close to the suction line. The degree of consolidation at a given elevation, $U(z)$, can be estimated as $(1 - \Delta u_f/80)\%$, in which Δu_f is the excess pore water pressure at the end of vacuum preloading and is measured as the difference between the final pore water pressure and suction line, as shown in Figs 8(a) and 8(b).

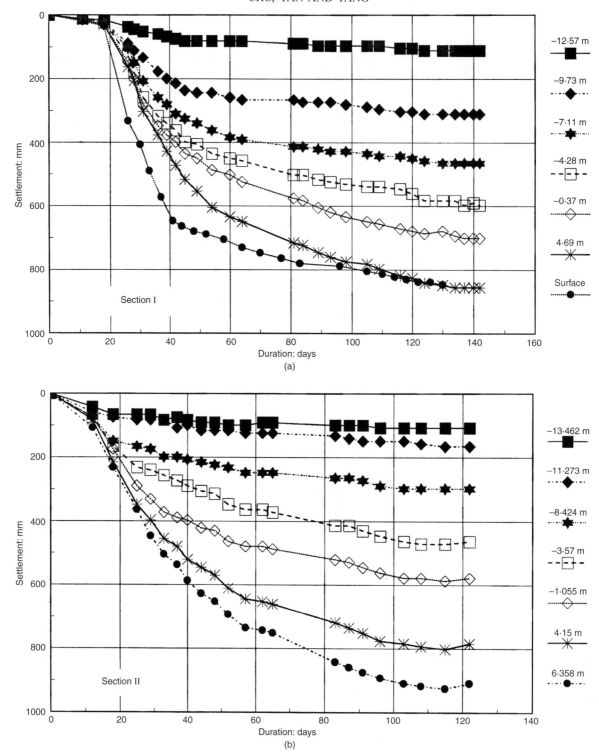

Fig. 5. Settlement plotted against duration of vacuum loading: (a) at Section I; (b) at Section II

Increase in undrained shear strength

Field vane shear tests were conducted before and after vacuum preloading in both Section I and Section II, and the results are presented in Figs 9(a) and 9(b). It can be seen that the undrained shear strength after vacuum preloading increased two- to three-fold for both sections. Using the field vane shear data, the ground bearing capacity was estimated to be greater than 80 kPa based on the Chinese Code for Soil Foundation of Port Engineering (JTJ 250-98, 1998). In this code, the bearing capacity, q_u, is calculated as $q_u = (\pi + 2)C_u$, in which C_u is the undrained shear strength of the soil.

DISCUSSIONS

The lateral displacement

As can be seen from Fig. 6, lateral displacement was greatest at the ground level and reduced sharply with depth. The lateral displacement at ground level is plotted against time in Fig. 10. It can be seen that the ground lateral displacements measured for Sections I and II on the north boundary are quite consistent. The displacement measured on the south boundary of Section I is higher than that on the north side. Unlike the surface settlement, the rate of the lateral displacement does not reduce rapidly with time. This indicates that the ground lateral displacement is affected not only by the primary consolidation, but

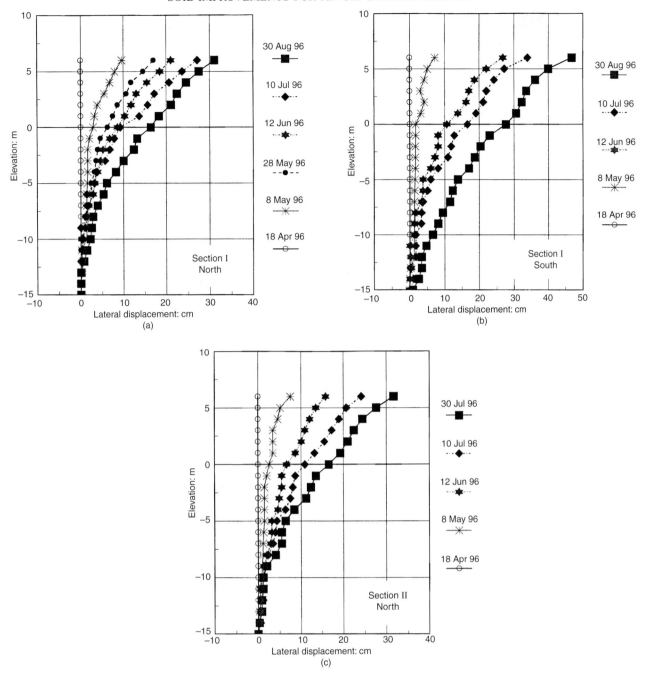

Fig. 6. Lateral displacement at different elevations: (a) at the north boundary of Section I; (b) at the south boundary of Section I; (c) at the north boundary of Section II

also by the shearing and secondary consolidation processes. At the end of the vacuum preloading period, the lateral displacements were up to 48·3 cm in Section I and 31·7 cm in Section II. Cracks had developed on the ground surface at about 10 m away from the edge of the preloaded area. As there were no adjacent buildings or facilities, the lateral displacement and cracks were tolerable. However, for sites where adjacent structures are present, lateral displacements can cause problems. The effect of lateral displacement and any risk of damage to adjacent structures is often an important factor that needs to be considered when using the vacuum preloading method for soil improvement work.

The depth of soil improvement

One of the common concerns in the use of vacuum preloading is that the vacuum is effective only up to depths of about 10 m. The field monitoring data obtained from this project have shown that vacuum preloading is still effective even for soils 20 m

below the ground surface. It needs to be pointed out that there are two processes involved in vacuum preloading. The first is pumping water out from the vertical drains, and the second is increasing the effective stress to consolidate the soil by reducing the pore water pressure in the soil. In the former, the effective depth cannot exceed 10 m. However, in the latter, the limit in the depth will depend on the well resistance. Another common concern is that the vacuum generated will reduce with depth. This is not the case, as Figs 8(a) and 8(b) show. The reduction in the pore water pressure was almost uniform throughout the whole 20 m depth of soft soils. With the use of PVDs, the well resistance is normally negligible (Chen, 1987). The vacuum pressure can be easily distributed to 20 m or even deeper.

CONCLUSIONS

The application of the vacuum preloading method to the improvement of a site with about 20 m thick of very soft and

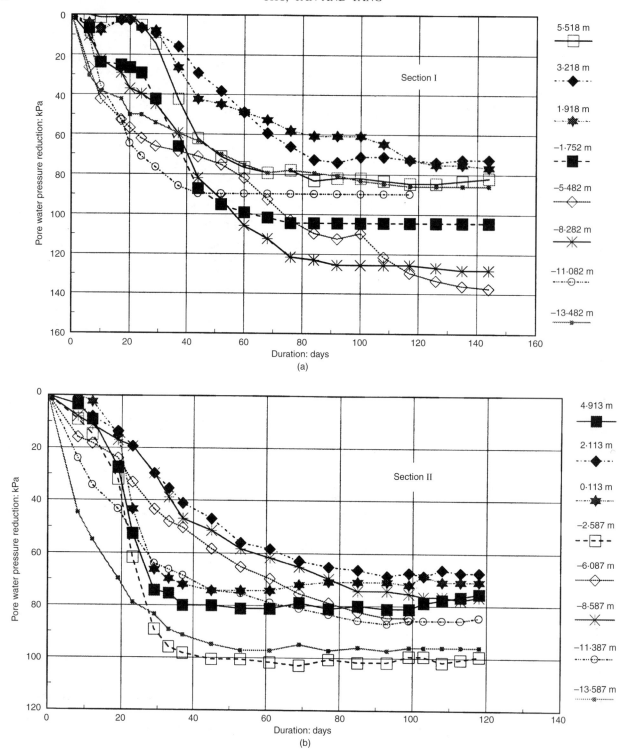

Fig. 7. Pore water pressure reduction plotted against duration of vacuum loading: (a) at Section I; (b) at Section II

soft soil was reported. Based on the study presented, the following conclusions can be made:

(a) The vacuum preloading method can be used effectively for the improvement of very soft clays, which could be difficult to treat using fill surcharge.

(b) The vacuum distribution system comprising PVDs at a square grid of 1·0 m together with horizontal 100 mm diameter corrugated flexible collector pipes was effective in distributing the vacuum pressure. A vacuum pressure of 80 kPa was maintained throughout the whole 20 m depth of soft clay.

(c) A vacuum pressure of 80 kPa plus 40 kPa of surcharge

achieved more than 80% average degree of consolidation within 4 months. The rate of settlement at all depths reduced to residual levels after about 100 days. The maximum ground settlement recorded was approaching 1 m.

(d) The application of the vacuum caused an inward lateral movement. The lateral displacement increased with time at almost the same rate until the end of the vacuum preloading period. The maximum lateral displacement measured at the site was 48·3 cm. Cracks on the ground surface were observed. The effect of lateral displacement needs to be considered when the vacuum preloading method is applied to soil improvement work.

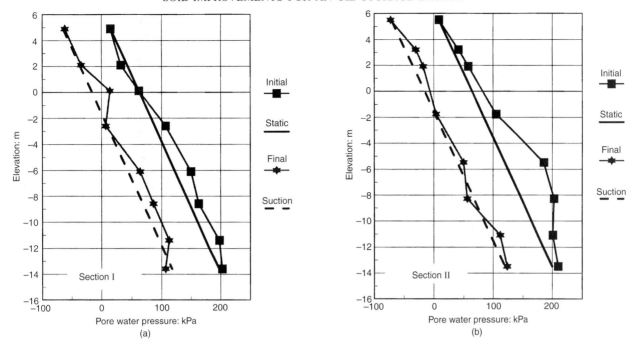

Fig. 8. Initial and final pore water pressure distributions: (a) at Section I; (b) at Section II

Table 1. Average degree of consolidation calculated for different layers at Section I

Layer	1	2	3	4	5	Whole
Elevation: m	+6·4 to +4·2	+4·2 to −1·1	−1·1 to −3·6	−3·6 to −11·3	−11·3 to −13·5	+6·4 to −15·0
Degree of consolidation: %	87	100	100	86·3	80	84

Table 2. Average degree of consolidation calculated for different layers at Section II

Layer	1	2	3	4	5	Whole
Elevation: m	+4·7 to −0·4	−0·4 to −4·3	−4·3 to −7·1	−7·1 to −9·7	−9·7 to −12·6	+6·4 to −14·0
Degree of consolidation: %	97	100	86·3	82·8	81·0	87

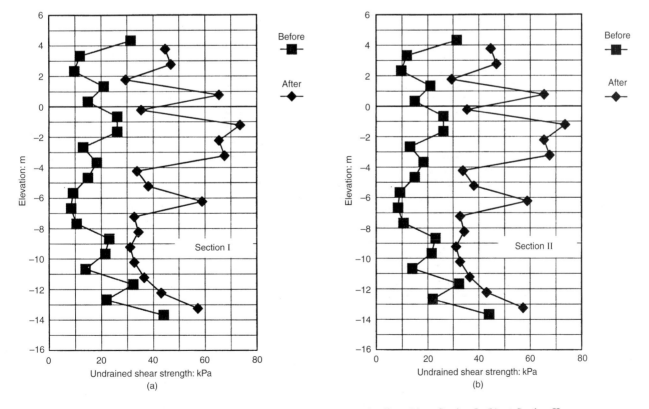

Fig. 9. Field vane shear test results measured before and after vacuum preloading: (a) at Section I; (b) at Section II

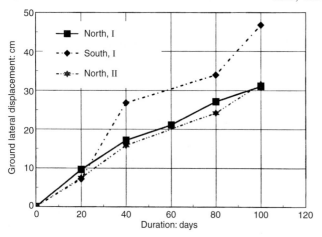

Fig. 10. Ground lateral displacement plotted against duration of vacuum loading

(e) Vacuum preloading achieved a two- to threefold increase in the undrained shear strength of the soft clays as measured by field vane shear tests. Vacuum preloading can be combined with a fill surcharge load to achieve further soil improvement if required.

REFERENCES

Chen, H. (1987). Analysis of the mechanism of vacuum preloading method. *The soft soil foundations of Tianjin* (eds Z. Hou *et al.*, 73–83). Tianjin Science and Tech. Publ. Hous.

Chen, H. and Bao, X. C. (1983). Analysis of soil consolidation stress under the action of negative pressure. *Proc. 8th European Conf. on Soil Mech. Found Eng., Helsinki* **2**, 591–596.

Choa, V. (1990). Soil improvement works at Tianjin East Pier project. *Proc. 10th Southeast Asian Geot. Conf., Taipei* **1**, 47–52.

Holtz, R. D. (1975). Preloading by vacuum: current prospects. *Transportation Research Record*, No. 548, 26–79.

JTJ 250-98 (1998). *Chinese Code for Soil Foundation of Port Engineering*. Ministry of Construction, China.

Kjellman, W. (1952). Consolidation of clayey soils by atmospheric pressure. *Proceedings of a conference on soil stabilization*, Massachusetts Institute of Technology, Boston, 258–263.

Qian, J. H., Zhao, W. B., Cheung, Y. K. and Lee, P. K. K. (1992). The theory and practice of vacuum preloading. *Comput. Geotech.* **13**, 103–118.

Sridharan, A. and Rao, S. (1981). Rectangular hyperbola fitting method for one dimensional consolidation. *Geotech. Testing J.* **4**, No. 4, 161–168.

Tan, S. A. (1993). Ultimate settlement by hyperbolic plot for clays with vertical drains. *J. Geotech. Engng Div., ASCE*, **119**, No. 5, 950–956.

TPEI (1995). *Vacuum preloading method to improve soft soils and case studies*. Tianjin Port Engineering Institute.

Yan, S. W. and Chen, H. (1986). Mechanism of vacuum preloading method and the numerical method. *Chin. J. Geotech. Engng* **8**, No. 2, 35–44.

Ye, B. Y. *et al.* (1983). Soft clay improvement by packed sand drain-vacuum-preloading method. *The soft soil foundations of Tianjin* (eds Z. Hou *et al.*), 126–131. Tianjin Science and Tech. Publ. House.

Almeida, M. S. S., Santa Maria, P. E. L. Martins, I. S. M., Spotti, A. P. & Coelho, L. B. M. (2000). *Géotechnique* **50**, No. 6, 633–643

Consolidation of a very soft clay with vertical drains

M. S. S. ALMEIDA*, P. E. L. SANTA MARIA*, I. S. M. MARTINS*, A. P. SPOTTI*, L. B. M. COELHO*

The paper analyses the performance of an embankment built on a very soft and extremely compressible organic clay where prefabricated vertical drains have been installed. Coefficients of horizontal consolidation, c_h, obtained from settlement analyses took into account the smear resulting from the installation of the drains and also the vertical drainage. These values of c_h are compared with the measurements obtained from special laboratory tests with drains installed in the sample, and from in-situ tests using the piezocone. The three sets of figures showed good agreement with each other, which suggests that the secondary consolidation was negligible compared with the primary consolidation in the case herein, in which vertical drains were used.

KEYWORDS: clays; consolidation; embankments; in-situ testing; laboratory tests; settlements.

Cette étude analyse la performance d'un talus construit sur une argile organique extrêmement compressible et très tendre où des drains verticaux préfabriqués ont été installés. Les coefficients de consolidation horizontale c_h donnés par les analyses de tassement ont pris en compte les salissures causées par l'installation des drains ainsi que le drainage vertical. Nous comparons ces valeurs c_h aux mesures données par des essais de laboratoire spéciaux avec les drains installés dans l'échantillon et aussi par les essais sur place utilisant le piézocône. Les trois lots de résultats sont en accord, ce qui laisse penser que la consolidation secondaire était insignifiante par rapport à la consolidation primaire, dans ce cas de drains verticaux.

INTRODUCTION

Although the vertical drain technique (e.g. Holtz *et al.*, 1991) is often adopted in order to accelerate settlements of soft soils, doubts still remain about the efficiency of the drains installed in very soft organic clay deposits and the influence of the drain installation in such clays, causing deformation in the areas surrounding the drains. On the other hand, doubts also arise with regard to the most suitable technique for estimating the coefficient of horizontal consolidation used when calculating the settlement rate.

The purpose of this paper is to clarify the above points by analysing the performance of an embankment built on a very soft and extremely compressible organic soil (Almeida, 1995, 1998) in which prefabricated vertical drains have been installed. The deposit is located on a site in the western zone of the city of Rio de Janeiro, where a training centre for the National Commercial Apprenticeship Service (SENAC) is being built on an almost square plot of land with an area of 92 000 m².

Coefficients of horizontal consolidation obtained from the analysis of field settlements are compared with the figures obtained from the piezocone test and laboratory using a special oedometer test. The data obtained are used to define suitable experimental techniques for defining the horizontal consolidation coefficient of soft organic soils. Conclusions are also made with regard to the performance of the prefabricated vertical drains in the gain in clay strength, by comparing the results of vane tests performed before and after the clay consolidation.

SITE DESCRIPTION

The site under study is about square with 300 m sides with a cut corner, as shown in plan view in Fig. 1. In this figure are also shown the square-based settlement plates with 0·50 m sides, used to measure settlements. These 20 steel settlement plates (SP01 to SP20) were installed before the embankment was built. The pattern of the settlement plates is not regular because they were not installed in the regions where piles were to be driven for buildings.

Ground conditions comprise very soft grey clay, of fluvial-marine origin and with shell fragments, with a top layer of peat.

Manuscript received 1 February 2000; revised manuscript accepted 7 September 2000.
*COPPE, Federal University of Rio de Janeiro.

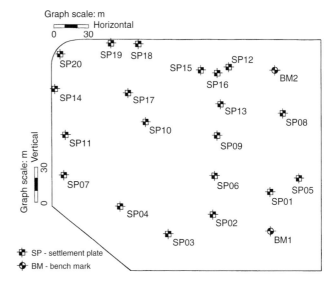

Fig. 1. **Overall view of the site, and location of the settlement plates**

Fig. 2 gives data of water content and Atterberg limits that permit the characterization of three distinct sublayers (Oliveira, 1997; Rodrigues, 1998): a first sublayer (depths 0–3 m), with water content of around 500%; a second sub-layer (depths 3–7 m), with water content of around 200%; and a third sublayer (depths 7–12 m), with water content of around 100%. The maximum thickness of the compressible soil is 12 m. A sandy soil of alluvial origin is found under the compressible soil, and is in turn underlain by residual soil. Results from chemical analyses of the soft clay have indicated that it has organic content between 40 g/l and 60 g/l, with evidence of an organic clay. The organic content of the underlying residual soil was 4·9 g/l. Other analyses indicate that the clay lies in a practically neutral (6·5 < pH < 8·1) and saline environment with a strong presence of calcium and magnesium.

A number of SPT boreholes have been performed in this site, which made it possible to obtain the contours of soft clay thickness shown in Fig. 3. Note that the clay thickness is 12 m over most of the site. SPT boreholes were also performed in October 1997 after embankment construction. Cross-section A–A indicated in Fig. 3 is presented in Fig. 4 for the situations before and after embankment construction. Fig. 4 shows quite

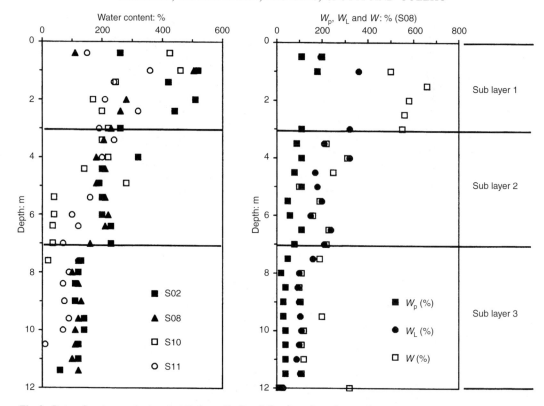

Fig. 2. Data of water content and Atterberg limits of the deposit under study

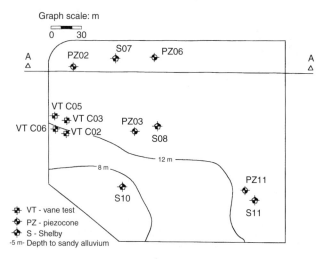

Fig. 3. Location of boreholes and thickness of soft clay strata

clearly that the compression of the peat layer was of great importance.

Figure 3 also shows the location of boreholes S07, S08, S10 and S11, where soil samples have been retrieved for consolidation and other laboratory tests, and boreholes PZ02, PZ03, PZ06 and PZ11, where piezocone tests have been performed. Consolidation tests were performed on samples 50 mm and 71 mm in diameter and 20 mm high, obtained from samples of 100 mm and 125 mm in diameter respectively. The quality assessment of the samples showed (Almeida, 1998) that 52% of the samples were of very good or excellent quality and 30% were good to fair, based on the guideline proposed by Lunne *et al.* (1997). Fig. 5 shows the variation in over-consolidation ratio, OCR, and compression ratio, $CR = C_c/(1 + e_0)$, with depth. It is found that the average value of OCR is 1·5 below 2·0 m depth, and that the average compression ratio of the clay deposit is 0·52. A detailed discussion of the properties of this deposit has been given by Almeida (1998).

Figure 6 presents the corrected point resistance, q_T, measured

at borehole PZ06, the location of which is shown in Fig. 3. The data for PZ06 are typical of a number of piezocones performed at the site. As expected, q_T increases with depth, which has also been noticed in vane tests (to be presented subsequently).

CONSTRUCTION METHOD
Figure 7 shows the typical cross-section of the completed embankment (Rodrigues, 1998). Immediately after instrumentation had been installed, a 0·60 m thick drainage blanket was placed; the top 0·30 m layer was crushed stone and the lower 0·30 m layer was sand. The construction of the drainage blanket began in December 1995, but was stopped for two months as a result of the heavy rains in Rio de Janeiro in February 1996. The drainage blanket was completed in June 1996, and about another three months were spent awaiting the invitation to bid and contracting of the company to build the embankment.

The drainage blanket provided the working platform for the equipment used to install the prefabricated vertical drains (MEBRA-DRAIN MD7407) in a 1·70 m triangular grid. Around 244 km of drains were installed on the site. After the vertical drains had been installed, a non-fabric geotextile was placed as a separation material between the drainage blanket and the compacted embankment. Vertical drains and geotextiles were not installed in 30% of the whole area, where structures were to be constructed on piles.

The compacted embankment was built in 300 mm thick layers with 95% degree of compaction. The control of field compaction of the embankment using around 700 tests has shown average values of the specific weights of the embankment and drainage blanket to be 17·43 kN/m³ and 18·50 kN/m³ respectively.

Figure 8 shows a number of settlement–time curves obtained from the readings of the settlement plates. The embankment height has also been recorded for each settlement measured, as shown in Fig. 8. The last construction stage should have been performed when the embankment achieved a degree of consolidation, U, of about 80%, but quite a long delay occurred owing to contractual issues between the owner and the subcontractor in charge of the building construction. The range of time corresponding to $U = 80\%$ is indicated in Fig. 8.

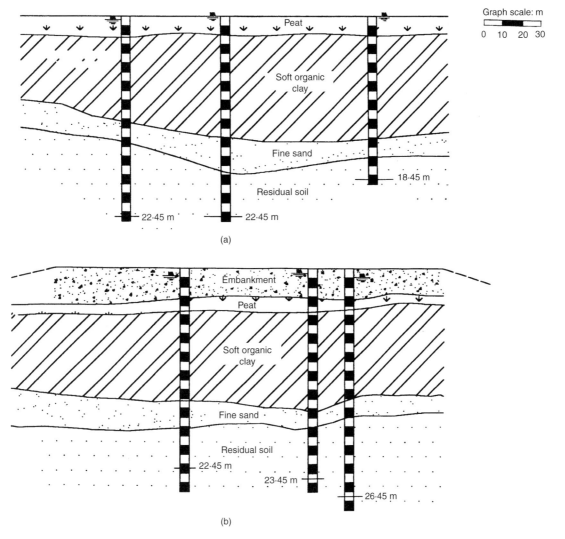

Fig. 4. Soil profiles (a) before and (b) after embankment construction

Electric and Casagrande piezometers were also installed in a number of boreholes to monitor pore-pressures. As shown in Fig. 9, the drainage blanket and the underlying soil acted as drainage interfaces, thus suggesting that vertical consolidation could be significant and should be added to the radial consolidation when analysing settlement data with the purpose of obtaining coefficients of horizontal consolidation.

SETTLEMENT ANALYSIS

The Asaoka (1978) method was used to obtain final settlements and coefficients of consolidation at each settlement plate. The settlement analysis assumed a negligible secondary consolidation given the primary consolidation, bearing in mind, on one hand, that the $\Delta\sigma'_v/\sigma'_{vo}$ ratio between the increase in vertical stress and the in-situ effective stress is generally greater than unity (Leonards & Altschaeffl, 1964) and, on the other hand, that the use of vertical drains accelerates the primary settlements (Leroueil, 1996).

The final settlements obtained by means of the Asaoka method for a period of up to 720 days are compared in Fig. 10 with the settlements calculated based on the results of laboratory tests. These calculations took into account the decrease in the applied embankment load caused by the settled and submerged part of the embankment. Good agreement is noted. The discrepancy found in some cases is justifiable by the use of the results of only four boreholes carried out on the 9 ha site. The average relative error of the group of plates was 6·3%. The good agreement between measured and calculated figures sug-

gests that secondary consolidation settlements were not important in the present case of vertical drains.

The horizontal coefficient of consolidation, c_h, for each settlement plate was calculated for the period of up to 720 days, using the Asaoka (1978) method modified by Magnan & Deroy (1980) in the case of combined radial and vertical drainage, using the equation

$$c_h = -\frac{\dfrac{\ln(\beta_1)}{\Delta t}}{\dfrac{8}{d_e^2 F_s(n)} + \dfrac{\pi^2}{4 H_d^2 r}} \tag{1}$$

where β_1 is the slope of the straight line in the Asaoka graph; $\Delta t = 40$ days is the time interval adopted; d_e is the diameter of the radial drainage cylinder; H_d is the distance from drainage to purely vertical drainage, assuming it is equal to half the layer thickness (see Fig. 3); and $r = c_h/c_v$, assumed here to be equal to 1·5 (Coutinho, 1976; Lacerda et al., 1977).

Considering the smear, but disregarding the hydraulic resistance of the drains, then (Hansbo et al., 1981):

$$F_s(n) = \ln\left[\frac{n}{s}\right] - 0.75 + \left[\frac{k_h}{k_s}\right]\ln(s) \tag{2}$$

where s is the ratio between the diameter of the area with smear around the drain and for prefabricated drains (Hansbo et al., 1981) $s = d_s/d_w = 1.5$ and $k_h/k_s = 2.0$ is used. It is easy to see (Rixner et al., 1986) that slight variations of these two

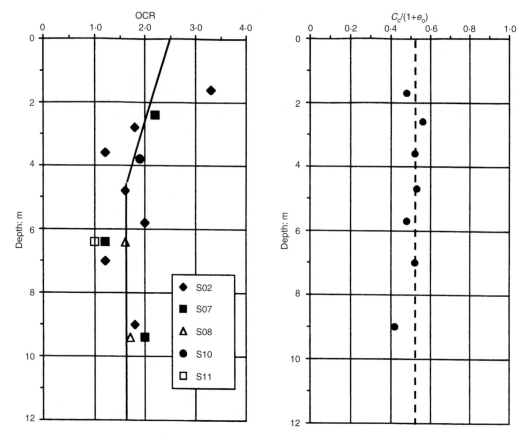

Fig. 5. History of stresses and compressibility parameter of the compressible soil

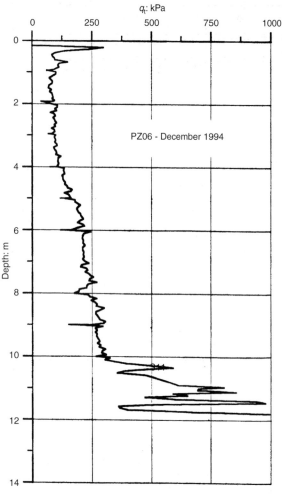

Fig. 6. Typical piezocone test

variables have almost no influence on the value of c_h, considering the variation band of this soil parameter. In equation 2, also

$$n = \frac{d_e}{d_w} \qquad (3)$$

d_w is the diameter equivalent to the drain $= 2(a + b)/\pi = 0.066$ m, where a and b are the dimensions of the prefabricated drain; $a = 0.1$ m and $b = 0.03$ m. $d_e = 1.05\ l$ for a triangular grid, and $l = 1.70$ m; then $d_e = 1.785$ m, and $n = 26.7$.

Substitution of the variables in equation (2) results in $F_s(n) = 2.94$. Table 1 shows the height of the embankment (drainage blanket and compacted embankment) for each settlement plate of the instrumentation, and the c_h values calculated for the maximum period of 720 days using equation (1). Table 1 also shows U values for the maximum period of 720 days, most of which are close to 90%. The calculations of degree of consolidation after 1440 days resulted in U values close to 100% for most plates.

COMPARISON OF c_h VALUES

The results of the coefficients of horizontal consolidation obtained from the settlement data will be compared here with those obtained from special oedometric tests and piezocone dissipation tests.

Determination of c_h by means of special consolidation tests

The c_h laboratory values were obtained (Coelho, 1997) by adopting a methodology (Seraphim, 1995) that uses conventional consolidation equipment and specimens in which vertical drains are inserted, thus causing pure radial drainage.

The samples tested were recovered from a depth of 9–9.5 m using the Osterberg stationary piston sampler, which has an inside diameter of 125 mm. Test samples were tested with 7, 19 and 37 drains, each with two drain diameters: 6.3 mm and 4.1 mm. The test samples were 100.8 mm in diameter

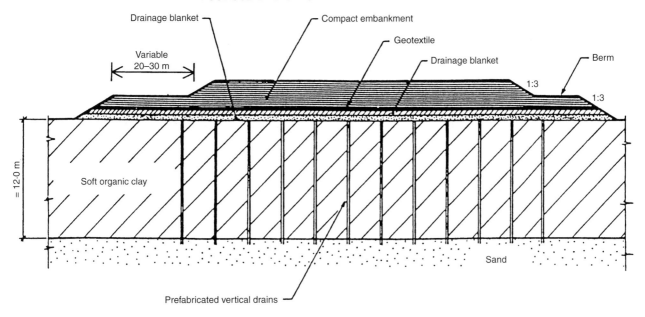

Fig. 7. Typical cross-section of the embankment

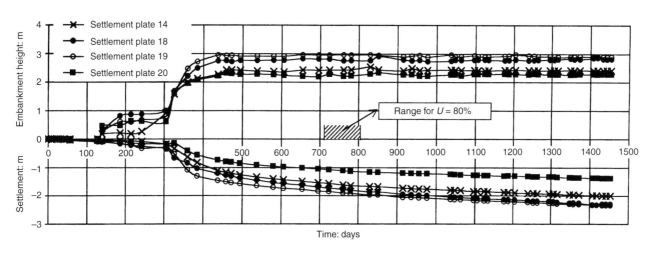

Fig. 8. Settlement–time curves measured in some settlement plates

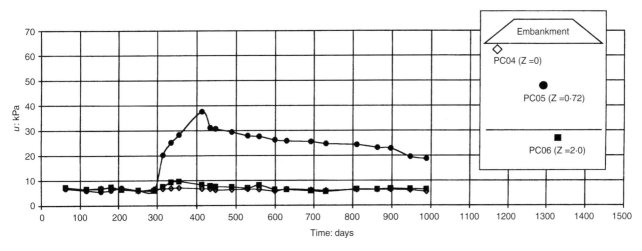

Fig. 9. Pore pressures measured in the bottom and top draining layers and in soft clay

(8000 mm² of base area) and 30 mm in height, resulting in tests with n ratio values of 11·46, 7·22, 5·73, 3·61, 3·82 and 2·41. Fig. 11 shows a diagram of these grids.

Each of the six grids was designed on Perspex plates, 3 mm thick, with the same diameter as the specimens. In the drain positions, the plates were perforated with 4·1 mm and 6·3 mm diameter holes. These plates were placed on top of the test sample, directly touching the soil, and acted as a gauge for the

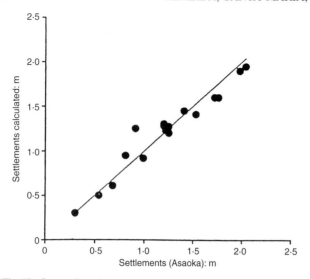

Fig. 10. Comparison between measured and calculated settlements

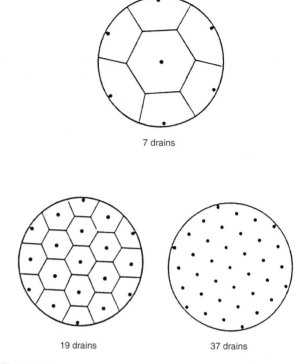

Fig. 11. Diagram of the grids with 7, 19 and 37 drains

correct placement of the drains. Vertical drainage in the soil was prevented at the top by the perforated Perspex plate and at the base by substituting the porous base stone of the conventional test with a blind Perspex plate, with a larger diameter than that of the ring. Therefore drainage occurred only in a radial direction to the drains. Fig. 12 shows a diagram of the oedometric consolidation cell adapted to the special consolidation tests.

The cavities in the drains, made by installing plastic straws, were filled with very fine beach sand, passing through the #100 screen, put in position with the help of a paper funnel. Loading was done in seven stages, and stresses of 3·13 kPa, 6·25 kPa, 12·5 kPa, 32 kPa, 80 kPa, 200 kPa and 500 kPa were applied.

The possible problems that may occur in this special oedometer test are (Martins, 1994) the relative rigidity between the drains and the clay, which interferes in the vertical compression; and the presence of spurious drainage—that is, undesirable drainage occurring at the soil–ring interface—which can adulterate the results obtained for c_h. Hence

$$c_h(\text{true}) = \frac{c_h(\text{calculated})}{I_r \times I_p} \qquad (4)$$

where I_r and I_p are factors of influence of the relative drain/

clay rigidity and spurious drainage respectively. Once the area ratio, R_a, is defined as being the ratio between the total area of the drain cross-sections and the area of the test sample cross-section, the variation of the factors of influence I_r and I_p with R_a can be represented for a certain drain diameter, as illustrated in Fig. 13. It is therefore found that the real c_h is closer to the c_h value calculated when the product $I_r \times I_p$ passes through a minimum. Table 2 gives the values of the area ratios for the drain grids used in the tests.

The consolidation coefficients were determined by using Barron's (1948) 'equal strain' solution applied to the curves of deformation against log time for each test loading stage, as shown in Fig. 14. The deformation for 100% of primary consolidation was obtained using Casagrande's recommendation. The point corresponding to 0% consolidation was based on

Table 1. Degree of consolidation and field consolidation coefficients for the period of up to 720 days

Plate	Height of drainage blanket: m	Embankment height: m	U consolidation degree: %	c_h: 10^{-8} m²/s
SP01	0·525	2·671	97·4	10·5
SP02	0·363	1·938	98·5	7·6
SP03	0·182	1·931	93·4	5·5
SP04	0·292	1·974	75·0	4·0
SP05	0·654	2·471	97·2	10·3
SP06	0·277	2·324	89·3	7·4
SP07	0·395	1·802	82·6	4·4
SP08	0·450	1·869	84·1	6·6
SP09	0·438	2·415	88·1	7·9
SP10	0·430	2·275	90·1	5·3
SP11	0·253	1·935	79·5	4·2
SP12	0·400	1·661	88·4	8·1
SP13	0·551	2·199	92·0	9·0
SP14	0·247	2·390	89·3	5·5
SP15	0·600	1·562	65·7	3·7
SP16	0·796	2·392	91·4	8·4
SP17	0·466	2·131	81·6	4·6
SP18	0·841	2·745	89·7	5·7
SP19	0·590	2·941	94·2	6·7
SP20	0·602	2·326	96·9	9·7
Average			88·2	6·8

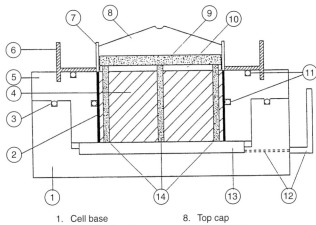

1. Cell base
2. Consolidation ring
3. Base O ring
4. Soil specimen
5. Cell top
6. Inundation base
7. Extension
8. Top cap
9. Porous stone
10. Perforated Perspex plate
11. O ring
12. Water outlet
13. Perspex plate
14. Vertical sand drains

Fig. 12. Diagram of the special oedometric consolidation cell

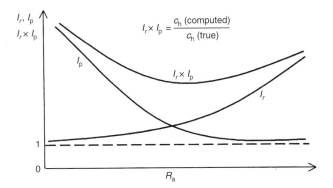

Fig. 13. Variation in factors of influence with area ratio

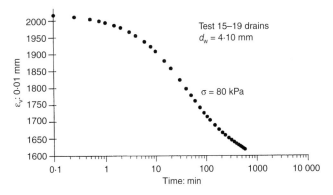

Fig. 14. Typical data of special oedometer test

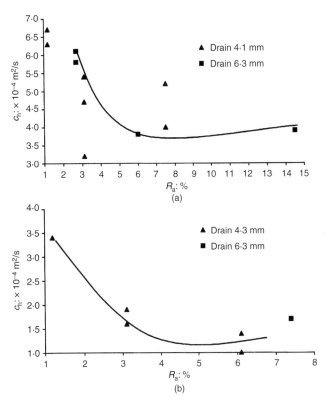

Fig. 15. Calculated c_h values versus R_a (a) soil understudy (b) soil reconstited in laboratory

finding that the $U \times T_h$ curve referring to Barron's solution is approximately a straight line for U values up to 10%, where T_h is the time factor as defined by Barron (1948).

Figure 15(a) shows the graphs of calculated c_h values against R_a for the clay under study, and Fig. 15(b) shows data for a soil reconstituted in laboratory, and therefore with a higher degree of similarity between the test samples. Although the results corresponding to the two tests basically show the same aspect, a wider variability is found in the tests performed in natural soil with 4·1 mm diameter drains. By analysing Fig. 15, it is found that the influence of the spurious drainage is minimum from approximately R_a equal to 7%, and the influence of the relative rigidity between sand drains and clay is not significant in the band of R_a values under study. It is, however, worth mentioning that, for the same clay, the relative drain/clay rigidity is also a function of the material that forms the drains, in addition to parameter R_a, and thus this last conclusion cannot be generalized for any kind of drain material.

The vertical effective stress, at the depth being tested, is $\sigma'_v = 56$ kPa for the whole embankment, which corresponds exactly to the average stress of one of the stages in the consolidation test. Table 3 shows c_h values for each test

performed when $\sigma'_v = 56$ kPa. The tests omitted in Table 3 correspond to the conventional consolidation tests, numbers 5, 8, 11, 14 and 17.

Discarding the results corresponding to the tests with seven drains ($R_a < 2·7$%) in Table 3, other tests are found in a narrow range with figures varying between 3·6 and $6·8 \times 10^{-8}$ m²/s, the average value being $c_h = 5·0 \times 10^{-8}$ m²/s.

Determination of c_h from dissipation tests using a piezocone

Figure 16 shows typical data for a piezocone dissipation test obtained at the site. The Houlsby & Teh (1988) method and the procedure described by Danziger et al. (1997) were used to obtain the c_h values from the piezocone dissipation tests, by

Table 2. Area ratios of drain grids used in tests

Grid	7 drains 4·1 mm dia.	7 drains 6·3 mm dia.	19 drains 4·1 mm dia.	19 drains 6·3 mm dia.	37 drains 4·1 mm dia.	37 drains 6·3 mm dia.
R_a	1·16%	2·73%	3·14%	7·42%	6·12%	14.45

Table 3. c_h values obtained from the special consolidation test (Coelho, 1997)

Test	Number of drains	Diameter of drains: mm	c_h: 10^{-8} m^2/s $\sigma_v = 57$ kPa
1	19	6·3	4·8
2	19	4·1	3·8
3	07	4·1	10·2*
4	07	6·3	8·8*
6	19	4·1	5·6
7	19	6·3	5·6
9	19	6·3	3·6
10	19	4·1	4·4
12	07	6·3	7·8*
13	07	4·1	8·4*
15	19	4·1	3·6
16	19	6·3	5·6
18	37	6·3	6·8
19	37	4·1	6·0
Average			5·0

*Values disregarded for averaging purposes.

Table 4. c_h values: dissipation tests using a piezocone

Test/depth	$c_{h\ piezocone}$: $\times 10^{-8}$ m^2/s	$c_{h(n.a.)}$: $\times 10^{-8}$ m^2/s
PZ02/6·21 m	36·0	3·6
PZ02/10·06 m	23·9	2·4
PZ03/2·10 m	50·9	5·1
PZ03/6·13 m	47·6	4·8
PZ03/9·15 m	30·0	3·0
PZ06/1·93 m	79·0	7·9
PZ06/6·00 m	68·9	6·9
PZ06/9·00 m	137·3	13·7
PZ11/1·99 m	208·7	20·9*
PZ11/10·42 m	134·2	13·4
Average	81·7	8·2

*Non-typical value, disregarded

Table 5. Summary of c_h values

Methods	Range of c_h: $\times 10^{-8}$ m^2/s	c_h average: $\times 10^{-8}$ m^2/s
Asaoka: whole layer	3·7–10·5	6·8
Piezocone	2·4–13·7	8·2
Special oedometer test (Coelho, 1997) $\sigma_v = 57$ kPa, z = 9·0 m	3·6–6·8	5·0

adopting a soil rigidity index $I_r = 93$ obtained by Oliveira (1997).

The stress-deformation condition imposed on the clay deposit during the penetration of the piezocone causes an over-consolidation process (Levadoux, 1980), and therefore the consolidation coefficient figures ($c_{h\ piezocone}$) calculated for these tests are for the over-consolidated condition of the consolidation curve. In order to obtain the c_h values of the normally consolidated band $c_{h(n.a.)}$, the formulation suggested by Baligh & Levadoux (1986) is followed, the ratio $C_s/C_c = 0·10$ being adopted for the calculations, where C_s is the swell index and C_c the compression index. The results of $c_{h\ piezocone}$ and $c_{h(n.a)}$ obtained for boreholes SP02, SP03, SP06 and SP11 are given in Table 4.

Discussion of c_h values

Table 5 and Fig. 17 compare c_h values obtained from the settlement analysis with the measurements in the special oedo-

meter consolidation tests and the dissipation tests using the piezocone. The results obtained show good agreement between all c_h values. These results suggest that the secondary consolidation is of little relevance in the case herein.

Note, however, that the consolidation coefficients determined by the Asaoka method diminish with an increase in the degree of consolidation (Almeida, 1996). The c_h values are also sensitive to the time interval, Δt, adopted. Notwithstanding this criticism, the Asaoka method is the simplest and easiest method to use.

A case history without vertical drains (Almeida *et al.*, 1993),

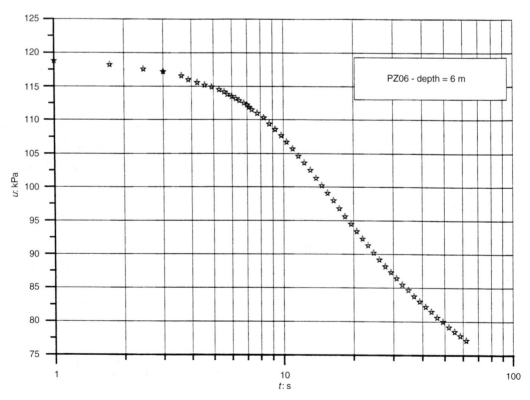

Fig. 16. Typical data of piezocone test

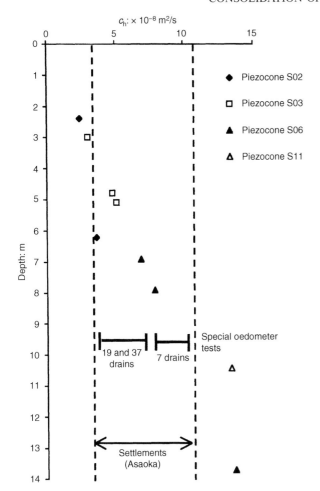

Fig. 17. Ranges in variation of c_h values obtained

was at the end of November 1997. Two borehole profiles were carried out in each test programme, and all were performed (Nascimento, 1998) using the same electric vane equipment and at the same location of the site. The results of both programmes are shown in Fig. 18. The location of these four boreholes is shown in Fig. 3. A number of other vane tests were performed along the border of the site in order to provide data for stability analyses.

A settlement of $\Delta h = 1.35$ m was assumed on the test location, based on the average figure for the nearby settlement plates PR11 and PR14. For $U = 90\%$ (see Table 1) in the test period, and as the embankment height on the site is equal to $h = 2.26$ m, there is a computed increase in effective stress equal to $\Delta\sigma'_v = (1.35 \times 8.6 + 18.4 \times 0.91) \times 0.90 = 25.6$ kPa due to the embankment.

For purposes of comparison, and bearing in mind the settlements that have occurred and the differences in depth in both test programmes, the depth data were normalized as a result of the thickness of the layer at the time of the test. Also, the S_u data were handled statistically to reduce them to the same normalized depths. Bessel's correction was used for sampling of less than 30 specimens in a population, and Student's distribution was used with $P = 95\%$. Table 6 gives the figures for an average undrained resistance, S_{um}, for each normalized depth for the two test programmes, which permits the calculation of the increased resistance, ΔS_u also given in Table 6.

In the design of embankments built in stages, it is common

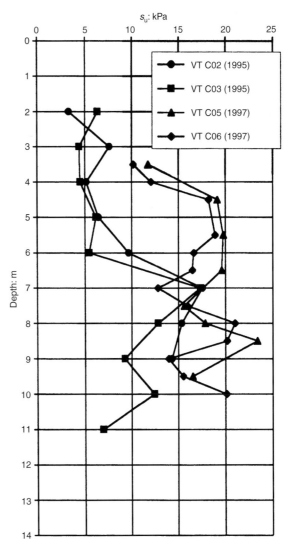

Fig. 18. Results of the two vane test campaigns, before and after embankment construction

where the secondary consolidation was negligible, bearing in mind the existing high stress ratio ($\Delta\sigma_v/\sigma_{vo}$), also showed good agreement between the three data groups: field (Asaoka), piezocone and laboratory. Another well-studied case, with and without vertical drains (Almeida *et al.*, 1992), where the secondary consolidation was not negligible, gave field consolidation coefficient values (Asaoka) higher than those measured using the laboratory and piezocone tests, particularly for the case of purely vertical consolidation, but with less difference in the field and laboratory radial consolidation.

The ranges of c_h values obtained, shown in Table 5, reflect the methodology and number of points used for these findings. For example, the c_h values measured in special consolidation tests show a small range, $3.6–6.8 \times 10^{-8}$ m²/s (see Table 3), consistent with the measurements taken in samples collected at the same depth (9·00–9·50 m). The settlement analysis, on the other hand, by the fact that it had been determined for 17 plates installed in terrain covering an area of 9 ha and in layers varying in thickness, showed a wider range, $3.7–10.5 \times 10^{-8}$ m²/s (see Table 1). The same may be said of the piezocone readings (see Table 4), with c_h between 2·4 and 13.7×10^{-8} m²/s (excluding the atypical c_h value = 20.9×10^{-8} m²/s), apparently because they were taken at different points on the site and at different depths. Because of their nature, piezocone measurements are nevertheless quite useful for design, particularly in their speed in obtaining the data.

ANALYSIS OF THE GAIN IN STRENGTH

Two vane test programmes were performed in the soft clay, specifically to analyse the gain in strength. The first was in early 1996, before the embankment was built, and the second

Table 6. Analysis of the gain in strength

Normalized depth: m	VT02 and VT03 $S_{u_{ave}}$: kPa	VT05 and VT06 $S_{u_{ave}}$: kPa	ΔS_u: kPa	$\Delta S_u / \Delta \sigma'_v$
0·35	4·84	11·34	6·5 ± 1·3	0·25
0·40	5·68	14·53	8·8 ± 2·1	0·34
0·45	6·49	18·74	12·2 ± 0·7	0·47
0·50	7·22	19·08	11·8 ± 2·3	0·46
0·55	10·77	19·09	8·3 ± 2·1	0·32
0·60	16·48	18·14	1·6 ± 2·2	0·06
0·65	15·86	17·30	1·4 ± 2·5	0·05
0·70	13·96	15·35	1·4 ± 2·4	0·05
0·75	12·62	16·89	4·3 ± 2·9	0·17

to calculate the increased strength, ΔS_u, as a result of the increase in effective stress, $\Delta \sigma'_v$. A conservative figure $\Delta S_u / \Delta \sigma'_v = 0.22$ may be adopted for the normalized increase in resistance. The $\Delta S_u / \Delta \sigma'_v$ values obtained for the present case are given in Table 6. More than half the top part of the clay layer, with high water content (see Table 6), was found to show $\Delta S_u / \Delta \sigma'_v$ above 0·22, as expected. The reason is not clear, however, for the extremely low $\Delta S_u / \Delta \sigma'_v$ value of the bottom half of the clay layer with small water content. However, what is clear is that the vertical drains have contributed to the improvement in the clay layer by increasing its shear strength, although this increase has varied along the clay layer.

CONCLUSIONS

A study was performed on the behaviour, in terms of settlements, of a very soft organic clay with prefabricated vertical drains. Settlements of up to 2·0 m for a 3·0 m high embankment were measured in this very compressible deposit. The settlements were observed over a four-year period, but the analysis covered a period of up to two years when the average degree of consolidation reached around 88%. The present analysis assumed the settlements for the secondary consolidation to be negligible when compared with the settlements through primary consolidation.

Good agreement was found between final settlements measured and calculated by means of oedometer consolidation tests. The coefficient of horizontal consolidation, c_h, obtained from settlement plates took into account the smear in the radial consolidation from the drain installation, as well as the vertical consolidation. Values of c_h were measured in special oedometer tests with drains installed in the sample, and in dissipation tests using the piezocone to compare with field values. The three sets of figures were found to be quite close, with little variation of the laboratory figures, followed by the field readings and lastly the piezocone data. The generally good agreement obtained suggests that secondary consolidation was in fact negligible compared with primary consolidation in the case herein using vertical drains.

The increase in clay shear strength, assessed by the vane tests performed before and after the construction of the embankment, showed a higher gain in S_u for the top half of the clay layer with high water content and, oddly enough, a negligible gain for the bottom half of the clay layer with lower water content.

ACKNOWLEDGEMENTS

The authors wish to thank the technical team of COPPE's Geotechnics Laboratory for its support in the work undertaken between 1995 and 1999 with respect to the site investigation, instrumentation installation and field monitoring. Francisco Bitencourt and Daniel Soares of SENAC are acknowledged for their ongoing interest and support in the studies described herein.

REFERENCES

Almeida, M. S. S. (1995). Engineering properties of regional soils: state-of-the-art-report soft clays. *Proc. 10th Panamerican Conf. Soil Mechanics and Foundation Engineering, Guadalajara* **4**, 161–176.

Almeida, M. S. S. (1996). *Embankments on soft clays: from design to performance evaluation* Federal University of Rio de Janeiro (in Portuguese).

Almeida, M. S. S. (1998). Site characterization of a lacustrine very soft Rio de Janeiro organic clay. *Proc. First Int. Conf. Site Characterization, ISC'98, Atlanta, Georgia*, 961–966.

Almeida, M. S. S., Danziger, F. A. B., Almeida, M. C. F., Carvalho, S. R. L. & Martins, I. S. M. (1993). Performance of an embankment built on a soft disturbed clay. *Proc. 3rd Int. Conf. Case Histories in Geotechnical Engineering, Missouri*, 351–356.

Almeida, M. S. S. & Ferreira, C. A. M., (1992). Field, 'in situ' and laboratory consolidation parameters of a very soft clay. *Predictive Soil Mechanics, Wroth Memorial Symposium, Oxford*, 73–93.

Asaoka, A. (1978). Observational procedure of settlement prediction. *Soils Found.* **18**, No. 4, 87–101.

Baligh, M. M. & Levadoux, J. N. (1986). Consolidation after undrained piezocone penetration II: Interpretation. *J. Geotech. Engng Div., ASCE* **112**, 727–745.

Barron, R. A. (1948). Consolidation of fine-grained soils by drains wells. *J. Geotech. Engng Div., ASCE* **113**, 718–742.

Coelho, L. B. M. (1997). *Considerations about an alternative test for the determination of the horizontal coefficient of consolidation on the soils*. MSc thesis, COPPE, Federal University of Rio de Janeiro (in Portuguese).

Coutinho, R. Q. (1976). *Radial consolidation of Rio de Janeiro very soft clay*. MSc thesis, COPPE, Federal University of Rio de Janeiro (in Portuguese).

Danziger, F. A. B., Almeida, M. S. S. & Sills, G. C., (1997). Piezocone research at COPPE/Federal University of Rio de Janeiro. *Proceedings of the international symposium on recent developments in soil and pavement mechanics, Rio de Janeiro, Balkema* **1**, 229–236.

Hansbo, S., Jamiolkowski, M. & Lok, L. (1981). Consolidation by vertical drains. *Géotechnique* **1**, 45–66.

Holtz, R. D., Jamiolkowski, M. B. R., Lancellotta, R. & Pedroni, S. (1991). *Prefabricated vertical drains, design and performance*. London: Heinemann-CIRIA.

Houlsby, G. T. & Teh, C. I. (1988). Analysis of the piezocone in clay. *Proc. 1st Int. Symp. on Penetration Testing* **2**, 777–783.

Lacerda, W. A., Costa Filho, L. M., Coutinho, R. Q. & Duarte, E. R. (1977). Consolidation characteristics of Rio de Janeiro soft clay. *Proceedings of the conference on geotechnical aspects of soft clays, Bangkok*, 231–244.

Leonards, G. A. & Altschaeffl, H. G. (1964). Compressibility of clay. *J. Soil Mech. Found. Div., ASCE* **90**, No. SM5, 133–135.

Leroueil, S. (1996). Compressibility of clays: fundamental and practical aspects. *J. Geotech. Engng Div., ASCE* **122**, No. 7, 534–543.

Levadoux, J. N. (1980). *Pore pressure generated during cone penetration*. PhD thesis, MIT, Cambridge, MA.

Lunne, T., Berre, T. & Strandvyk, S. (1997). Sample disturbance effects in soil Norwegian clays. *Proceedings of the international symposium on recent developments in soil and pavement mechanics, Rio de Janeiro, Balkema* **1**, 81–102.

Martins, I. S. M. (1994). General report session 2: Site investigation, laboratory and instrumentation. *Proc. 10th Brazilian Cong. Soil Mechanics and Foundation Engineering*.

Nascimento, I. N. de S. (1998). *Development and use of an electric vane borer equipment*. MSc thesis, COPPE, Federal University of

Rio de Janeiro (in Portuguese).

Oliveira, O. C. (1997). *Measurement of the undrained strength of clays by means of in situ tests*. MSc thesis, COPPE, Federal University of Rio de Janeiro (in Portuguese).

Rodrigues, A. S. (1998). *Performance appraisal of prefabricated vertical drains in compressible organic soil*. MSc thesis, COPPE, Federal University of Rio de Janeiro (in Portuguese).

Seraphim, L. A. (1995). Coefficient of consolidation in radial drainage. *Proceedings of the international symposium on compression and consolidation of clayey soils*, Vol. 1, 165–170.

INFORMAL DISCUSSION

Session 1

CHAIRMAN DR ANGUS SKINNER

Dr Angus Skinner, *Chairman*

The main purpose of this meeting is to try and stimulate discussion between theoreticians and practising engineers. Both of our presenters Julie Shang and Marcio Almeida are in fact a combination of both: they are both practical engineers as well as theoreticians. So I open the field now to those who wish to ask questions about either technique or interpretation

Mike Rogers, *IGES*

A question for Dr Shang. When comparing consolidation after loading with consolidation after the vacuum extraction, did it take longer to get to the same consolidation or did you get less consolidation?

Dr Julie Shang

Basically, the time of consolidation is really a function of soil hydraulic conductivity and, as recorded in the field tests, the consolidation time is about the same. The vacuum loading saves time because the vacuum pressure is applied quickly. Figure 6 shows that the surcharge must be applied in increments to avoid shear-capacity failure or bearing-capacity failure, but vacuum-loading pressure can be applied in one go.

Peter Felton

What are the energy costs of applying a vacuum for such a long period? How does it compare with the energy cost of constructing an embankment?

Dr Julie Shang

According to the project engineer, the typical vacuum pumps used are 7·5 kW and one pump will control an area of about 40 m by 60 m. If you calculate the energy cost - the depth of a prefabricated vertical drain insertion and the treatment area covered by each vacuum pump – it was quite cost-effective, and is affordable even in China.

David Nash, *University of Bristol*

What is the justification for using the constant values of coefficient of consolidation given that in soft soils the stiffness is very non-linear and the permeability is likely to decrease as the clay consolidates? Would the author please expand on his conclusion that creep was not occurring during primary consolidation.

Dr Marcio Almeida

Perhaps it is not justified not to take into account the variation of the coefficient of consolidation with depth or even with effective stress. However, the variation is usually low, an order of magnitude of 10. It is not uncommon to be uncertain about which coefficient of consolidation to adopt. I appreciate that it does vary, but I have doubts about which value to use, so it is completely justified to use a constant value for practical purposes. It is therefore possible to simply use the Terzaghi theory or Barron theory or whatever.

If you want to take all factors into account, you have to include combined drainage and smear - I suppose those are more important. If you do take into account combined drainage and smear, it is more important to use variation of the coefficient of consolidation with time.

Regarding creep, there is no reason why secondary consolidation or creep does not start from the early stages of consolidation – of course it does. It is more important if you do not accelerate consolidation. If consolidation is accelerated using vertical drains, what you actually get in the field is mainly primary consolidation, which is what occurred here. That is the reason for not taking account of the later stage, in the later stages of consolidation the presence of creep will be more important.

However, there are some case histories where vertical drainage occurred and no acceleration, and for those instances the agreement was no good at all. The reason was that whatever method was used for calculating primary consolidation (Asaoka, Terzaghi or Barron), in the field you have secondary consolidation to take into account, which means you do not get good agreement between field and lab data, or even between field and piezocone data.

Dr Angus Skinner, *Chairman*

I am going to disagree with you. I would like to take up this point that David Nash just asked and talk a little about variable compressibility and variable permeability, and stress dependent parameters.

I tried to choose a soil in between the soils used by Dr Shang and Dr Almeida with a voids ratio of 3. This is a highly non-linear soft soil and has a very non-linear compressibility characteristic, but you can actually fit a compressibility characteristic to that.

The consolidation data is really all the data people have on which to prepare a design, and very often time is of the essence in a practical problem like this. So for each of the test data you can fit the small strain theory and obtain the values of coefficient of consolidation.

For the data for the stress range that the authors have presented, the coefficient of consolidation from the lab data is essentially constant. You can use those data and develop from the compressibility characteristic the coefficients of consolidation and the coefficient of permeability (which varies with stress). You can then put both of those into the small-strain theory. Both of the authors have used the Barron small-strain theory to analyse their data. If you take both the stress dependent-permeability and the stress-dependent compressibility into account with the large-strain analysis you get a factor in terms of delay of about 1·7 in this soil.

It is not really the compressibility characteristics or the settlement characteristics that are important here. I have taken three stages of the pore-pressure development at different times. The first one is at 80 days after consolidation and there is not much difference between all three interpretations. After 320 days into the consolidation process there is a distinct difference. The analysis taking into account the variation of the permeability with effective stress is very significant. The pore pressure is pinched off by the material changing its permeability later on, and towards the end of consolidation the fact that the non-linear permeability or stress-dependent permeability is coming into affect is very important.

Professor Almeida used Asaoka's construction to interpret his data and this is where you are cutting the progress of the consolidation in terms of settlement. The small-strain theory directly from the data gives a coefficient of consolidation of 0·523. However, if you take the non-linear characteristics into account you get an apparent coefficient of consolidation of 0·3.

Now if you go to the pore-pressure data you get a completely different picture. The interpretation using small-strain theory gives a coefficient of consolidation from the pore-pressure data of 0·08, a difference of a factor of 7.

I think that the interpretation of field data needs some further input in terms of characteristics of the soil itself and we must begin to realize that we have got the ability to analyse our results, measurements and test results using a much better fit for our soil parameters. If we take those into account then I think we can in fact make some progress, both the practitioner and the person who is doing the research and understanding the behaviour of the soil.

Dr Marcio Almeida

I have also found that depending on the time you interpret your data, you may see a decrease of coefficient of consolidation with time. Even when using Asaoka's method, it is clearly seen that field data C_h or C_v decrease with time.

Dr Angus Skinner, *Chairman*

Field measurements with interpreted coefficients of consolidation usually give larger values. Apparent coefficients of consolidation from the laboratory data, which are taken from samples that are disturbed, give longer times than you experience in the field. However, with these very very soft soils, where vacuum techniques are being used and where very soft materials are being compressed, I think you will get the reverse. It is about time some of our researchers did some work in terms of getting into the field and really measuring what is happening in the consolidation process in these very very soft and highly compressible soils.

Richard Jardine, *Imperial College, London*

With vacuum consolidation, it of course is not the same sort of K_o process. It is effectively superimposing some isotropic stress changes on top of the stress field that exists in the ground. That is one reason why shrinkage, rather than outward displacement of the inclinometers, is observed.

One question in connection with that: have you had a look at how that will modify the soil properties with regard to such factors as stability or bearing capacity? We are used to the idea of having patterns of anisotropic behaviour and there are a variety of theoretical or empirical procedures for looking at that. However, this isotropic type of consolidation will modify that pattern and I wonder whether that has been looked at.

A second question in connection with the case history. You formed clay walls around the side of the runway to give a seal. I wonder if that will give you any problems in the long term by actually trapping water inside, in the more permeable layers at the surface, or whether you actually try to do something to disrupt the hydraulic barrier that you have below the runway? Or would that be an engineering problem?

And a third point, to reinforce Dr Skinner's intervention. The third *Géotechnique* symposium in print, on vertical drains, which was in 1981, included a number of case histories that reinforced the importance of non-linear parameters. I was a second author on one of those case histories, which included some reinterpretation of the Barron theory to allow you to look at piezometric decay, and from that, by a little bit of manipulation, you could see how the rates of consolidation were changing, or how the consolidation parameters were changing.

We considered a foundation that went from a lightly over-consolidated state to consolidation with a multi-stage structure and relatively higher effective stresses. We found that C_v values, just on a simple interpretation, tumbled by a factor of about 30 between the start of the job and the end. We found that was fairly predictable by doing in-situ permeability testing, where we modified the effective-stress regime during in-situ permeability testing and combined that with non-linear compressibility data from oedometer tests. With a relatively structureless soil we did not find the point that you remarked upon, where the laboratory gave poor indications of field behaviour – in fact they tied up quite nicely.

Dr Julie Shang

Vacuum pressure is basically applied isotropically in soils so that the general change is in the soil pore pressure. Basically, the analyses were done looking at the effective stress change after the negative pore pressure is developed, so the immediate observation is the settlement. Looking at the three dimensional conditions for the surcharge preloading case, the soil is also squeezed out. In the case of vacuum preloading, there is an inward lateral displacement in the soils. Chinese engineers have been handling this problem from a practical viewpoint. There is a design code, which requires that any vacuum treatment area must be at least 30 m from any adjacent structure. It is found that soil cracks develop due to the nature of vacuum preloading. In terms of calculation, they use a mass balance method, basically considering the mass conservation assuming the soil is 100% saturated. One of the reasons I got interested in this project was to look at numerical simulation of this technique. So far we have finished the one-dimensional case but are still working on the three-dimensional case. I hope to have something more to say soon.

Regarding the cut-off wall, I do not believe there is a problem. For the case reported, a vertical prefabricated drain is installed to serve as the vertical drainage layer. At the surface, there is a sand drainage layer. I believe these provided sufficient drainage channels. For treatment, the runway was divided into sections with slurry cut-off walls. I still need to check, but I do not see there is a problem. The project was constructed more than 10 years ago and no problems have been reported.

Considering the point about in-situ permeability testing. You are quite right – it can be seen from our paper that in terms of calculation the in-situ monitored settlement is close to two times higher than predicted from the one-dimensional, basic Terzaghi's theory. It is obvious that the one-dimensional constant coefficient of consolidation cannot predict this case due to the large strain involved in the case. For this case study, we are really reporting on the design and the estimate conducted by Chinese engineers in the 1980s and we want to give a true report on the state of art at that time. There is some other work currently on-going.

Dr Marcio Almeida

My question to Dr Shang relates to PhD research at Laval University using vacuum preloading. It has been found that when the water level is significantly below the ground level, when the vacuum is applied, much higher at ground level, there is a rise in the water table which decreases the effective stress and so entirely changes your effective-stress situation. It may be a surprise to see this.

We are hoping to go much beyond the yield stress when starting from a lower stress level. Do you appreciate the situation that when you have vacuum preloading you have got to have your drainage blankets where you apply your vacuum, close to the water level? You may have to dig a hole and put your drainage layer at the bottom there. If you do not do that you will raise the water table and then you change entirely the effective stress pattern. That was my conclusion of vacuum preloading from this particular study in Quebec.

Dr Julie Shang

For the water level, the groundwater table, that has been quite true for all the cases that I know of where vacuum dewatering and consolidation was conducted on very soft, saturated soils.

The first case history we worked on was the Tianjing Island reclamation project for the Xingang Port, which was published by Shang, Tang and Miao in the *Canadian Geotechnical Journal* in 1998 [Vol. 35, No. 5, 740–749]. That was a fully saturated material. For the case study on the Yaoqiang Airport Runway presented at this symposium, the groundwater was about 2·5 m below the ground surface, which caused major problems. At first, the vacuum pressure could not be obtained. A leakage was found in the unsaturated zone by the sand layer – this is why cut-off walls were installed. The data at the surface sand layer show there was no significant increase in shear strength. This can be expected because the soil below the base of the sand layer during the consolidation process becomes saturated, which would affect the shear strength and the effective stress status would change. The main focus here is really for a soft fine grained soil. My presentation showed a major increase in shear strength which was observed in the clay layer at a depth of 3 m to 11 m, a soft under-consolidated clay layer. That is really the focus of this technology, vacuum preloading and consolidation is mainly for fine-grained material with low hydraulic conductivity and, I would say, is not applicable for sands, or silty sands, or similar kinds of soil.

Tang, M., Shang, J. Q., Chu, J., *et al.*, Almeida, M. S. S., *et al.* (2002). *Géotechnique* **52**, No. 2, 148–154

WRITTEN DISCUSSION

Vacuum preloading consolidation of Yaoqiang Airport runway

M. TANG and J. Q. SHANG

Soil improvement by the vacuum preloading method for an oil storage station

J. CHU, S. W. YAN and H. YANG

Consolidation of a very soft clay with vertical drains

M. S. S. ALMEIDA, P. E. L. SANTA MARIA, A. P. SPOTTI, L. B. M. COELHO and I. S. M. MARTINS

A. E. Skinner, *Skinner & Associates*

The authors of these three papers treat consolidation, due to the flow of water to the vertical drains, assuming the soil has both constant compressibility m_v and constant permeability k, which generate a constant value for the coefficient of radial or horizontal consolidation, c_h. Would not the authors agree that, because the soils involved were soft and very compressible, they could have had significant changes in stiffness and permeability, due to the effective stress changes imposed during consolidation? If this were the case, analyses that took into account these stress-dependent parameters should yield additional insight into the observed responses of the soils involved.

The discusser does not have sufficient detail of the soils encountered by the authors to reanalyse the results of the field studies that were presented. However, as an illustration, analyses for one very soft soil are given, showing the different soil responses generated by the different assumptions.

The compression and swelling data, shown in Table 1 and Fig. 1, are from each stage of a one-dimensional consolidation test on a sample of soft 'intertidal' marine sediment.

If it is assumed that, because the stress increments in this test are small, m_v and k are reasonably constant during each stage, then the time–settlement responses can be compared with theoretical predictions. Fig. 2 shows the comparisons, fitted at 50% consolidation, and the values for c_v that this fit generates. The loading stage c_v values are plotted in Fig. 3. For the stress range considered in the subsequent analyses an average 'lab' value of $c_v = 0.523$ m²/yr is obtained.

Table 1. Data for test sample

Stage	Stress in kN/m²		Height: mm		Voids ratio, e	m_v for each stage of loading m²/kN	c_v: m²/year	k: 10^{-9} m/s	k: 10^{-3} m/yr
	From	To	At start	At end					
1 (bedding)	?	7·545	10·94	10·13	3·134				
2	7·545	15·09	10·13	9·60	2·919	0·006 91	0·550	1·205	38·001
3	15·09	30·18	9·60	9·03	2·687	0·003 93	0·499	0·623	19·647
4	30·18	60·36	9·03	8·21	2·351	0·003 02	0·550	0·527	16·620
5	60·36	30·18	8·21	8·30	2·389				
6	30·18	15·09	8·30	8·39	2·426				
7	15·09	30·18	8·39	8·36	2·412				
8	30·18	60·36	8·36	8·15	2·326				
9	60·36	120·72	8·15	7·41	2·025	0·0015	0·511	0·243	7·663
10	120·72	241·44	7·41	6·47	1·641	0·001 06	0·493	0·165	5·203
11	241·44	120·72	6·47	6·57	1·681				
12	120·72	60·36	6·57	6·64	1·710				
13	60·36	120·72	6·64	6·55	1·673				
14	120·72	241·44	6·55	6·36	1·596				
15	241·44	482·88	6·36	5·77	1·357	0·000 38	0·359	0·0434	1·369
16	482·88	965·76	5·77	5·14	1·100	0·000 23	0·498	0·0357	1·126
17	965·76	482·88	5·14	5·28	1·156				
18	482·88	241·44	5·28	5·46	1·230				
19	241·44	120·72	5·46	5·64	1·304				
20	120·72	60·36	5·64	5·83	1·380				
21	60·36	30·18	5·83	6·02	1·457				
22	30·18	15·09	6·02	6·18	1·526				
23	15·09	3·018	6·18	6·48	1·645				

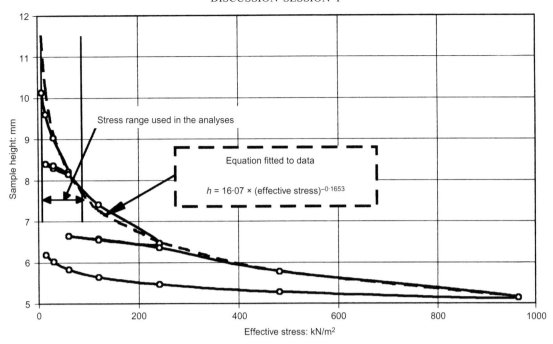

Fig. 1. Sample: test data for each stage

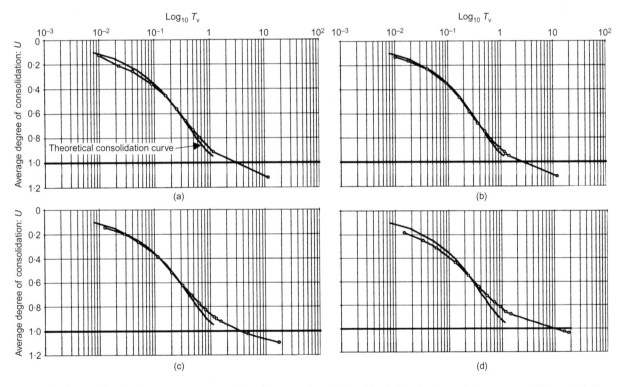

Fig. 2. Settlement data for four separate stages of loading; c_v values obtained by fitting the theoretical curves at 50% consolidation. (a) Stage 2: data fitted to theoretical solution for $c_v = 0.55$ m2/yr. (b) Stage 3: theoretical solution fitted for $c_v = 0.5$ m2/yr. (c) Stage 4: data fitted to theoretical solution for $c_v = 0.55$ m2/yr. (d) Stage 9: data fitted to theoretical solution for $c_v = 0.51$ m2/yr

The measured m_v and derived c_v values for each stage were then used to derive stress-dependent permeability k values. It is possible, with some consolidation apparatus, to obtain the stress-dependent k values directly.

The data for sample height and permeability against effective stress were fitted with trend lines in the form of power functions, as shown in Fig. 4. The equations for the trend lines for the sample height of the normally consolidated branch, and for the permeability, were used to generate settlement and pore pressure values from numerical iterative calculations. The calcu-

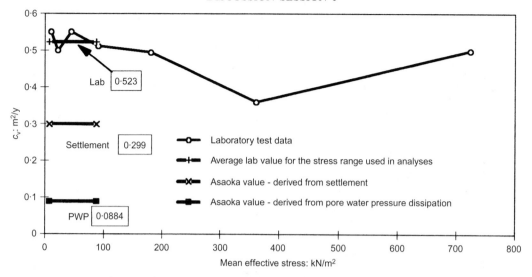

Fig. 3. Coefficient of consolidation, c_v: laboratory values compared with those from the Asaoka construction applied to the numerical solution for stiffness and permeability varying with effective stress

lations were performed in a spreadsheet, with various constraining assumptions, and the results are shown in Figs 5 and 6. The influence of the stress-dependent compressibility and permeability, depicted by curves E in these figures, is very clear. The early reduction in permeability, close to the well, has the effect of slowing the dissipation of the excess pore water pressure and settlement, as compared with the analyses for which compressibility, m_v, and permeability, k, were taken as constant.

In the paper 'Consolidation of a very soft clay with vertical drains' the authors have used the Asaoka method to back-calculate values of the coefficient of horizontal consolidation, c_h. The Asaoka (1978) method is a graphical inversion of Barron's (1948) equal strain solution, for which the compressibility, m_v, and permeability, k, are taken as constant. The authors also took into consideration the modifications to Asaoka's method suggested by Magnan & Deroy (1980). It is worthy of note that Magnan *et al.* (1983) suggest that the inverted Barron formula can be used to evaluate c_h from pore pressure dissipa-

tion data as well as from the settlement data. In Figs 7(a) and Fig. 7(b) Asaoka's construction is shown, using the results from the numerical analyses. It is clear that the plots for the output data, when stress-dependent compressibility and permeability are allowed for, are very different from the results generated using the constant 'lab' c_v value.

The back-calculated c_h values from the Asaoka construction are given in Table 2. Widely different values are obtained from consideration of settlement as compared with pore water pressure dissipation, and both are different from the appreciably constant value seen in the laboratory tests.

In light of these numerical evaluations, the variations in the value of the fitted consolidation coefficient, c_h, is a function of the stress-dependent soil behaviour. Would not the authors of the three papers agree, as most soft compressible soils exhibit quite strong stress-dependent compressibility and permeability, that forecasts made by the soils engineer should take these soil responses into account? The numerical analyses presented here

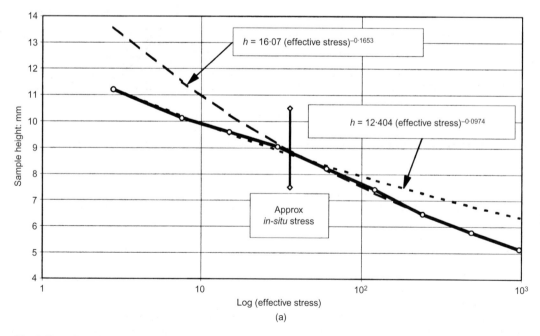

Fig. 4. Raw data for PK test sample, with two power function relationships fitted to the overconsolidated and normally consolidated branches: (a) sample heigfht; (b) permeability

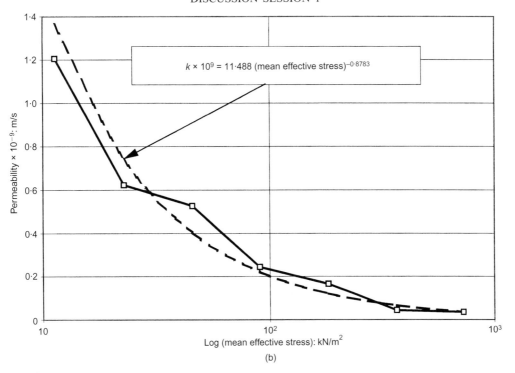

$$k \times 10^9 = 11{\cdot}488 \text{ (mean effective stress)}^{-0{\cdot}8783}$$

Fig. 4. (*continued*).

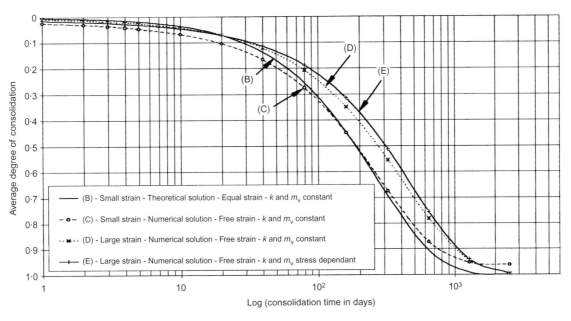

Fig 5. **Various solutions for a drain well. Drain diameter 0·1 m; effective diameter of drained material 1·3 m; well ratio 13**

indicate that, if the soils engineer does not take stress-dependent compressibility and permeability into account, 'forecasts' for drain well performance are unlikely to coincide with the response of the soil to the drainage processes in the field.

Authors' reply (Chu *et al.*)

The authors agree with the discusser that neither the compressibility, m_v, nor the permeability, k, is constant. In our paper, none of these parameters was assumed constant. In fact, we also recognise that the consolidation parameters of soft clay vary with the consolidation process (Chu & Choa, 1997). Two of our PhD students had just completed their projects trying to

model this problem. One (Xiao, 2000) carried out large-scale model tests to study the way in which the consolidation properties of soil around a vertical drain vary with the consolidation process. Another (Nie, 1999) has developed a computer program to model how the variation of consolidation parameters affects the consolidation of soft clay around vertical drains.

Authors' reply (Almeida *et al.*)

We should like to express our thanks to the discusser for his comments concerning the stress dependence of the compressibility and permeability of soft soils. Although he is right,

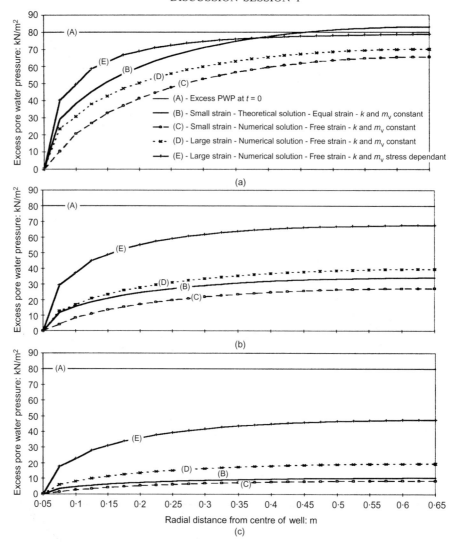

Fig. 6. Various solutions for a drain well. Drain diameter 0·1 m; effective diameter of drained material 1·3 m; well ratio 13. Effective pore water pressure after (a) 80 days, (b) 320 days and (c) 640 days of consolidation

strictly speaking, the compressibility and the permeability vary in the same direction, so that the ratio k/m_v remains reasonably constant and so does c_v, as shown in Table 1, where it can be observed that c_v varies within a narrow interval of $0.52 \pm 5\%$ m²/yr for an effective stress range from 7·5 to 240 kN/m². Besides, the assumption of taking c_v as a constant does not imply that k and m_v are both constant, but is equivalent to assuming that, as the soil particles are moved closer together, the decrease in permeability is proportional to the decrease in compressibility.

In fact the main sources of mismatch in predictions of consolidation analysis are non-linearity in stress–strain behaviour, large strains and secondary compression. Martins & Lima (1996) performed a comparative study of linear and non-linear consolidation theories, and showed that although the average degree of consolidation, $\bar{U}$, may be accurately predicted from linear theory, the pore pressure is always underestimated in this case. The errors are related to the ratio between final and initial effective stresses σ'_f/σ'_i and, for $T_v = 2$, vary from about 17% for $\sigma'_f/\sigma'_i = 1.5$ to about 39% for $\sigma'_f/\sigma'_i = 3.0$. For large σ'_f/σ'_i ratios, pore pressure predictions could be seriously in error if the chord AB in Fig. 8 is taken as an approximation for the arc AB, the true relationship between void ratio and vertical effective stress. The case study presented by Almeida *et al.* (1994) of a 20 m thick embankment placed over a soft clay layer with a σ'_f/σ'_i ratio of the order of 30 illustrates such a case. The

non-linear relationship between void ratio and vertical effective stress had to be taken into account in order to estimate the average degree of consolidation concerning settlements and excess pore pressures, which were quite different. Notwithstanding the strong non-linear relationship between void ratio and effective stress, a constant-valued c_v was used to predict the settlements and pore pressures that were still to occur. Last, but not least, the question raised by the discusser about the validity of using a constant c_v value with both k and m_v stress dependent, which he argues against, can be answered by remembering the non-linear theory of consolidation developed by Davis & Raymond (1965), where one of the basic hypotheses is exactly that which is discussed herein.

The inadequate use of small strain analysis when large strains take place also accounts for the departure of field measurements from theoretical predictions, as illustrated in Fig. 5 provided by the discusser. This figure also points out the minor influence of considering k and m_v to be stress dependent compared with the influence of taking into account large strains in the computation of the average degree of consolidation.

Finally, the authors believe that secondary compression (used herein in the sense of strain-rate-dependent settlements) is always present in the consolidation of soft soils, and occurs together with primary consolidation. Settlements due to secondary compression are frequently a source of error leading to failure in accurate predictions. Although the authors recognise

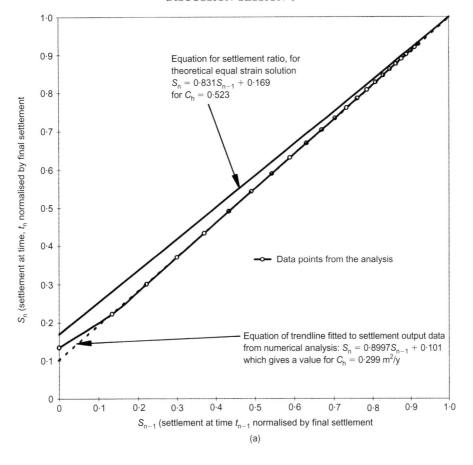

(a)

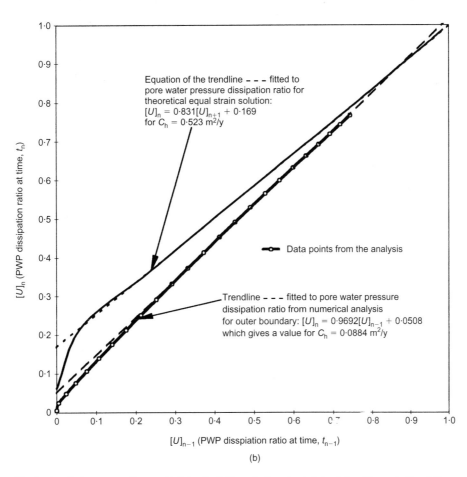

(b)

Fig. 7. Asaoka's construction applied to the PK analysis output data: (a) settlement, S_n, at time t_n normalised by the final settlement; (b) $[U]_n$, dissipation ratio at time t_n

Table 2. Comparison of measured and derived values of the coefficient of consolidation for the isotropic test soil.

	Coefficient of consolidation (measured or derived), c_v or c_h: m^2/yr
Average value from laboratory tests	0·523
Asaoka's construction applied to numerical settlement results (stress-dependent soil parameters)	0·299
Asaoka's construction applied to numerical pore water dissipation results (stress-dependent soil parameters)	0·088

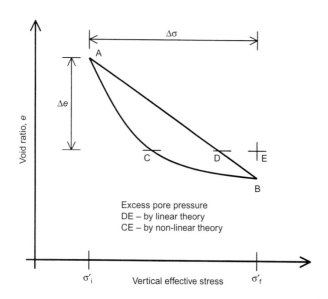

Fig. 8. Linear and non-linear void ratio plotted against effective stress relationships and associated excess pore pressures

the importance of secondary compression, its effect was disregarded in the presented analysis of field data for, in the authors' opinion, soil mechanics has not yet provided a satisfactory approach to handling such types of settlement. As the discussion of secondary compression is beyond the scope of this discussion reply, some authors' ideas about the phenomenon can be found in Martins & Lacerda (1985) and Martins *et al.* (1997).

Author's reply (Tang and Shang)

We agree with Dr Skinner's comments and have no further comment to make.

REFERENCES

Almeida, M. S. S., Danziger, F. A. B., Almeida, M. C. F., Carvalho, S. R. F. & Martins, I. S.M. (1994). Performance of an embankment built on a soft disturbed clay. *Solos e Rochas* **17**, No. 1, 67–73.

Asaoka, A. (1978). Observational procedure of settlement prediction. *Soils Found.* **18**, No. 4, 87–101.

Barron, R. A. (1948). Consolidation of fine-grained soils by drain wells. *J. Geotech. Engng Div., ASCE* **113**, 718–742.

Chu, J. and Choa, V. (1997) Characterization of Singapore marine clay at Changi East. *Proc. 14th Int. Conf. Soil Mech. Foundation Engng, Germany,* 65–68.

Davis, E. H. & Raymond, G. P. (1965). A non-linear theory of consolidation. *Géotechnique* **15**, No. 2, 161–173.

Magnan, J. P. and Deroy, J. M. (1980) Analyse graphique des tassements observés sousles ouvrages. *Bull. Liaison Lab. P .et Ch.* **109**, 45–52.

Magnan, J. P., Pilot, G. and Queyroi, D. (1983). Back analysis of soil consolidation around vertical drains. *Proc. 8th Eur. Conf. Soil Mech. Found. Engng, Helsinki* **2**, 653–658.

Martins, I. S. M. & Lacerda, W. A. (1985). A theory of consolidation with secondary compression. *Proc. 11th Int. Conf. Soil Mech. Found. Engng, San Francisco* **1**, 567–570.

Martins, I. S. M. & Lima, G. P. (1996). A comparative study of linear and non-linear consolidation theories. *Solos e Rochas* **19**, No. 1, 29–43 (in Portuguese).

Martins, I. S. M., Santa Maria, P. E. L. & Lacerda, W. A. (1997). A brief review about the most significant results of COPPE research on rheological behaviour of satured clays subjected to one-dimensional strain. *Proceedings international symposium on recent developments in soil and pavement mechanics,* Rio de Janeiro, pp. 255–265.

Nie, X. Y. (1999). *Consolidation of soft clays around vertical drain.* PhD thesis, Nanyang Technological University, Singapore.

Xiao, D. P. (2000). *Consolidation of soft clay using vertical drains.* PhD thesis, Nanyang Technological University, Singapore.

Session 2

Alonso, E. E., Gens, A. & Lloret, A. (2000). *Géotechnique* **50**, No. 6, 645–656

Precompression design for secondary settlement reduction

E. E. ALONSO*, A. GENS* and A. LLORET*

A large water treatment plant is to be built on soft deltaic deposits. Precompression has been selected as the method for achieving the required ground improvement. An instrumented preload test has been carried out to obtain reliable information on precompression performance. Distributions of displacements throughout the foundation depth, obtained using sliding micrometer extensometers, have proved extremely useful for identifying the mechanisms of behaviour controlling ground deformation. The magnitudes of displacements are largely dependent on the overconsolidation state of the soil. As primary consolidation settlements take place rather quickly, they can be largely controlled by applying a preload over a limited period. The main design criterion therefore concerns secondary settlements. Laboratory and field data indicate clearly that overconsolidating the soil, even in moderate amounts, significantly reduces the secondary compression rate. The performance of an unloading stage in the preload test provides crucial information in this regard. Therefore applying a preload surcharge larger than the final structure load is quite effective in controlling the magnitude of subsequent secondary settlements. The information collected during the site investigation and in the preload test provides the basis for the development of a ground deformation model that can be used for computing settlement histories. The model is used as a design tool for the final proposal of the precompression treatment required for the various structures of the plant.

KEYWORDS: case history; compressibility; consolidation; field instrumentation; ground improvement; settlement

Une grande usine de traitement des eaux doit être construite sur des dépôts deltaïques tendres. On a choisi la méthode de précompression pour améliorer le sol de manière adéquate. Nous avons fait un essai de précharge instrumenté afin de récolter une information fiable sur la performance de précompression. Les répartitions des déplacements à la profondeur de fondation, obtenues en utilisant des extensomètres micromètres à coulisse se sont révélées extrêmement utiles pour identifier les mécanismes de comportement contrôlant la déformation du sol. La magnitude des déplacements dépend fortement de l'état de surconsolidation du sol. Etant donné que les tassements de consolidation primaires se produisent assez rapidement, ils peuvent être largement contrôlés en appliquant une précharge sur une période limitée. Le principal critère conceptuel porte donc sur les tassements secondaires. Les données obtenues sur le terrain et en laboratoire indiquent clairement qu'en surconsolidant le sol, même modérément, on réduit de manière significative le taux de compression secondaire. La performance d'un stade de décharge dans l'essai de précharge donne une information cruciale à cet égard. Donc, le fait d'appliquer une surcharge de précharge plus grande que la charge de structure finale est assez efficace pour contrôler la magnitude des tassements secondaires subséquents. L'information rassemblée pendant l'investigation du site et dans l'essai de précharge, sert de base au développement d'un modèle de déformation du sol qui pourra être utilisé pour calculer les historiques de tassement. Nous utilisons ce modèle comme outil conceptuel pour la proposition finale du traitement de précompression nécessaire pour les différentes structures de cette usine.

INTRODUCTION

The use of precompression to improve the foundation characteristics of soft, fine-grained soils has been widely used in geotechnical engineering. Generally, design for settlement reduction has focused mainly on primary consolidation movements using well-established soil mechanics principles and procedures (e.g. Aldrich, 1965; Johnson, 1970; Mitchell 1981, Stamatopoulos & Kotzias, 1983). If the time to reach the end of primary consolidation is relatively short, primary consolidation settlements can be largely eliminated by applying a preload over limited periods of time. In those cases, the time-dependent settlement of the ultimate structures is basically controlled by the secondary consolidation (drained creep) of the soil. Although secondary settlements are certainly smaller than primary consolidation settlements, sometimes they may become a significant design issue, especially when the structure is sensitive to ground movements.

The construction of a large water treatment plant on soft deltaic deposits near Barcelona has required the design of a preloading operation with the objective of minimizing subsequent settlements. The plant will occupy about 45 000 m² close to the mouth of the Llobregat River. Fig. 1 shows a plan view of the current design. The various structures will generally apply moderate loads, but a number of them are quite sensitive to total and differential settlements. As the firm ground appears at great depths and the structures are very extensive, deep

foundations have been discarded in favour of ground improvement by means of precompression.

The plant is located in the Llobregat delta region, the general geology of which is well known since it has been an area intensively occupied in the past. Three main stratigraphic units are distinguished. From top to bottom, they are: (a) a permeable unit made up mainly of sands (upper aquifer); (b) a low-permeability intermediate unit of soft clays and silts with frequent granular partings; and (c) a deep permeable unit composed of gravels and sands (lower aquifer). The intermediate aquitard effectively separates the upper aquifer from the lower aquifer, making them independent from a hydrological point of view. Whereas the quality of the water in the upper aquifer is low, the lower aquifer has been intensely exploited in the past. Piezometric levels 25 m below hydrostatic have been measured in the deep aquifer, although water pressures have been steadily rising in the last two decades owing to reductions in water extraction.

As will be shown later, primary consolidation is in fact quite rapid, so secondary deformations are the main source of concern regarding future plant settlements. This requires the development of a practical and reliable method to estimate secondary settlements and to assess the effectiveness of preloading in reducing them. Generally, the design of precompression to reduce future secondary settlements has received comparatively less attention (Jamiolkowski *et al.*, 1983; Magnan, 1994). A number of authors (e.g. Ladd, 1971; Mesri, 1973; Koutsoftas *et al.*, 1987; Yu & Frizzi, 1994) have remarked on the significant reduction of secondary compression deformation when the soil is overconsolidated even to a modest degree, indicating that the use of a preload surcharge larger than the final structure load is an effective method for reducing secondary settlements.

Manuscript received 4 February 2000; revised manuscript accepted 12 May 2000.
* Technical University of Catalunya, Barcelona, Spain.

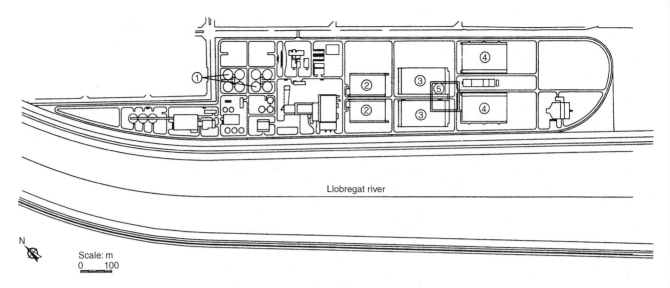

Fig. 1. Plan view of the water treatment plant: 1, anaerobic digesters; 2, primary clarifier; 3, activated sludge reactor; 4, secondary clarifier; 5, preload test embankment

Bjerrum (1972) analysed qualitatively the effect of a preloading on subsequent secondary settlements using his conceptual creep model.

In this paper, precompression design will be based on the establishment of a direct relationship between overconsolidation ratio (OCR) and secondary compression coefficient, C_α. Such an approach has some limitations from a theoretical point of view, but it provides a good base for achieving useful results in practice. Advantage will be taken of the field measurements from a preload test in order to develop a ground deformation model that can be used for prediction and design. The precise measurement of the distribution of settlements in the ground will play a central role in this respect. An unloading stage is included in the preload test with the specific aim of observing directly the effects of overconsolidation on secondary compression deformations.

SITE INVESTIGATION AND GROUND CHARACTERIZATION

The site investigation was carried out in stages. The first stage was directed towards achieving a general knowledge of the stratigraphic sequence of the area where the plant is to be built and of the properties of the materials encountered. It involved drilling of eight deep boreholes (up to 80 m long) and six piezocone (CPTU) tests; 18 vibrating wire piezometers were also installed to observe the hydraulic conditions of the site. The second stage was concentrated on the area where the preload test was going to be carried out, and was focused mainly on the determination of the specific ground properties required for precompression design. Seven additional boreholes were drilled and four static penetration tests (two of them CPTU tests) were performed. A cross-hole seismic test was also carried out to determine the elastic properties of the various strata. Detailed soil profiling was undertaken from the visual observations of the soil samples and from the cone penetration tests. An extensive programme of laboratory tests was carried out on the samples obtained from the boreholes. A third site investigation stage covering in detail the whole of the plant has just been carried out, but the results are outside the scope of this paper.

The approximate ground sequence at the site of the preload test is depicted in Fig. 2. It consists of:

(a) 0–3 m. Upper silt. This material is of little significance as it will generally be excavated and removed to build the foundations of the various structures.

(b) 3–10 m. Sands (upper aquifer). Medium to fine sands, grey

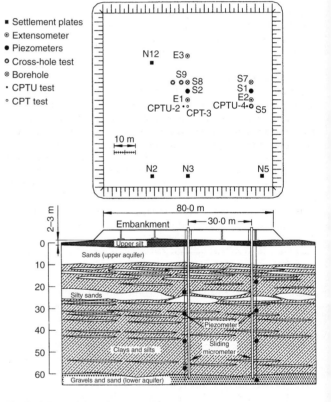

Fig. 2. Soil profile and layout of the preload test

and brown in colour; occasionally small silt inclusions are found. Most values of SPT (corrected for depth) are in the 20–30 range.

(c) 10–60 m. Clays and silts (intermediate layer). This constitutes the compressible soft fine-grained deposits that give rise to most settlements. They are dark in colour. This layer contains a significant numbers of sand and silty sand partings, the proportion of which decreases from the top of the layer towards the bottom. In fact, it is not usually feasible to identify precisely the limit between this layer and the upper aquifer; the transition is quite gradual. As depth increases, the amount of sand and silty sand becomes very small. The organic matter content of the fine-grained material is low, about 0·7% in average and never higher than 1·5%.

(d) 60–70 m. Gravels and sands (lower aquifer). Sandy gravels predominate, although occasionally they are replaced by dense medium-size sands of limited thickness. This layer provides a quite rigid substratum to the previous materials and can be considered incompressible in practice.

Some of the geotechnical indices and parameters obtained in the site investigation are shown in Fig. 3. The fine-grained materials classify mainly as CL (low plasticity clays), and the ratio between C_s and C_c is about 0·12 on average. The coefficient of consolidation has been determined both in the laboratory (oedometer tests) and in the field (CPTU dissipation tests). The field values are in general significantly higher, reflecting the presence of draining sublayers. Sometimes, however, large values of the coefficient of consolidation are obtained in the laboratory, corresponding to specimens with a high sand content.

A water table about 1 m above sea level is present in the Upper aquifer. The piezometer readings (Fig. 4) show that the distribution of pore pressure in the intermediate fine-grained layer is not fully hydrostatic; pore pressures fall below hydrostatic as the lower aquifer is approached, reflecting the reduced piezometric levels there. Hydrostatic conditions prevail only up to a depth of approximately 15–20 m, where the presence of granular materials is more widespread.

As indicated above, an important soil behaviour feature for design is the reduction of the secondary compression coefficient with overconsolidation ratio. A limited series of special oedometer tests on the natural soil have been carried out. The tests were performed as follows. First the specimens were consolidated (in steps) to their estimated in-situ preconsolidation stress stress. Afterwards an additional load was applied and left for a sufficient period of time, typically about 3 days, so that the secondary compression coefficient could be reliably determined. Then the applied vertical load was reduced in order to give the sample a certain degree of overconsolidation. The new value of the secondary compression coefficient was then determined from the new slope of the vertical strain plotted against log time. Examples of the results obtained in three different samples are shown in Fig. 5a. Fig. 5b presents the results of the various tests in terms of the ratio between the value of C_α in the overconsolidated state and the value of the normally consolidated C_α. Note that the reduction of C_α is very sharp; small increases of overconsolidation ratio reduce secondary compression very significantly. The results obtained are in the same range as those determined by Ladd (1971).

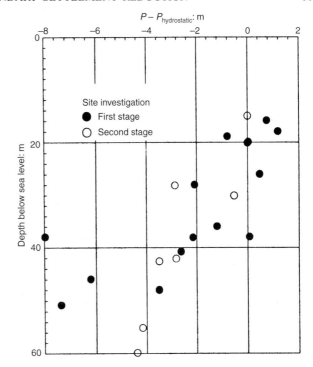

Fig. 4. Excess pore water pressure with respect to hydrostatic conditions measured in piezometers installed during site investigation

PRELOAD TEST

As is often the case in precompression treatments concerning large projects, a preload test was performed in order to obtain more reliable information for design. Apart from the usual objective of checking consolidation times under field conditions, the preload test was also devised to provide information on the effects of overconsolidation and on the relative contributions of the various soil layers to the total settlements. This led to the measurements of movements throughout the soil column and to the inclusion of an unloading stage.

The preload test was performed by building an embankment 4 m high (effectively applying a distributed load of 80 kPa) over an area of 80 m × 80 m (Fig. 2). The construction lasted for two months, from the end of October to the end of December 1996. The test was monitored continuously for a period of two years, until December 1998. From 10 to 22 December 1998, the

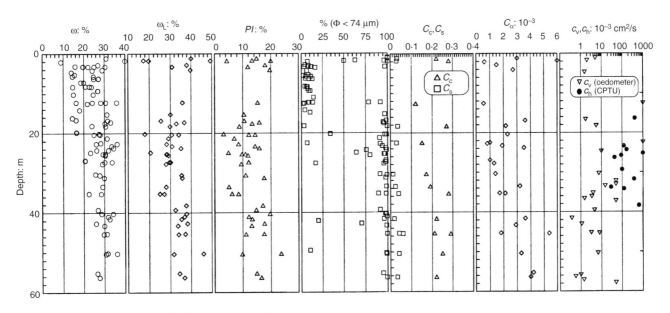

Fig. 3. Soil properties determined in the site investigation

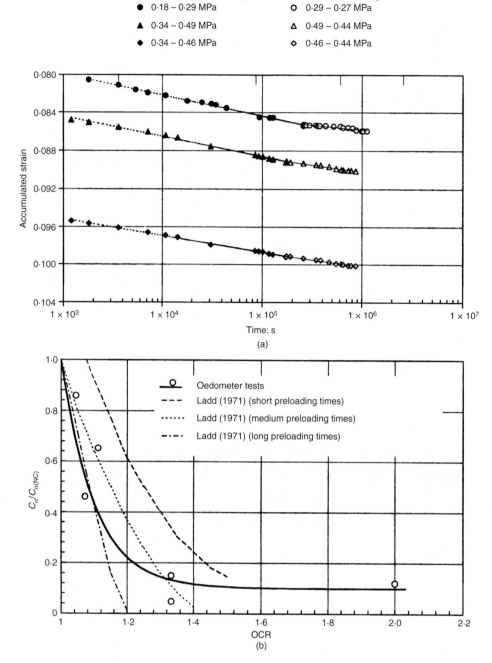

Fig. 5. (a) Vertical strains measured in long term oedometer tests on natural clay. (b) Reduction of secondary compression coefficient with overconsolidation ratio observed in oedometer tests

top 1·5m of embankment was removed to check on the effects of unloading on displacements, which were monitored for an additional eight months, to August 1999. At the end of August 1999, the first construction operations started, disturbing the patterns of displacements, and the preload test was concluded.

To understand the settlement mechanisms properly it was necessary to measure precisely the distributions of displacements throughout the foundation soil. Two sliding micrometers (Kovari & Amstad, 1982, 1983), capable of monitoring vertical displacements at 1 m intervals, were installed at the locations shown in Fig. 2. The micrometer type was specially selected so that it could measure large movements (up to 10 cm/m) with an adequate accuracy (0·03 mm/m). Adjacent to each sliding micrometer, four vibrating wire piezometers were installed at different depths, as shown in Fig. 2. In addition, vertical displacements at the base of the embankments were measured by means of settlement plates laid out in a grid of 20 m × 20 m. For clarity, Fig. 2 indicates only the location of

the settlement plates for which the results are given in this paper. The starting date for monitoring data is taken to be 21 October 1996.

The development of settlements at several points at the base of the embankment is shown in Fig. 6(a). The sequence of embankment loading/unloading is also indicated. Naturally, maximum settlements are observed in the centre of the loaded area, reaching a value of about 23 cm. The associated variations of piezometric heads at three different depths are shown in Fig. 6(b). Considering the piezometers placed in the intermediate clay and silt layer (depths 18 m and 45 m), it is apparent that the pore pressure response to loading is very suppressed, indicating a rapid dissipation of pore water pressure. Primary consolidation times are short and comparable to embankment construction times. The measurements at a depth of 63 m correspond to a piezometer placed in the gravels and indicate the pore water pressure changes in the lower aquifer. It can be observed that piezometric levels are not constant. This causes

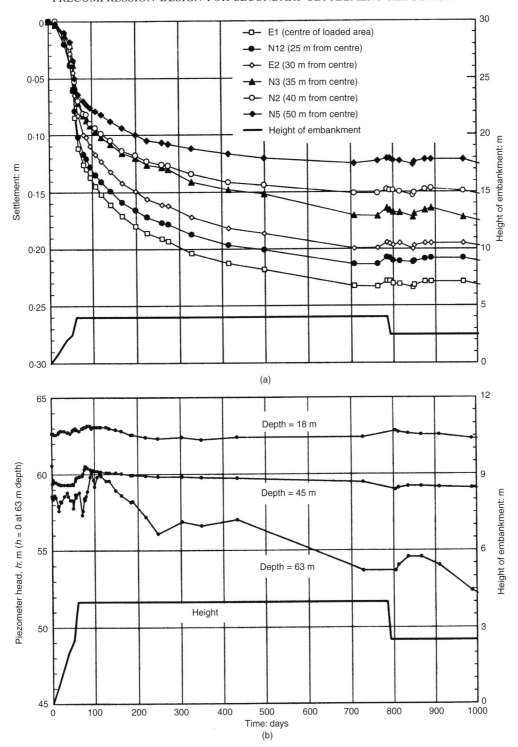

Fig. 6. (a) Vertical settlements at the base of the embankment during the preload test. Embankment height is indicated. (b) Piezometric heads measured during the preload test at various depths. Embankment height is indicated

some difficulties in the interpretation of measurements, fortunately limited to the deeper part of the intermediate layer.

The essential information for subsequent design is provided by the distributions of vertical displacements in depth (Fig. 7). They are obtained from the readings of the two sliding micrometers during the loading stage of the test. The measurements are given in both differential and accumulated form. The results are very similar in the two extensometers.

As expected, the contributions of the upper silt layer and the sand (upper aquifer) layer are small; most settlements occur in the intermediate layer where fine-grained materials are generally

predominant. Observation of settlement distributions within the intermediate levels yields very important information. Most deformations occur in the upper part of the layer, between 14 and 20 m deep, in spite of the fact that there the proportion of granular, less compressible material is higher. In the deeper part of the layer the deformations are moderate or small, although the proportion of clay is very high. This reflects the fact that in the upper part of the layer the material is likely to be normally consolidated, because the pore pressures have been basically controlled by the upper aquifer, not subjected to extractions, and have remained very close to sea level throughout. In

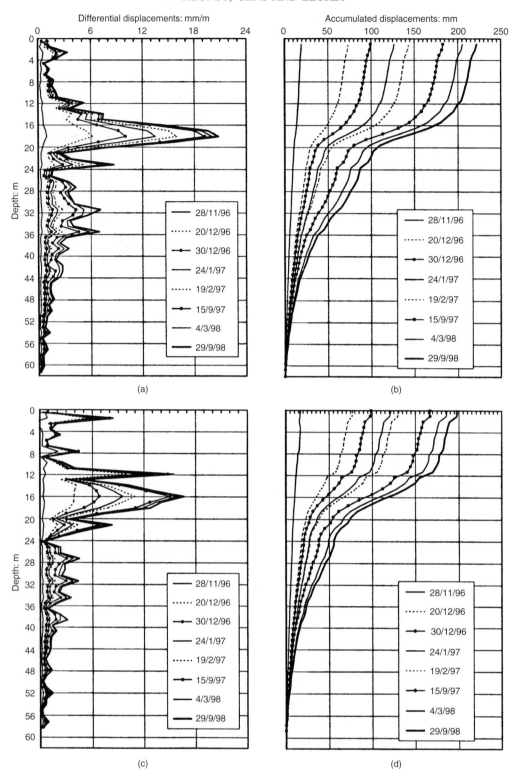

Fig. 7. (a) Distributions of differential vertical displacements measured in extensometer 1. (b) Distributions of vertical displacements measured in extensometer 1. (c) Distributions of differential vertical displacements measured in extensometer 2. (b) Distributions of vertical displacements measured in extensometer 2. Date of origin for displacement measurements: 23 Oct. 1996

contrast, in the deeper part of the layer, the material has become overconsolidated by the past reduction of piezometric levels of the lower aquifer.

The extensometer data permit the analysis of the foundation considering it as a series of slices 1 m thick. If the vertical stress increments are computed, it is possible to estimate the equivalent values of compression index every metre using the observed differential displacements due to primary consolidation. The vertical stress increments caused by the embankment

load have been obtained from an elastic finite element analysis using a realistic distribution of stiffness throughout the soil profile. The results for the locations of the two extensometers are shown in Fig. 8(a). The compression index values obtained in this way exhibit significant differences (Fig. 8(b)). In the upper part of the layer compression indices are high, typical of materials in a normally consolidated state. In the medium and deeper levels the values of compression index are much lower, characteristic of overconsolidated soils.

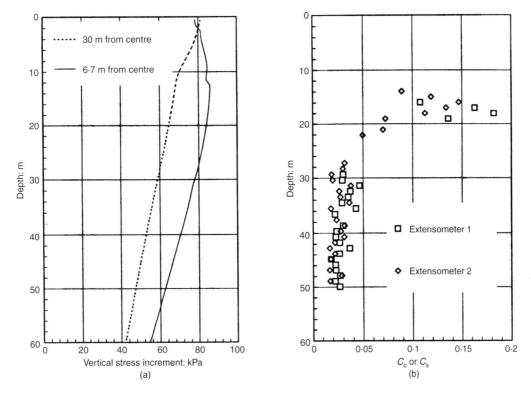

Fig. 8. (a) Computed vertical stress increment due to the applied preload surcharge. (b) Distribution of primary compression indices with depth

Differential vertical displacements corresponding to different ground depths are plotted in Fig. 9(a) against time in a semi-log scale. In this representation the evolution of settlements at long times is approximately linear, typical of secondary compression phenomena. It is also very interesting that, plotting the evolution of strain rates with time, the field data appear to be a natural extrapolation from laboratory oedometer data to longer times.

After two years of observations an unloading stage was performed in the test, reducing the applied stress from 80 kPa to 50 kPa. Fig. 6(a) shows that, after a small rebound, settlements continued but at a smaller rate. The behaviour of the ground during unloading, however, is best examined by referring to distributions of displacements (Fig. 10(a)). Now, the development of vertical displacements is quite uniform throughout the entire layer. For comparison, Fig. 10(b) is presented alongside. It contains the distribution of vertical displacements for approximately one year before unloading (note the different displacement scale). The contrast is very clear. By providing overconsolidation to the soil, the unloading stage has removed the large deformation peaks that existed and has resulted in a quite uniform settlement development occurring at a reduced rate. The same observation can be made by examining the results in terms of secondary compression coefficient, C_α. The values of C_α at each point can be obtained from observation of the individual deformation readings against log time. As Fig. 11 shows, all the large values of C_α have disappeared after unloading. The results for depths below 50 m are not plotted because they are affected by the reduction of water pressures in the lower aquifer.

GROUND DEFORMATION MODEL

The transfer of the results of the preload test to the design of the precompression treatment under a variety of situations requires the development of a ground deformation model. For use in practice this model should be simple, but it must incorporate all relevant behavioural features. Given the large extent of the preload area in relation to the deformable ground

thickness, a one-dimensional model has been deemed acceptable.

The basic information required for a particular location is:

(a) a detailed soil profile achieved by way of a close examination of the borehole samples and the CPTU records
(b) an estimation of the initial value and the development of effective stresses and preconsolidation pressure throughout the soil column.

In this section the model for the preload test zone will be described. It will be calibrated against field observations so that it can contribute to a more quantitative interpretation of the test. The soil profile derived from the site investigation data is presented in Fig. 12. The most important part is the distinction, in the intermediate layer, between free-draining sandy sublayers and the fine-grained intervals where transient consolidation phenomena occur. As Fig. 12 shows, the free-draining granular materials are far more widespread in the upper part of the intermediate layer.

The initial stress in the ground column prior to the performance of the test is shown in Fig. 13(a) together with the estimated maximum past stress (preconsolidation pressure). This maximum past stress has been computed from the expected effects of the known 25 m water level reduction of the lower aquifer. The effect is considered to extend until the granular materials become more predominant towards the upper part of the intermediate layer, where hydraulic conditions are likely to be controlled by the upper aquifer water levels. Therefore it was expected that the material has remained normally consolidated up to a depth of about 18–20 m. This is consistent with the estimations of primary consolidation indices shown in Fig. 8(b). Subsequent changes in vertical effective stress during embankment loading and unloading are computed from the elastic finite element analysis. The successive vertical stress distributions are plotted in Fig. 13(a). Neglecting possible ageing and bonding effects, the distribution of overconsolidation ratios can now be computed at various stages of the preload test. They are presented in Fig. 13(b). It can be observed that after embankment loading a large part of the intermediate layer

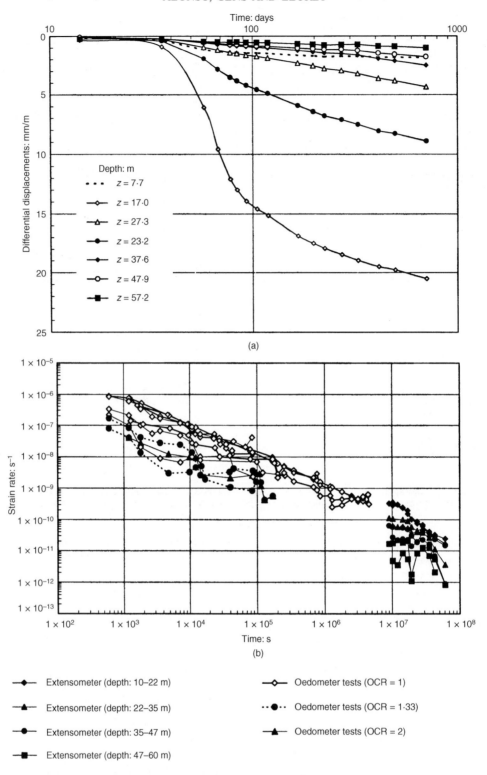

Fig. 9. (a) Variation of differential vertical displacements at various depths with time. (b) Comparison of measured secondary compression strain rates measured in the laboratory and in the field

is normally consolidated, so that large deformations both in primary and secondary compression can be expected there. However, after unloading, all the foundation soil has at least some degree of overconsolidation.

Using this estimation, the data from the extensometers can be used to relate the observed reduction in the value of the secondary compression coefficient, C_α, with overconsolidation ratio. The results are shown in Fig. 14, where, in spite of some scatter, several interesting features can be noted. In the first place there is a strong reduction in the value of C_α with very moderate increases of overconsolidation ratio. Therefore

preloading beyond the final structure load should be quite effective in reducing secondary settlements. The reduction trend is quite similar to that obtained by Ladd (1971) for long preloading times. The results obtained from oedometer tests are also included in the same figure. They tend to plot above the field results, possibly reflecting the significantly shorter preloading times. It is also worth noting that the points obtained in the loading and unloading stages appear to follow the same trend. Therefore there is no need to distinguish between those two situations in the computation of secondary settlements.

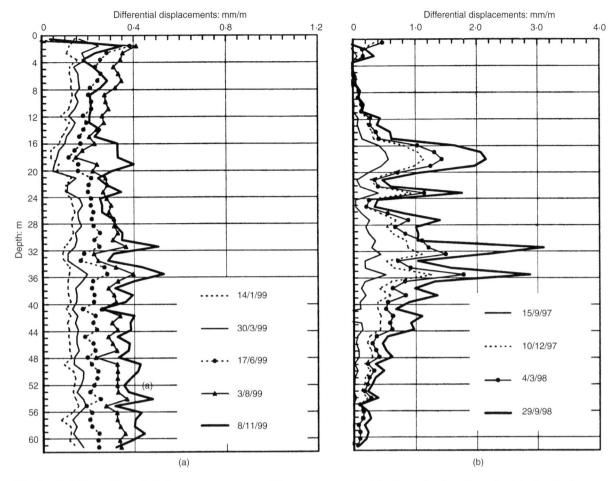

Fig. 10. (a) Differential vertical displacements measured in extensometer 1: (a) after unloading (date of origin for displacement measurements: 22 Dec. 1998; (b) before unloading (date of origin for displacement measurements: 27 July 1997

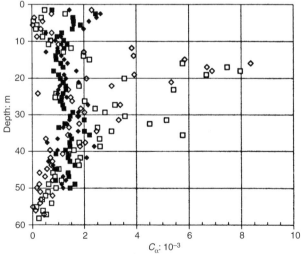

□ Extensometer 1 (before unloading)
◇ Extensometer 2 (before unloading)
■ Extensometer 1 (after unloading)
♦ Extensometer 2 (after unloading)

Fig. 11. Observed values of the secondary compression coefficient, C_α, before and after unloading

The ground deformation model is now constructed as follows:

(a) The soil column is divided into a number of individual intervals based on the soil profile obtained from site investigation.

(b) The development of effective vertical stresses and preconsolidation pressures throughout the loading history is computed as indicated above.

(c) The time history of settlements is calculated as the sum of the contributions of the vertical displacements for each soil interval. The total calculation time is divided into a large number of time increments, and the development of settlements is computed by means of a time-advancing scheme.

In the fine-grained soils the deformations are the sum of the immediate elastic deformation (computed from elastic expressions), the primary consolidation deformation (computed using the oedometer parameters C_c and C_s) and the secondary compression deformation. Secondary compression strain occurring in a time interval $(t_{i-1} - t_i)$, at a point with vertical coordinate z, $\Delta\varepsilon_{vs}(z)$, is computed as

$$\Delta\varepsilon_{vs}(z) = C_\alpha(P_c(z, t_i)/\sigma_v'(z, t_i))\log(t_i/t_{i-1}) \qquad (1)$$

where $p_c(z, t_i)$ and $\sigma_v'(z, t_i)$ are the preconsolidation stress and the effective vertical stress respectively at time t_i.

The value of secondary compression depends on the overconsolidation ratio, OCR, according to the empirical expression derived from the preload test results (Fig. 14):

$$C_\alpha = 0.008[0.1 + 0.9 \exp(-13(OCR - 1))] \qquad (2)$$

For consistency with Bjerrum's (1972) conceptual model, the preconsolidation pressure increases during secondary compression in accordance with

$$(p_c(t_i)/p_c(t_{i-1})) = (t_i/t_{i-1})^{C_\alpha/(C_c-C_s)} \qquad (3)$$

For the granular sublayers only instantaneous elastic deformations and drained creep deformations are computed. Their contribution to total settlement is small.

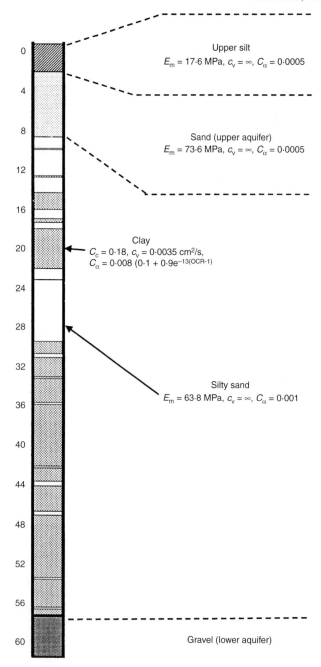

Fig. 12. Soil profile representing ground conditions in the preload test area

ments that the various structures can safely undergo. The basic criteria concern, therefore, the reduction of settlements expected to occur after construction of the structure, i.e. the time-dependent settlements. Because of the planned construction schedule, preload times are limited to six months approximately. Therefore the main design parameter to be determined is the height of the precompression embankment to be adopted for each particular structure.

The design process for each structure of the plant follows the following steps:

(a) adoption of a soil profile based on the detailed examination of the site investigation data (borehole cores and samples, CPTU results)
(b) evaluation of the history of stresses and preconsolidation pressures taking into account all the stages of preloading, construction and operation
(c) application of the ground deformation model outlined above to compute the expected settlement history of the structure.

Steps (a)–(c) are carried out for various heights of the preload embankment. From the analysis of the time-dependent settlements of the finished structure at various times, the optimum embankment height for each case is selected.

As an example, the results of the calculations carried out for the anaerobic digesters are presented. The location of those structures is shown in Fig. 1. They are the heaviest structures of the plant, and apply a pressure on the ground of 124·5 kPa. Of this total, 24·5 kPa corresponds to the self-weight of the structure and the rest, 100 kPa, to the load applied when filling the tank. Prior to construction, 1 m of ground is excavated to house the foundation.

The loading sequence considered for analysis includes the following:

(a) application of the precompression embankment load (variable)
(b) preload maintained for six months
(c) preload removal
(d) ground excavation
(e) structure construction
(f) filling of the structure.

A subsequent emptying/filling of the tank after the one-year operation has also been included in the calculations. The computed history of displacements for four different preload embankment heights is shown in Fig. 16(a). The relevant variable for design is, however, the magnitude of the time-dependent settlements occurring after construction. They are plotted as a function of the height of preload embankment in Fig. 16(b). It can be noted that the additional reduction of settlements for embankment height values larger than 8 m becomes small. Indeed, in this particular case, a preload embankment 8 m high has been adopted in the final design.

The ground deformation model can now be applied to the preload test. The parameters used are indicated in Fig. 12. Since the model has been calibrated using some of the preload test data, the close agreement between global settlement computations and observations is not especially significant (Fig. 15). It should be pointed out that the agreement also extends to the details of the settlement variations with depth. The model can therefore be used to provide a more quantitative insight into the results of the test. For instance, in the same Fig. 15, the contribution of primary and secondary consolidation settlements to total movements can be readily examined. The deformation ground model calibrated in this manner can be used as a tool for precompression design.

PRECOMPRESSION DESIGN

The design of the precompression required for building the water treatment plant is controlled by the magnitude of settle-

CONCLUDING REMARKS

When precompression is applied to a relatively free-draining material, consolidation times are short, and therefore primary settlements can be effectively controlled by using a preload equal to the final load over a limited period of time. The subsequent movements associated with unloading/reloading will be small, and will occur rapidly. In these cases, the main source of long-term displacements is likely to be secondary compression settlements.

Using laboratory and field data gathered in relation to the precompression design for a large water treatment plant, it has been shown that preconsolidating the soil is an effective way to reduce secondary settlements to acceptable levels. Both laboratory and field observations show that there is a sharp reduction in the value of the secondary compression coefficient with only small degrees of overconsolidation. The performance of an unloading stage at the end of the preload test has provided particularly useful data in this respect.

The deployment of extensometers to obtain precise measure-

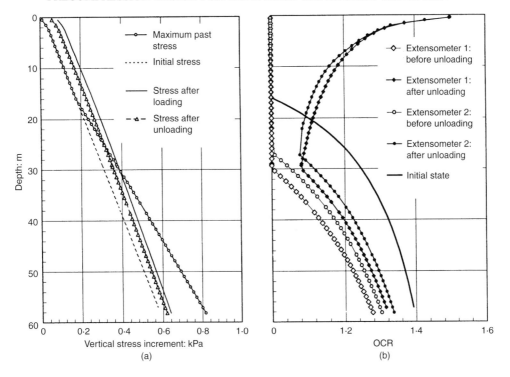

Fig. 13. (a) Initial preconsolidation stress and vertical stress distributions at various times of the preload test. (b) Distribution of overconsolidation ratio (OCR) with depth at various times of the preload test

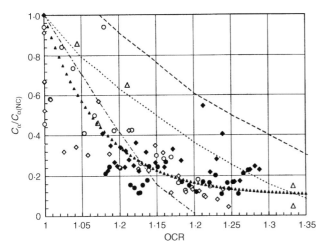

Fig. 14. Reduction of secondary compression coefficient with over-consolidation ratio derived from field measurements. Oedometer test results have been added for comparison

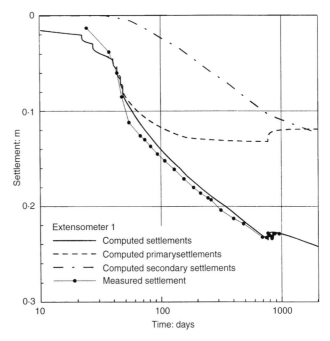

Fig. 15. Global settlements of the preload test computed using the ground deformation model. Observed settlements are plotted for comparison

ments of the displacement distributions at depth has provided very valuable information towards achieving a good understanding of the various phenomena involved and their relative importance. This information has provided the basis for the development of a simple but comprehensive ground deformation model that can be used for computing and predicting future settlements. The final result is a practical design tool firmly based on the observed field behaviour of the ground.

ACKNOWLEDGEMENTS
The authors acknowledge the support of Depurbaix S.A. and EMSSA in the development of this work. Field work was carried out by Eurogeotécnica S.A. The authors are also grateful for useful discussions held with the site engineer, Jordi Vila.

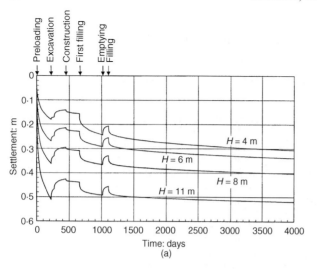

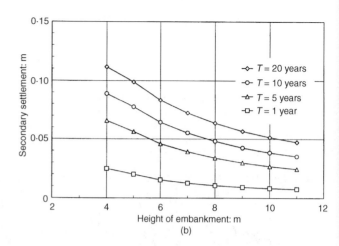

Fig. 16. (a) Settlement histories computed for the anaerobic digesters using different preload embankment heights, *H*. (b) Variation of computed secondary settlements at various times with preload embankment height, *H*

NOTATION

ω	Water content
PI	Plasticity index
C_c	Compression index
C_α	Secondary compression coefficient
c_v	Vertical consolidation coefficient
c_h	Horizontal consolidation coefficient
ω_L	Liquid limit
$\Phi < 74\ \mu m$	Grain size smaller than 74 microns
C_s	Swelling index

REFERENCES

Aldrich, H. P. (1965). Precompression for support of shallow foundations. *J. Soil Mech. Found. Div., ASCE* **91**, No. 2, 5–20.

Bjerrum, L. (1972). Embankments on soft ground. *Proceedings of the specialty conference on performance of earth and earth-supported structures*, India Vol. 2, 1–54.

Jamiolkowski, M., Lancellotta, R. & Wolski, W. (1983). Precompression and speeding up consolidation. General report. *Proc. 8th Europ. Conf. Soil Mech. Found. Engng*, Helsinki **3**, 1201–1226.

Johnson, S. J. (1970). Precompression for improving foundation soils. *J. Soil Mech. Found. Div., ASCE* **96**, No. 1, 111–144.

Koutsoftas, D. C., Foott, R. & Handfelt, L. D. (1987). Geotechnical investigations offshore Hong Kong. *J. Geotech. Engng. Div., ASCE* **113**, No. 2, 87–105.

Kovari, K. & Amstad, Ch. (1982). A new method of measuring deformations in diaphragm walls and piles. *Géotechnique* **22**, No. 4, 402–406.

Kovari, K. & Amstad, Ch. (1983). Fundamentals of deformation measurements. *Proceedings of international symposium on Field measurements in Geomechanics*, Zurich, Vol. 1, 219–239.

Ladd, C. C. (1971). *Settlement analysis of cohesive soils*, Research Report R71-2. Cambridge, MA: MIT.

Magnan, J. P. (1994). Methods to reduce the settlement of embankments on soft clay: a review. In *Vertical and horizontal displacements of foundations and embankments*, ASCE Geotechnical Special Publication No. 40, Vol. 1, 77–91.

Mesri, G. (1973). Coefficient of secondary compression. *J. Soil Mech. Found. Div., ASCE* **99**, No. 1, 123–137.

Mitchell, J. M. (1981). Soil improvement. State of the art report. *Proc. 10th Int. Conf. Soil Mech. and Found. Engng*, Stockholm **4**, 509–565.

Stamatopoulos, A. C. & Kotzias, P. C. (1983). Settlement-time predictions in preloading. *J. Geotech. Engng. Div., ASCE* **109**, No. 6, 802–820.

Yu, K. P. & Frizzi, R. P. (1994). Preloading organic soils to limit future settlements. In *Vertical and horizontal displacements of foundations and embankments*, ASCE Geotechnical Special Publication No. 40, Vol. 1, 476–490.

Gohl, W. B., Jefferies, M. G., Howie, J. A. & Diggle, D. (2000). *Géotechnique* **50**, No. 6, 657–665

Explosive compaction: design, implementation and effectiveness

W. B. GOHL,* M. G. JEFFERIES,† J. A. HOWIE‡ and D. DIGGLE§

Although used for over 70 years, Explosive Compaction (EC) has not attained widespread acceptance despite the attraction of low cost and ease of treating large depths. Lack of familiarity with the method, and an empirical design approach unrelated to theory, appear the primary cause for reticence in adopting EC. To alleviate these concerns, practical design considerations for EC based on detailed experience from nine applications and trials are presented here to illustrate the predictable and repeatable effectiveness of EC. Design is based on cavity expansion theory. EC readily gives volume changes 2–3 times larger than might occur under large earthquake motions, with final average relative densities often greater than 70%. Further, environmental and vibration control issues do not constrain the use of EC provided that appropriate explosives and delayed detonation sequences are used. As pronounced post-blast time effects are evident in penetration testing, evaluation of the effectiveness of EC should be based on a combination of pre- and post-blast penetration testing and volume change measurements.

Bien qu'utilisée depuis plus de 70 ans, la méthode de compactage aux explosifs (CE) n'est pas encore acceptée de manière universelle malgré son faible coût et son efficacité pour les grandes profondeurs. Cette méthode est mal connue et la cause principale de cette réticence à l'adopter semble être son caractère empirique, dissocié de la théorie. Pour apaiser les craintes, et pour prouver son efficacité prévisible et reproductible, nous présentons ici les fondements pratiques de la méthode CE en nous basant sur une expérience détaillée dérivée de neuf applications et essais. Le principe repose sur la théorie de l'expansion de cavité. La méthode CE produit des changements de volume de 2 à 3 fois plus importants que ceux qui peuvent se produire pendant les forts mouvements sismiques, avec des densités relatives moyennes finales souvent supérieures à 70%. De plus, les questions d'environnement et de contrôle des vibrations ne limitent pas l'utilisation de la méthode CE du moment qu'on emploie des explosifs appropriés et des séquences de détonation à retardement. Etant donné que des effets de temps prononcés post-explosion sont évidents dans les essais de pénétration, l'évaluation de l'efficacité de la méthode CE devra être basée sur une combinaison d'essais de pré et post explosion et de mesures des changements de volume.

KEYWORDS: compaction; ground improvement; liquefaction; sand.

INTRODUCTION

Explosive compaction (EC) has been used in various projects throughout the world over the last 70 years. EC involves placing a charge at depth in a borehole in loose soil (generally sands to silty sands or sands and gravels), and then detonating the charge. Several charges are fired at one time, with delays between each charge to enhance cyclic loading while minimizing peak acceleration. Often several charges will be stacked in one borehole with gravel stemming between each charge to prevent sympathetic detonation.

EC is attractive, as explosives are an inexpensive source of readily transported energy and allow densification with substantial savings over alternative methods. Only small-scale equipment is needed (e.g. geotechnical drill or wash boring rigs), minimizing mobilization costs and allowing work in confined conditions. Compaction can be carried out at depths beyond the reach of conventional ground treatment equipment.

Most EC has been driven by concerns over liquefaction, and has been on loose soils below the water table (and to depths of nearly 50 m). However, compaction also increases ground stiffness and strength, and EC has wide application for general ground improvement.

Like many other geotechnical processes, explosive compaction has been designed largely on experience rather than theory. It is common to carry out a trial before starting full-scale treatment. This empirical design basis appears to be an obstacle to the widespread use of an otherwise inexpensive and effective compaction method: owner's review boards are often reticent in approving proposals, contractors are unsure of risk factors when bidding work, and consulting geotechnical engineers lack familiarity with the method. The aim of this paper is to present detailed experience from several applications to counter these concerns, explaining the methodology used. EC can be systematically designed, and can achieve repeatable and consistent results. We also consider environmental, vibration, and other incidental effects to illustrate that these will not usually constrain EC.

TYPICAL GROUND RESPONSE DURING EXPLOSIVE COMPACTION

Similar ground response was observed at the sites discussed in this paper: compaction does not occur concurrently with explosive detonation. Following detonation of the charges, some immediate ground heave or small settlements occurs around individual blast holes. Nothing is then apparent at ground surface for at least several minutes or, in the case of fine sand, tens of minutes; then the ground starts to settle, and continues to settle for upwards of an hour. Compaction induced by explosives is not over the few seconds of explosive detonation but rather is an induced consolidation over several hours, even with sandy gravels. Explosives generate residual excess pore water pressure that dissipates and results in consolidation-related compaction.

Where the water table is near the surface, and there is no overlying impermeable layer, sand boils usually develop as the residual excess pore water migrates at depth and forms preferential pathways to the surface. The vertical seepage pipes associated with these sand boils can be 100 mm or more in diameter (we have seen them as large as 600 mm), and large quantities of water escape to the surface through them. Pore water escapes for perhaps as long as 2 h. If the water table is at depth, or if there is an overlying impervious layer, then sand boils develop around the blast holes.

The ground settles concurrently with the escape of pore water. Settlements may continue at a slow rate depending on soil permeability and drainage conditions. Once an area of ground has been shot and pore pressures have largely dissipated, blasting from adjacent areas causes additional settlement de-

Manuscript received 7 February 2000; revised manuscript accepted 28 June 2000.
* Pacific Geodynamics Inc., Canada.
† Golder Associates & UMIST, UK.
‡ University of British Columbia, Canada.
§ Foundex Explorations Ltd, Canada.

pending on soil density and stiffness. The first detonation sequence of blast holes (first pass) destroys any bonds existing between sand grains due to ageing and other geologic processes, and causes the majority of settlement within the soil mass. Subsequent passes cause additional settlement by cyclic straining.

Penetration resistance of the compacted sands shows pronounced time dependence. Some case histories show no increase or even a decrease in penetration resistance immediately after blasting, while other case histories show only a modest increase in resistance. However, the penetration resistance two weeks after blasting is often double the pre-compaction value. Delayed strength gain must be allowed for in developing an explosive compaction project if penetration resistance is the basis for the work. An example of delayed strength gain and achievable penetration resistance is shown in Fig. 1.

THEORY

Historically, charge weights and spacing have been selected based on empirically derived correlations between explosive charge density and induced post-blast settlement. Vibrations (peak particle velocities) are controlled on the basis of empirically derived relationships between observed ground velocities, distance from the blast and explosive charge mass per delay. Blast effectiveness and the magnitude of vibrations vary with the soil conditions, type of explosive, charge length, blast hole layout and the sequence of detonation. A more theoretical basis for design is desirable.

Dimensional analysis

Detonation of an explosive produces two forms of energy release; the shock wave from the detonation front, and the work done by the high-pressure gas formed in the explosion as it expands. About one third of the charge energy is available for work in expanding the cavity containing the charge, and it is this component that is of interest for compacting the soil. The shock wave is predominantly a compression wave, although

shear waves are formed at the corners of cylindrical charges. However, the cavity expansion is pure shear in the elastic phase, and largely shear in the elastic–plastic stage.

Most work on explosive compaction has relied on similitude in examining trends in the experience record, adopting Hopkinson's number (HN) to estimate empirically the radius of influence around a blast hole. HN is often taken to be

$$HN = W^{0.33}/r \qquad (1)$$

where W is the charge mass delay (kg), and r is the distance from the charge to the point of interest. But equation (1) is not dimensionless. The characteristic size of the explosive charge is

$$z = (W/\rho)^{0.33} \qquad (2)$$

where ρ is the mass density of the explosive, and z is the face length of an equivalent cubical source. Although adopting W without regard to explosive type might seem too simple given the differing bulk strength of various explosives, it is difficult to discern an effect of explosive type in the experience record because of the variation in the proportion of energy radiated in the blast wave from one explosive and charge arrangement to another. Thus W is used at face value without regard to explosive heat content or detonation pressure.

A consistent form of HN is

$$HN = z/r = (W/\rho)^{0.33}/r \qquad (3)$$

Using dimensional analysis (Chadwick et al., 1964; Higgins et al., 1978), the effectiveness of an explosion, E, can be expressed as

$$E = f(HN, t', h/z, gh/V_s^2, \varphi', \lambda, \ldots) \qquad (4)$$

where t' is dimensionless time, h/z is the burial depth ratio, and the other parameters relate to the elastic/plastic properties of the soil. Thus the blast effectiveness is a function of charge density, time, the strength and stiffness of the soil, and two terms related to the depth of the charge. Because stiffness is strongly dependent on the square root of the effective stress, a first approximation of charge effectiveness in a given soil type is

$$E = k(W/\rho)^{0.5} h^{-0.5} R^{-1} \qquad (5)$$

where E is the fraction of maximum achievable vertical strain, R is the radius of a circle of area equal to the rectangular/triangular region compacted by a blast hole, and k is a site-specific coefficient related to the soil properties and damping. The time term has been dropped, as interest is in the induced plastic strains, not the ground motion during the blast. Equation (5) implies that for maximum effectiveness the charge weights must increase with depth. Design using a constant HN or constant loading factor will give less than optimum compaction.

Cavity expansion

Although equation (5) provides a first estimate for design (given knowledge of k, some values of which are provided later), improved understanding can be gained from treating detonation of an explosive charge as the rapid expansion of a cavity. Cavity expansion is primarily a shearing phenomenon, with the expansion and then collapse of a cavity inducing a cycle of shear strain within the soil. The magnitude of the strain diminishes with distance from the blast, and the radius of influence of a particular charge detonation depends on the size and geometry of the charge and the rate of energy release relative to the properties of the soil.

Wu (1995, 1996) developed a non-linear, spherically symmetric finite element program that assumes that a charge detonation may be idealized by assuming a blast pressure–time input applied normal to the surface of a spherical cavity. Soil is represented by a hyperbolic constitutive model, using the Masing criterion to represent hysteretic effects. The high strain rates with blast-induced cavity expansion require a viscous component of strength in the model for realistic predictions: a simple

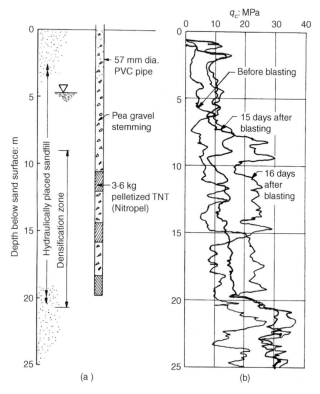

Fig. 1. Example of (a) explosive compaction loading (after Stewart & Hodge, 1988) and (b) achieved change in CPT resistance (after Rogers et al., 1990)

linear (Newtonian) dependence on shear strain rate is used. The program outputs dynamic shearing strains, ground accelerations and velocities, plastic volume change potential and residual pore water pressures for both single and multiple charge detonations. The approach used in the Wu model is by no means a rigorous representation of soil physics, but captures an engineering approximation given appropriate calibration.

Soil behaviour observed in laboratory testing has shown that a minimum threshold shear strain is necessary to initiate pore pressure generation (e.g. Dobry *et al.*, 1982), and that multiple cycles of shear strain are more effective than single cycles. Superposition of spherically symmetric models, allowing for the relevant distances, simulates 3-D arrays of blast holes with reasonable results obtained provided that the model is first calibrated by analysis of test blasts at the site in question.

The effect of the detonation of multiple charges in fine-grained sand is shown in Fig. 2. The residual pore pressure increases after each detonation and in the expected way. Also shown is a multi-blast simulation using the Wu model, showing that pore water pressures are reasonably modelled.

Implications of the theory

Summarizing, theory indicates that:

(*a*) The initial induced pore pressure during blasting reflects the increase in total stress due to the detonation shock wave.

(*b*) The accumulated shear strains govern the amount of settlement achieved, and are reflected in the residual excess pore water pressure. Multiple cycles of a given shear strain level will be more effective than single cycles.

(*c*) The zone of influence of a given charge detonation increases as the size of the cavity increases, but the radius of the zone of disturbance caused by a given charge will depend on the mean confining stress and the strength and stiffness of the soil surrounding it.

(*d*) To create the same radius of zone of disturbance, it will be necessary to increase the charge weight as the depth

increases, as stress, strength and stiffness generally increase with depth. Dimensional analysis suggests that a charge mass should increase proportionally to the square root of the depth.

In practice this theoretical guidance requires explosive charges distributed and timed to maximize the magnitude and number of cycles of shear strain of the soil in the zone being densified. But there is little point in detonating charges in a zone that has liquefied because of adjacent charges; pore pressures should be allowed to dissipate first. Wave theory can be used (e.g. the Wu model) to estimate the peak shear strains and corresponding

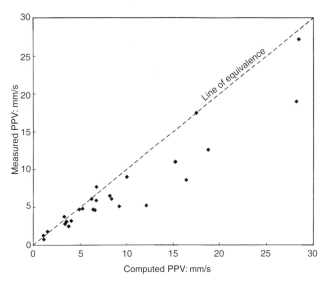

Fig. 3. Blast wave simulation: measured PPV plotted against computed PPV

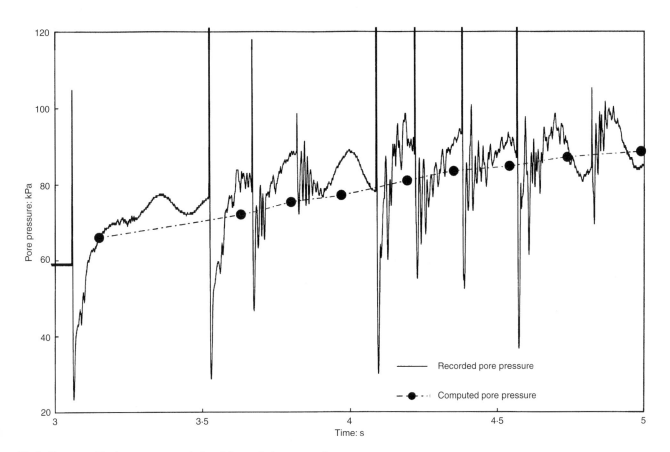

Fig. 2. Excess residual pore pressure induced by explosive compaction

Table 1. Summary of case histories

Project and year	Volume: m³ Max. depth: m	Soil type	Site factor k, (eqn (5))	Initial avg. D_r: %	Charge density: g/m³	Volume change: %	Final avg. D_r: %	Key considerations
Molikpaq I, Amauligak (1986)	67 000 21	Clean dredged medium fine sand	117	45	Nitropel 36	6·4	75	Limited headroom for drilling equipment, rapid EC execution, on-board safety, blast pressures on hull of structure, meeting CPT specification
Molikpaq II, Sakhalin (1998)	169 000 41	Clean dredged medium coarse sand	143	35	Super Blastex 52	8·0	75	
Trans-X (1993)	65 000 15	Medium fine alluvial sand with silt layers	100	50	Iremite TX 37	3·7	70	Limiting PPVs in urban environment, adequate densification for seismic liquefaction control
Quebec HQ SM-3 Dam (1995)	200 000 20	Clean alluvial fine to coarse sand	99	45	Emulsion 38	6·2	75	Limiting hydrodynamic water pressures on ice sheet, meeting CPT specification
Coldwater Creek (1992)	90 000 40	Silty sand, gravel, cobbles	81	50	Nitropel 42	3·6	65	Minimizing slope instability adequate densification for seismic liquefaction control
Kitimat Hospital (1998)	27 000 12	Layered silty sand and sand	101	> 45 (Variable)	Super Blastex 48	3·5 overall 4·7 in sands	> 70 in cleaner sand layers	Vibration control, reducing potential earthquake settlements, maximizing CPT resistances
Sato Kogyo* (1997)	1125 12	Alluvial silty sand and n.p. silt	—	~45	Emulite 53	7·0	2–3 × increase in N_{sw}	Vibration control, maximizing post-blast settlements and penetration resistances
Kelowna* (1991)	8000 10	Sand to silty sand	104	25	Extragel (75) 9	5·5	50	Ground improvement against liquefaction
Elliot Lake* (1980)	1000 6	Silt tailings	—	~35 equivalent	Kinestick 5	4·2	q_c increased by 2–3 times	Behaviour of silt under earthquake loading

*Denotes field trial.

residual pore water pressures, with the site-specific factors determined by analysis of a small trial blast.

PRACTICAL CONSIDERATIONS
Vibration control

Induced vibrations on nearby structures need to be controlled where blast densification is carried out in developed areas. Blasting within 30–40 m of existing structures requires a reduction in the charge weights per deck (involving a reduction in blast hole spacings), and in the number of holes detonated at any one time. Also, when blasting is carried out on or adjacent to slopes, blast patterns are adjusted to restrict the zone of residual pore water pressure build-up and minimize the risk of slope instability.

For the above reasons, the number of charges detonated sequentially is often restricted to minimize the duration of shaking. Longer-duration shaking causes more damage to structures and increases residual pore water pressure build-up. Individual charge delays are also selected so that destructive interference in the frequency range of interest occurs between ground waves from the sequence of blasts. Design of the appropriate charge delays between adjacent decks in each borehole and between adjacent boreholes is carried out using the following process:

(a) Ground vibration patterns (peak particle velocities and frequency content) are determined at a particular location of concern remote from the blast point due to a single charge. This is best done using field measurements, but can also be carried out theoretically.

(b) The frequency range of potentially damaging vibrations is selected based on structural vibration theory or other considerations.

(c) The effects of sequential charge detonation from a decked array of boreholes are assessed by a simple linear combination of the single charge wave trains in which time delays between decks and between adjacent boreholes are varied. Optimum blast delays are then determined to minimize the peak particle velocity or, alternatively, the vibrational energy content in the frequency range of interest.

Based on vibration measurements recorded during detonation of a single charge at various distances from a blast source at an alluvial site in Japan, the linear wave superposition model described above was applied to compute the likely peak particle velocities (PPV) at the ground surface for a particular direction resulting from multiple detonations. The computed and measured peak velocities are plotted against each other in Fig. 3 and indicate that the use of linear combination of wave motions (incorporating appropriate time-shifts in the waveforms based on the prescribed detonation sequence and waveform scaling depending on charge weight–distance relationships determined for the site) generally leads to a conservative over-prediction of PPV. This is particularly true for small source–site distances, where non-linear effects caused by soil liquefaction around a blast point would be expected to reduce near-field motions.

Blast hole layout and detonation sequencing

Blast patterns generally use a staggered rectangular grid of boreholes at spacings of 4–9 m. Staggering is used to provide a pattern of two (or more) passes within a uniform grid. Boreholes are drilled over the full depth of soil deposit to be densified, and 75–100 mm diameter plastic casing installed (this casing size is convenient with the drills normally used). The casing is then loaded with explosive at one or more levels in the borehole (decks). A series of boreholes, each containing one or more decks, is then sequentially detonated. The number of blast holes detonated in any shot depends on vibration control considerations and on concerns about the effect of liquefaction and settlement on adjacent slopes and structures. Subsequent

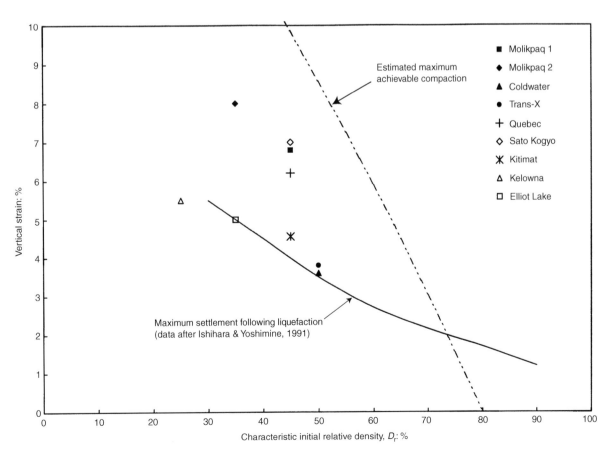

Fig. 4. Summary of settlements induced by explosive compaction

passes are detonated once pore water pressures generated by the previous pass have dissipated sufficiently (typically 1 to 2 days).

Using multiple blast passes promotes increased settlement and more uniform densification. Because local soil loosening can occur immediately around a charge, subsequent passes of blasting from surrounding boreholes are designed to re-compact these initially loosened zones. At least two passes are usually required.

The explosive types have varied, and are selected based on safety, suitability for underground work, and handling convenience. Most recent projects have involved 1·5 kg cartridges (chubbs) of blasting emulsion. Detonators are invariably the non-electric (nonel) system for safety. It is also important to use delays to limit peak ground accelerations, to optimize the frequency content of the vibration, and to gain an adequate number of distinct cycles. The first goal is achieved by avoiding decks in adjacent boreholes detonating simultaneously: typically delays of tens of milliseconds will be used between holes. The second goal is achieved by using long-period delays between the various decks in a hole – often 500 ms (which is usually close to the site response frequency). The third goal is achieved by shooting multiple decks in a blast: at least three, and as many as seven (the practical limit with convenient casing sizes). Stemming, typically 1–2 m of pea gravel, is used between decks to avoid sympathetic detonation.

TYPICAL DENSIFICATION RESULTS ACHIEVED

The authors have been involved in a total of six production EC projects and several additional field trials where EC has been carried out to examine the effectiveness of the densification process (judged from settlements or change in penetration resistance). These case histories are summarized in Table 1, with Fig. 4 showing achieved settlements plotted in terms of initial relative density. The effectiveness of EC is readily apparent from the settlements, with average volume changes over the densification zone in the range 4–10%. As expected, the settlement depends on the initial relative density of the deposit, soil gradation, and the blast design used.

Sand settlement could be viewed as plastic compression following a rather extreme initial excess pore pressure distribution, equivalent to a very large load increment ratio. But sands become progressively stiffer as they approach their minimum void ratio (Jefferies & Been, 2000). Large excess pore pressure ratios do not mean large strains with granular materials. As a first approximation, for vertical stress levels of less than about 1 MPa, it is difficult to compact sands to a relative density of more than about $D_r \approx 0·8$. The maximum vertical strain to be expected is then

$$\varepsilon_v = \frac{0·8 - D_r}{(1 + e_{max})/(e_{max} - e_{min}) - D_r} \qquad (6)$$

where D_r is the initial in-place relative density. For typical sand properties, equation (6) can be approximated as

$$\varepsilon_v = \frac{0·8 - D_r}{4 - D_r} \qquad (7)$$

Equation (7) is plotted in Fig. 4 and compared with the case histories. The trend in the data is captured, with the data suggesting that about two thirds of the maximum possible settlement is obtained in practice. Also shown are expected settlements following earthquake-induced liquefaction based on Ishihara & Yoshimine (1991).

For initial relative densities in the range 30–50%, volume changes in the range 4–10% were achieved, corresponding to final relative densities in the range 65–80%. As these strains have been distributed over substantial depths, induced settlements can be large: slightly more than 3 m settlement is the largest achieved to date. These settlements need to be considered when volumes of fill are being calculated for final site grading. Significant volumes of water will also escape where free-draining soils exist at the soil surface, and this may

necessitate the construction of soil berms to prevent offsite flooding.

The site specific factors k used for initial sizing of charges with equation (5) are also shown in Table 1. Sites with high attenuation (e.g. Coldwater Creek with its large boulders) show lower values of k than do sites with more uniform sands (e.g. Trans-X and Quebec), while the two sites with energy reflection back into the fill (the Molikpaq cases) have apparently high k values. Interestingly, the two Molikpaq cases had markedly different explosive charge layouts and yet give comparable k values; this gives some confidence in the adequacy of equation (5) as a first approximation.

ASSESSMENT OF DENSIFICATION AND TIME EFFECTS

Penetration resistance (measured with the standard penetration test, SPT, or the more reliable cone penetration test, CPT) is typically used to assess the success of ground improvement, but this can be misleading – at least initially. Penetration testing may indicate little effect of blasting, while the large settlements indicate that considerable density increase has indeed been achieved. Time effects complicate the assessment of the success of EC.

Time effects can be at least partially explained by consideration of ageing in sands. Skempton (1986) presented data suggesting that the SPT resistance $(N_1)_{60}$ in normally consolidated sands increased by about 60% over a period of 100–200 years from the time of deposition although there was little increase in D_r. Prior to blasting, the $(N_1)_{60}$ obtained would include the effects of ageing. Immediately after blasting, these

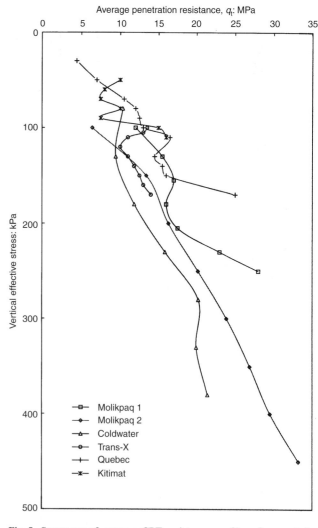

Fig. 5. Summary of average CPT resistance profiles after explosive compaction

effects will have been destroyed, and SPTs measured after the blast could be lower than those existing before the blast, even though settlement indicates that the D_r has increased. The reduction in $(N_1)_{60}$ for several days to weeks following blasting is poorly understood, but is probably due either to the destruction of particle-to-particle bonds caused during sand diagenesis by secondary compression or chemical reactions at sand grain contacts, or to changes in soil structure and effective stress states in the soil mass caused by blasting. Given sufficient time following blasting, $(N_1)_{60}$ increases with time owing to the re-establishment of these ageing effects with little increase in D_r. Time effects will also influence soil strength and stiffness, and this should be borne in mind during preparation of densification specifications. Time effects are also a factor in the assessment of other methods of densification (Mesri *et al.*, 1990).

Achieved average cone penetration tests resistance (q_t) profiles are shown on Fig. 5 (the Coldwater Creek data have been transformed from the Becker test data at the site using a representative q_t/N ratio). Fig. 5 shows the average at each site without regard to the length of time for ageing: simply, this is the available data. These q_t profiles are thus not equivalent, and there are further differences in soil type so that equal q_t values do not imply the same density at the different sites. Nevertheless, the data illustrate the magnitude of average penetration resistance readily achieved by EC: in terms of stress-normalized CPT resistance the data indicate that on average it is straightforward to compact ground to better than $Q > 70$, but equally it is

difficult to achieve $Q > 140$. For comparison, the critical state of sands corresponds to $Q \approx 35$ (see Been *et al.*, 1987), so these achieved resistances indicate substantially dilatant behaviour of the compacted sands.

An important aspect of soils is their natural variability. Also, penetration resistance is affected by soil type (penetration resistance is a behaviour not a property). The effect of both of these aspects is illustrated in Fig. 6 for three of the case histories, showing the range of penetration resistance (at 95% confidence – both low and high resistance spikes have been filtered) before and after EC. The Quebec case illustrates the situation with an alluvial sand, while the Coldwater Creek case shows the effects in a silty sand with cobbles and boulders (this site is in the Mount St Helens debris flow). The Molikpaq case shows what can be achieved with a hydraulically placed dredged sand. As can be seen, penetration resistances have been more than tripled by EC (judged by comparing minimum before with minimum after, average before with average after, etc.). The natural variability of penetration resistance (or density) appears to be unchanged or slightly increased by EC.

While settlement measurements should provide a more direct measure of compaction than penetration resistance, the use of surface settlement to evaluate EC is complicated by the difficulty of assessing initial D_r and by the need to ensure that the increase in density is consistent throughout the layer treated. Compaction uniformity with depth can be monitored by using

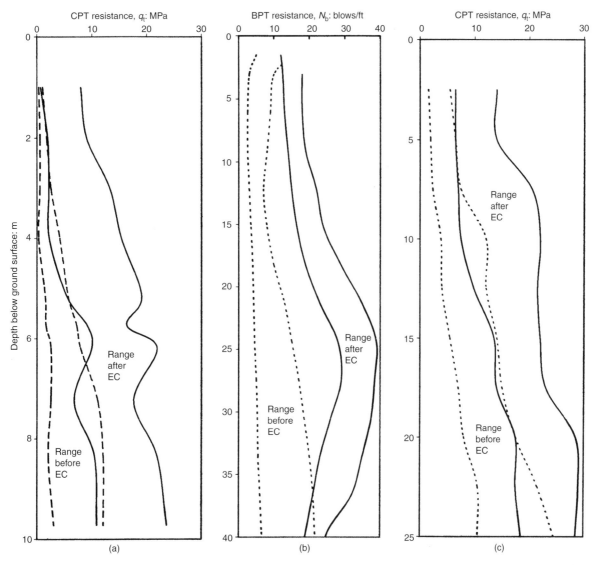

Fig. 6. Comparison of penetration resistance before and after EC showing range of data: (a) Quebec SM3 Dam (alluvial fine-coarse sand): (b) Coldwater Creek (silty sand and gravel); (c) Molikpaq Amauligak (dredged medium-fine sand)

deep settlement gauges placed at selected depths within the layer. Experience at reasonably homogeneous clean sand sites where the distribution of settlement with depth has been measured has shown that uniform settlement can be achieved (Fig. 7).

It seems likely that a combination of the use of settlement measurement and penetration testing will continue to be necessary to measure the performance of EC. Experience indicates that considerable increases in penetration resistances can be achieved in clean sands. These resistances are consistent with the increase in density from achieved settlements. In more silty soils, considerable settlements can still be achieved but penetration resistances will not be as great as at sand sites. All densification methods experience this effect, which is commonly misinterpreted as inadequate compaction.

ENVIRONMENTAL ISSUES

Below-ground explosive detonations result in large amounts of gas being released into the soil–water system, in the form of nitrogen oxide, carbon monoxide and carbon dioxide. Nitrogen oxide is inert in terms of environmental effects on groundwater. Release of carbon dioxide may lower the pH of the groundwater temporarily, while ammonia levels may also be temporarily elevated. But both nitrogen oxide and carbon monoxide are poisonous in air and venting is necessary if blasting is carried out within confined spaces.

In general, the chemical make-up of a particular explosive and its by-products should be reviewed for every project in order to assess its suitability for use at a particular site. Recently, the authors were involved at a trial EC project where extensive groundwater monitoring was carried out to assess the impacts of blasting on groundwater. While short-term spikes in ammonia levels were noted in close proximity to the blast zone, measurable increases diminished a short distance from the blast.

Where blasting is carried out under bodies of water, the impact of the blast-induced percussive wave on resident fish populations needs to be assessed. Blast effects can be minimized through the installation of air bubble curtains around the blast zone or by using special sonic devices to drive fish away from an area.

CONCLUSIONS

Explosive compaction is effective and predictable. Explosive weighting follows from laws of similitude, although site attenuation factors may require further consideration. Current numerical simulators are adequate for blast design, although test blasts remain very desirable at any site before production blasting. Induced settlement (densification) depends strongly on the initial density and less strongly on the soil properties. Penetration resistances should not be the only measure of improvement unless time and soil type effects are taken into consideration.

ACKNOWLEDGEMENTS

The projects whose data are reported here were carried out for various companies including Gulf Canada Resources, Washington Dept of Transport (WASHDOT), Quebec Hydro, and Sakhalin Energy Investment. We are grateful for the support of these companies and the opportunity for such innovative work. We also acknowledge the support of our former colleagues at Agra Earth & Environmental and Golder Associates. In particular, we should like to express our appreciation to Herb Hawson (Golder Associates) and Brian Rogers (formerly Gulf Canada, now Klohn Crippen) without whose support and encouragement much of this work would not have happened. Ron Elliot (Pacific Blasting) was the registered blaster in charge of explosive handling for these projects.

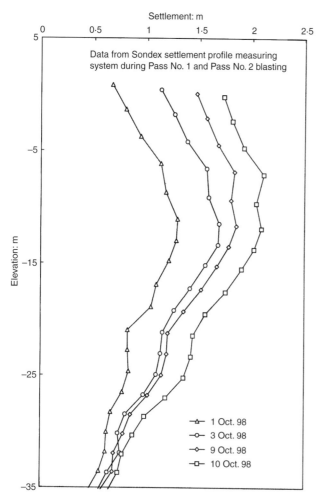

Fig. 7. Example of settlement profiles induced by explosive compaction

NOTATION

D_r relative density
e void ratio
g gravitational acceleration
h burial depth to centre of charge
HN Hopkinson's number
k site-specific attenuation factor
$(N_1)_{60}$ SPT blow count normalized to 100 kPa and 60% of maximum potential energy
N_b Becker penetration test resistance blow count
Q stress-normalized CPT resistance, $(q_t - \sigma_v)/\overline{\sigma}_v$
q_t cone penetration resistance corrected for tip unequal area effects
r radial distance
R radius of a circle of equal area to the rectangular/triangular region compacted by a blast hole
t time
V_s soil shear wave velocity
W mass of explosive charge (kg)
z equivalent cube side dimension of explosive charge
ε_v vertical strain
λ slope of critical state line in e–$\ln p'$ space
ρ density of explosive charge
$\sigma_v, \overline{\sigma}_v$ vertical stress and vertical effective stress respectively
φ' friction angle

REFERENCES

Been, K., Jefferies, M. G., Crooks, J. H. A. & Rothenburg, L. (1987). The cone penetration test in sands. Part 2: General inference of state. *Géotechnique* **37**, No. 3, 285–299.

Chadwick, P., Cox, A. D. & Hopkins, H. G. (1964). Mechanics of deep underground explosions. *Proc. R. Soc. London Ser. A* **256**, No. 1070, 235–300.

Dobry, R., Ladd, R. S., Yokel, F. Y., Chung, R. M. and Powell, D. (1982). *Prediction of pore pressure buildup and liquefaction of sands during earthquakes by the cyclic strain method*, NBS Building Science Series **138**. Washington, DC: National Bureau of Standards.

Higgins, C. J., Johnson, R. L. & Triandafilidis, F. E. (1978). *Simulation of earthquake-like ground motions with high explosives, Final report*. Albuquerque, NM: University of New Mexico, Department of Civil Engineering.

Ishihara, K. & Yoshimine (1991). Evaluation of settlements in sand deposits following liquefaction during earthquakes. *Soils Found.* **32**, No. 1, 173–188.

Jefferies, M. G. and Been, K. (2000). Implications for critical state theory from isotropic compression of sand. *Géotechnique* **50**, No. 4, 419–429.

Mesri, G., Fang, T. W. & Benak, J. M. (1990). Post-densification penetration resistance of clean sands. *J. Geotech. Engng Div. ASCE* **116**, No. 7, 1095–1115.

Rogers, B. T., Graham, C. A. & Jefferies, M. G. (1990). Compaction of hydraulic fill sand in Molikpaq core. *Proc. 43rd Can. Geotech. Conf.*, Canadian Geotechnical Society **2**, 567–575.

Skempton, A. W. (1986). Standard penetration test procedures and the effects in sands of overburden pressure, relative density, particle size, ageing and overconsolidation. *Géotechnique* **36**, No. 3, 425–447.

Stewart, H. R. and Hodge, W. E. (1988). Molikpaq core densification with explosives at Amauligak F-24. OTC 5684, Proc. 20th Annual Offshore Technology Conf, Houston.

Wu, G. (1995). *A dynamic response analysis of saturated granular soils to blast loads using a single phase model*. Research report submitted to Natural Sciences and Engineering Research Council (Canada), December 1995.

Wu, G. (1996). *Volume change and residual pore water pressure of saturated granular soils to blast loads*. Research report submitted to Natural Sciences and Engineering Research Council (Canada), December 1996.

Feng, T.-W., Chen, K.-H., Su, Y.-T. & Shi, Y.-C. (2000). *Géotechnique* **50**, No. 6, 667-674

Laboratory investigation of efficiency of conical-based pounders for dynamic compaction

T.-W. FENG,* K.-H. CHEN,* Y.-T. SU* and Y.-C. SHI*

The pounder for dynamic compaction is conventionally built to have a flat bottom. An innovative idea of using a conical-based pounder to improve the efficiency of dynamic compaction has been investigated, and the results are presented in this paper. Both a conical-based pounder and a flat-based pounder were used in an extensive programme of laboratory dynamic compaction tests. An air pluviation technique was used to prepare dry sand specimens, 60 cm in diameter and 50 cm high, to have initial relative densities of 30% and 40%. Wet sand specimens with an initial relative density of 60% were prepared from the dry sand specimens. A total of 14 dynamic compaction tests were performed. The results of each test include dimensions of the crater, heave around the crater, and cone penetration resistance at 13 penetration locations. It is found from the test results that the efficiency of dynamic compaction with the conical-based pounder depends on both the grain size characteristics and the volume change behaviour of the sand. The conical-based pounder is more effective in the Mai-Liao fine sand with 8% fines than in the clean Ottawa medium sand.

KEYWORDS: compaction; sands; ground improvement; laboratory tests; dynamics.

Le pileur conventionnel servant au compactage dynamique est à fond plat. Nous avons étudié la possibilité d'un pileur à base conique qui améliorerait l'efficacité du compactage dynamique et nous présentons ici les résultats de cette recherche. Nous avons utilisé un pileur à fond plat et un pileur à fond conique dans un vaste programme d'essais de compactage dynamique en laboratoire. Nous avons utilisé une technique de pluviation à l'air pour préparer des spécimens de sable sec de 60 cm de diamètre et de 50 cm de hauteur qui avaient des densités relatives initiales de 30 et 40%. Nous avons préparé des spécimens de sable mouillé ayant une densité relative initiale de 60% à partir des spécimens de sable sec. Nous avons effectué un total de 14 essais de compactage dynamique. Les résultats de chaque essai montrent les dimensions du cratère, le rejet autour du cratère, la résistance à la pénétration du cône en 13 endroits. Nous avons trouvé, d'après les résultats de ces essais que l'efficacité du compactage dynamique utilisant un pileur à base conique dépend de la taille des grains et du comportement de changement volumique du sable. Le pileur à base conique est plus efficace dans le sable fin de Mai-Liao avec 8% de fins que dans le sable moyen propre d'Ottawa.

INTRODUCTION

Densification of loose sandy soils by dynamic compaction (DC) has been found to be successful in many projects (e.g. Menard & Broise, 1975; Leonards *et al.*, 1980; Lukas, 1980, 1986; Jessberger & Beine, 1981; Gambin, 1983; Mayne *et al.*, 1984; Senneset & Nestvold, 1992; Chow *et al.*, 1994). Improvement of hydraulic reclaimed lands is one of the major applications of this technique. The dynamic compaction is executed by lifting a pounder to a specified height for release in free fall. Upon impact on the ground surface, the pounder induces high contact stress and produces a crater, which results in an increase in the density of the loose soils. Ground heave around the crater is sometimes observed (e.g. Charles *et al.*, 1981; Welsh, 1983; Mayne *et al.*, 1984). Furthermore, densification of loose soils at greater depths is achieved owing to both compression and shear waves generated by the impact of the pounder.

Compression waves are useful in reducing normal stresses between sand particles, and shear waves are helpful in creating shear displacement and thus rearrangement of sand particles: the higher the amplitude of both compression waves and shear waves, the greater the densification of loose sandy soils that is expected. So far, there is no information about the partition of dynamic compaction energy between compression waves, shear waves, and Rayleigh waves. Miller & Pursey (1955) show that for flat-base shallow foundations undergoing steady-state vibration, 67%, 27% and 6% of the energy is transmitted by the Rayleigh waves, compression waves and shear waves respectively. A conventional pounder also has a flat base, like that of a shallow foundation. Therefore it may be assumed that the partition of the dynamic compaction energy between the Ray-

leigh waves, compression waves and shear waves is similar to that for the steady-state vibration of shallow foundations (Smoltczyk, 1983; Van Impe, 1989). The energy imparted by the shear waves is a very small portion of the total compaction energy, and this suggests that modifying the base of the pounder might increase the shear wave energy.

Generation of shear waves at the ground surface can be accomplished by hitting a hammer on an inclined plane, such as that executed during the seismic cone penetration test for measurement of the shear modulus of soils. Therefore a new design of pounder with a conical base may be expected to induce a higher proportion of shear waves than that induced by pounders with flat bases. Accordingly, a laboratory study was carried out to investigate the use of a conical-based pounder for dynamic compaction of loose sands. This paper presents the results of this investigation.

Laboratory dynamic compaction tests have been carried out by a number of researchers. Orrje & Broms (1970) used a reinforced wooden box, 130 cm × 50 cm in plan area, to hold soils for laboratory dynamic compaction tests. Circular flat-based pounders with diameters of 5 cm, 10 cm and 15 cm were used. Ellis (1986) performed dynamic compaction tests in a metal drum, 45·7 cm in diameter and 57·6 cm high, on both dry and saturated sand, clay, and fly ash specimens. Four flat-based pounders of the same weight but with diameters of 3·1 cm, 5·1 cm, 6·5 cm and 10 cm were used. Poran & Rodriguez (1992) built a cubic steel tank 122 cm × 122 cm × 122 cm to conduct laboratory dynamic compaction model tests on dry sand specimens. Three circular flat-based pounders with diameters of 10·2 cm, 15·2 cm and 22·9 cm were used. The influence zones resulting from the densification were determined by using a portable nuclear instrument in the sand specimens. It was found from their tests that the influence zone has a bowl shape, with a width at the top equal to three times the diameter of the pounder and a depth equal to four times the diameter of the pounder. This experience was used to determine the relative dimensions of the pounder and the tank used in this investigation.

Manuscript received 31 January 2000; revised manuscript accepted 18 July 2000.
* Department of Civil Engineering, Chung Yuan Christian University, Taiwan.

LABORATORY DYNAMIC COMPACTION TESTS
Soils tested

Both natural and commercial sands were used in this investigation. The natural sand samples were obtained from the Mai-Liao reclaimed land in Taiwan, where the dynamic compaction technique has been used extensively. The commercial sand is Ottawa sand (ASTM C778). Grain size distribution curves of these sands are shown in Fig. 1. As can be seen from Fig. 1, the Mai-Liao sand is poorly graded with 8% fines, and the Ottawa sand is also poorly graded but without fines. The average grain size (D_{50}) is 0·16 mm for the Mai-Liao sand and 0·32 mm for the Ottawa sand. The specific gravity of the solid is 2·68 for the Mai-Liao sand and 2·65 for the Ottawa sand. The maximum void ratio (ASTM D4254-83) and minimum void ratio (ASTM D4253-83) are 1·11 and 0·74 respectively for the Mai-Liao sand, and 0·77 and 0·54 respectively for the Ottawa sand. The primary mineral of the sand samples is quartz, as determined from X-ray diffraction analyses. The sand particles are predominantly sub-angular and sub-rounded for the Mai-Liao sand and the Ottawa sand respectively.

Test equipment

A flat-based cylindrical pounder and a conical-based cylindrical pounder were used in this investigation. The dimensions of these pounders are shown in Fig. 2. Cheney (1983) concluded that, in dry sand, the diameter of the model footing must be greater than 15 times the average grain size. The diameter of the pounders is at least 250 times the average grain size. The pounders are made of steel, and each has a mass of 3 kg. An electromagnetic system (capacity 20 kg) was used to release the pounder from the pre-specified drop heights. Other major equipment used included a stainless steel tank, a soil sample container with a hoist system, an electric cone penetrometer, and a data acquisition system. The stainless steel tank has an inside diameter of 60 cm and a height of 60 cm (Fig. 3), designed according to space and operational conditions in the laboratory. Therefore both the diameter and the height of the tank are 7·5 times the diameter of the pounder. These are all larger than the dimensions of the influence zone from the experiments by Poran & Rodriguez (1992). Furthermore, Poran & Rodriguez (1992) verified that a pounder of 22·9 cm in diameter, 33 kg in weight, and with a maximum drop height of 2 m in their experiments was free from the problem of impact wave reflections from the tank boundary. For the pounder of 8 cm in diameter, 3 kg in weight, and with a maximum drop height of 60 cm in this study, the impact energy per unit contact area is smaller, and the pounder should thus be free from the problem of impact wave reflections from the tank boundary. A hole 1 cm in diameter was drilled at the centre of the base plate of the tank to allow drainage or measurement of pore water pressure, if desired. The pore water pressure was measured by using a pressure transducer (Kyowa, capacity 1000 kPa). An O-ring was installed at the interface between the bottom of the tank wall and the base plate of the tank to prevent leaking of water. The soil sample container is shown in Fig. 3. It includes two porous plates, an extension shaft below the porous plates, and two wire meshes (US sieve #10, opening 2 mm). Each porous plate (Fig. 4) has a total of 293 openings, each 1 cm in diameter and spaced at 3 cm centre to centre, uniformly distributed inside the test tank. The upper porous plate of the soil sample container is welded to the upper half of the container for holding sand

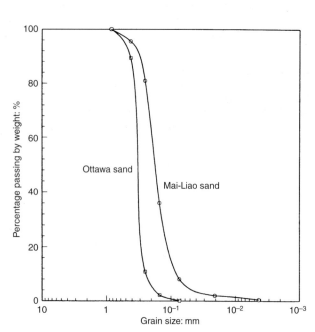

Fig. 1. Grain size distribution curves of sands tested

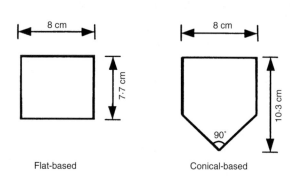

Fig. 2. Side view of cylindrical pounders

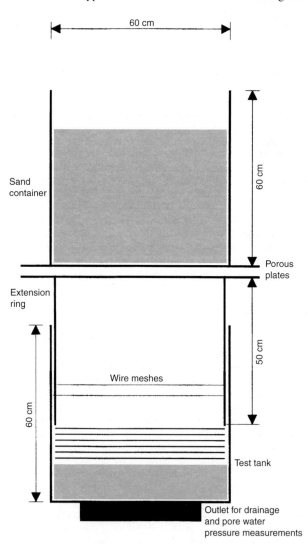

Fig. 3. Side view of test tank and soil sample container

Openings, 1 cm in diameter, 3 cm centre to centre

Fig. 4. Top view of upper porous plate

samples. The lower porous plate is placed immediately below the upper porous plate, and is guided by two tracks for sliding to control the flow rate of the sand. Two wire meshes are welded 10 cm above the bottom of the extension ring to separate sand particles dropped through the porous plates. A hoist system was used to lift the soil sample container. An electric cone penetrometer (Hogentogler, shaft diameter 9 mm, tip angle 60°, load capacity 600 kg) was used to measure the tip resistance for specimens with and without densification. A load sensor is installed immediately behind the tip of the electric cone. The data acquisition system consists of a personal computer and software to record the tip resistance of the electric cone as it penetrates into the specimen.

Test procedure

An air pluviation technique was used to prepare loose dry sand specimens with an initial thickness of 50 cm. Key factors of air pluviation are flow rate and drop height of sand particles. The soil sample container was used to execute the air pluviation. At the beginning of the specimen preparation, the openings in the upper porous plate were shut and the pre-weighed sands were poured into the sample container. The lower porous plate was then slid to a preset position so that the openings in the upper porous plate became partially open, thus allowing the sand sample to flow through the opened area. A constant drop height of sand particles was obtained by lifting the container at a constant speed equal to the rate of increase in the thickness of the sand specimen. Wet sand specimens were prepared from the air-pluviated dry sand specimens by allowing water to seep into the tank from the drainage inlet. The water table in the wet sand specimens was kept at a depth of 25 cm until dampness appeared on the surface of the specimens, so that the water table was always below the crater.

Pounder drop heights of 20, 30 and 60 cm were used for a parametric study. The pounder made an impact on the specimen surface at the central point only. Upon impact, the pounder partially penetrated into the specimen. The pounder was then carefully retrieved by hand and attached to the electromagnetic unit for the next release. A crater appeared after the removal of the pounder. Depth, width and heave were measured to characterize the crater. Trial DC tests were carried out to obtain the relationship between the crater depth and the drop number to determine the number of drops for the production laboratory DC test.

The electric cone penetrometer was driven downward by a motor at a constant speed of 0·5 cm/s. The tip resistance of the penetrometer was measured and was considered as an index, rather than as a calibrated penetration resistance. Penetrations were made at 13 locations on each specimen to determine the influence zone of densification. A Plexiglass plate was used to mark the locations for the cone penetration tests, as shown in Fig. 5. For each specimen, the penetration always started from the central point (i.e. location 0) and followed by locations 1, 2, 3, ..., 12, sequentially. The results of the cone penetration tests on two specimens without dynamic compaction show that the repeatability of the specimen preparation technique is satisfactory, since the penetration resistance profiles from the two specimens are in good agreement with each other.

For the tests without dynamic compaction, cone penetration tests were carried out as soon as the specimens were obtained. For the tests with dynamic compaction, the dimensions of the craters and heave were measured during and after the dynamic compaction, followed by the cone penetration tests. For the tests on wet sand specimens with dynamic compaction, the cone penetration tests were carried out as soon as the induced excess pore water pressure reached zero. By this procedure, the effect of ageing on cone penetration resistance is avoided.

Test results and discussions

In this investigation, eight DC tests were carried out on dry Mai-Liao sand specimens, four DC tests were carried out on wet Mai-Liao sand specimens, and two DC tests were carried

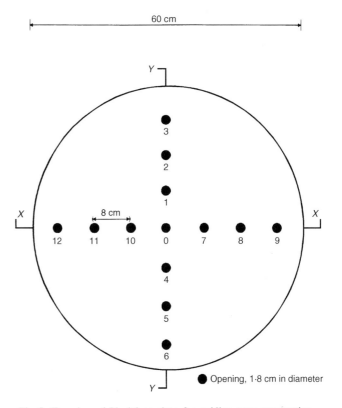

Fig. 5. Top view of Plexiglass plate for guiding cone penetration

out on dry Ottawa sand specimens. For each of the three types of specimens mentioned above, two DC trial tests were carried out with the flat-based pounder to determine the number of drops at which the increase in the depth of crater per drop became insignificant. It was found that the number of drops was 12 and 20 for the Mai-Liao sand and the Ottawa sand specimens respectively. They were then used for the laboratory production DC tests.

THE CRATERS

The craters produced in the different types of specimen are different in shape. The craters in the dry Mai-Liao sand specimens all have a bowl shape, as illustrated in Fig. 6. The craters in the wet Mai-Liao sand specimens are nearly cylindrical in shape, as illustrated in Fig. 7. Interestingly, heave around the craters was absent in these tests. Table 1 lists the width at the top, the depth and the volume of these craters. The crater volume was calculated with cylindrical coordinates r and z as follows:

$$\int_0^{2\pi} \int_0^{r_1} \int_{z=0}^{z=f(r)} r \, dz \, dr \, d\theta \tag{1}$$

The parameter r_1 in equation (1) is defined in Fig. 8 The relation $z = f(r)$ represents the variation of magnitude of

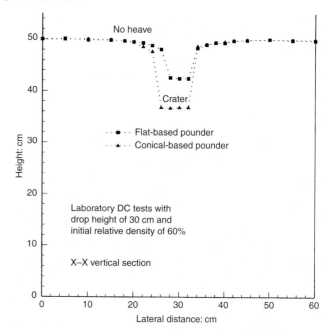

Fig. 7. Crater characteristics of wet Mai-Liao sand specimens

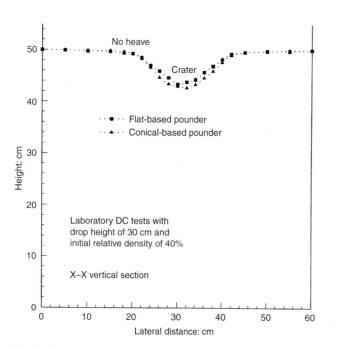

Fig. 6. Crater characteristics of dry Mai-Liao sand specimens

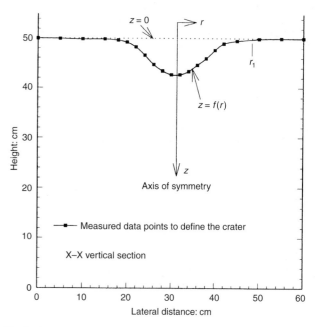

Fig. 8. Parameters used for calculating crater volume of Mai-Liao sand specimens

Table 1. Dimensions and volumes of craters in Mai-Liao sand specimens

Test no.	Specimen condition	Initial D_r: %	Pounder base	Drop height: cm	Crater width: cm	Crater depth: cm	Crater volume: cm^3	Ratio of crater volume
1	Dry	30	Flat	20	20·1	6·2	662	1
2	Dry	30	Conical	20	23·2	8·3	1170	$1170/662 = 1·77$
3	Dry	30	Flat	30	21·0	8·0	924	1
4	Dry	30	Conical	30	25·5	10·5	1802	$1802/924 = 1·95$
5	Dry	40	Flat	30	19·6	6·9	694	1
6	Dry	40	Conical	30	21·6	7·4	904	$904/694 = 1·30$
7	Dry	40	Flat	60	23·0	9·2	1274	1
8	Dry	40	Conical	60	25·2	9·9	1646	$1646/1274 = 1·29$
9	Wet	60	Flat	30	10·0	7·4	427	1
10	Wet	60	Conical	30	11·7	13·9	740	$740/427 = 1·73$
11	Wet	63	Flat	60	12·5	11·0	990	1
12	Wet	60	Conical	60	12·1	15·0	1320	$1320/990 = 1·33$

depression in the radial (r) direction, and can be obtained by carrying out regression analysis on the measured data points in Fig. 8.

The craters in the dry Ottawa sand specimens have a bowl shape, as shown in Fig. 9. In contrast to that observed from the Mai-Liao sand specimens, heave is clearly appearing around these craters. This observation deserves some discussions. First, one might suspect that the initial relative densities of 30% and 40% for the dry Mai-Liao specimens are too low, but the pounders did not submerge into the dry Mai-Liao specimens; the depths of the craters (Table 1) were only about the same as the height of the pounders. Second, the wet Mai-Liao specimens were probably liquefying, so that heaving of the sands was arrested. Third, the dry Mai-Liao sand specimens contain 8% fines and have high initial void ratios (around 1), which should have a contraction response upon dynamic compaction, whereas the dry Ottawa sand specimens have low void ratios (around 0·68), contain no fines, and are coarser than the Mai-Liao sand, which would have a dilative response with increasing drop number of the pounder. In fact, it was observed on the dry Ottawa sand specimens that heaving around the crater became significant after a few drops, when the initially loose Ottawa sand had become denser than its initial state. It appears from the above discussions that heave around the crater should not always be expected. Table 2 lists the width at the top, the depth and the volume of the craters in the Ottawa sand specimens. The crater volume is the difference between the depression volume and the heave volume. After adding a line to represent the specimen surface before DC, the heave and the depression due to DC can easily be distinguished, as shown in Fig. 10. The depression volume was then calculated by using equation (1). The heave volume was calculated by using the following equation:

$$\int_0^{2\pi} \int_{r_1}^{r_2} \int_{z=0}^{z=f(r)} r \, dz \, dr \, d\theta \qquad (2)$$

The $z = f(r)$ relation for the heave starts from r_1 and ends at r_2, and can be obtained from a regression analysis on the

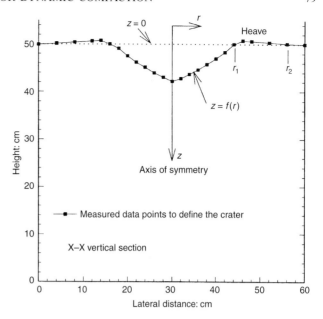

Fig. 10. Parameters used for calculating crater volume of Ottawa sand specimens

measured data points in Fig. 10. It can be calculated from Table 2 that the net depression volume for the crater induced by the flat-based pounder is about 3–4% larger than that for the crater induced by the conical-based pounder. Furthermore, Table 1 and Table 2 show that for the same initial relative density of 40% and drop height of 60 cm, but for different drop numbers of 12 and 20 for Mai-Liao sand and Ottawa sand respectively, the crater is larger in Mai-Liao sand than that in Ottawa sand. This is probably due to different grain characteristics in the Mai-Liao sand and the Ottawa sand.

For specimens of the same initial density, the larger the crater the higher the increase in specimen density, and thus higher the efficiency of dynamic compaction. It can be seen from Table 1 for the Mai-Liao specimens that the crater volume induced by the conical-based pounder is 29–95% larger than that induced by the flat-based pounder. It appears from the above data that the conical-based pounder is more efficient in densifying the Mai-Liao sand specimens than the flat-based pounder. On the other hand, the craters of the dry Ottawa sand specimens by different pounders are nearly identical. Since there was no liquefaction in these specimens, the shear waves generated by the impact of the pounders did very little in densification. This is expected for sands exhibiting a dilative response upon dynamic compaction. In this context, the densification is obtained mainly from compression rather than from shearing, and the conical-based pounder is probably inferior to the flat-based pounder. However, while impacting on the sand specimen, the conical-based pounder pushes more neighbouring sand particles aside laterally, and thus induces more shearing than that induced by the flat-based pounder, but its effect on densification is difficult to quantify.

The test results as listed in Table 1 show some further interesting relations. For the same initial relative density, the depth of the crater in all tests increases with increase in drop height (i.e. increase in compaction energy). The results of tests 3–6 show that, for the same drop height, the depth of the crater decreases with increasing initial relative density (i.e. decreases in compressibility). The widths of the craters of dry specimens

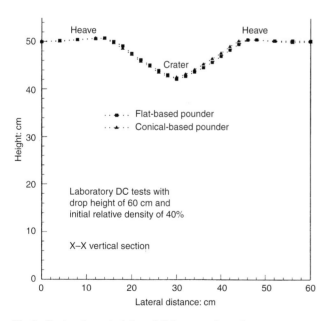

Fig. 9. Crater characteristics of Ottawa sand specimens

Table 2. Dimensions and volumes of craters in dry Ottawa sand specimens

Test no.	Initial D_r: %	Pounder base	Drop height: cm	Crater width: cm	Crater depth: cm	Heave volume: cm³	Depression volume: cm³	Crater volume: cm³	Ratio of crater volume
1	40	Flat	60	28	7·7	837	1803	966	1
2	40	Conical	60	27·8	7·6	812	1739	927	$927/966 = 0·96$

are larger than those of the wet sand specimens, whereas the depths of the craters of dry specimens are smaller than that of wet specimens. Note that the craters in dry specimens have sloping sides (Fig. 6), whereas the craters in wet specimens have nearly vertical sides (Fig. 7). There is zero cohesion in dry sand, so that the sides of the crater stabilize with a slope. The upper half of the wet specimens is partially saturated (see test procedure), and therefore an apparent cohesion develops, which results in the nearly vertical sides. In wet specimens, water appeared in the bottom of the crater owing to excess pore water pressure induced by the pounding. In this case, the next drop was conducted after the water had escaped from the bottom of the crater. Fig. 11 shows the pore water pressures measured at the bottom of a wet specimen with DC by the flat-based pounder and of a wet specimen with DC by the conical-based pounder. It can be seen that the pore water pressures increase rapidly with each pounding and dissipate less rapidly with time. For both specimens the maximum induced pore water pressures are around 1·1 kPa, which corresponds to a pressure head of 11 cm; but the induced pore water pressure near the crater must have been much greater than that measured at the bottom of the specimen, and must have led to a liquefaction condition so that the depths of the craters of wet specimens are larger than those of dry specimens.

INFLUENCE ZONE AS DETERMINED BY THE CONE PENETRATION TEST

The efficiency of dynamic compaction may also be expressed in terms of the extent of the influence zone in both the vertical and lateral directions. For specimens with the same initial conditions and the same compaction energy, the larger the influence zone the higher the efficiency. Cone penetration tests were carried out at 13 locations (see Fig. 5) in each specimen with or without DC to determine the influence zone. The result is illustrated by drawing contours of a given cone penetration resistance of six Mai-Liao dry sand specimens on the X–X vertical plane (see Fig. 5). Curve a–a in Fig. 12 represents the 300 kPa contour of two specimens without dynamic compaction. Curves b–b and d–d in Fig. 12 represent the 300 kPa contours of two specimens densified by the flat-based and the conical-based pounders respectively, with the same drop height of 30 cm and drop number of 12. Curves e–e and g–g represent the 300 kPa contours of two specimens densified by the flat-based and the conical-based pounders respectively, with the same drop height of 60 cm and drop number of 12. The cone penetration resistance of 300 kPa was selected to draw the contours since it was about the highest measured value for the specimens without DC. It can be seen from Fig. 12 that the

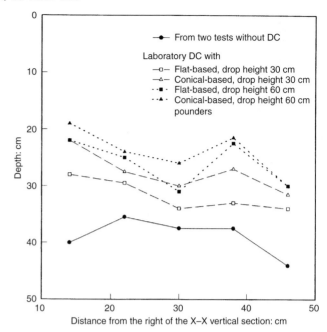

Fig. 12. Contours of penetration resistance of 300 kPa on vertical section X–X of six dry specimens of Mai-Liao sand, initial relative density 40%

specimens with DC have the same penetration resistance within 18–34 cm depth as that of the specimens without DC within 36–45 cm depth. This clearly indicates an effective densification by DC. Furthermore, curves b–b and e–e in Fig. 12 are in general below curves d–d and g–g, which indicates a larger influence zone and therefore a better efficiency of the conical-based pounder than of the flat-based pounder.

Cone penetration tests on two specimens of Ottawa sand show that the efficiency of DC with the conical-based pounder was about the same as with the flat-based pounder. This can be illustrated by showing the cone penetration resistance measured at location 0 (see Fig. 5) in Fig. 13, and the cone penetration resistance measured at locations 1, 4, 7 and 10 (see Fig. 5) in Fig. 14. Cone penetration tests were carried out after filling up the craters carefully with sands. It can be seen from Fig. 13 that the post-DC cone penetration resistance within one pounder diameter of depth underneath the crater is about 20–100% larger with the flat-based pounder than with the conical-based pounder. This is expected since the flat-based pounder induces a

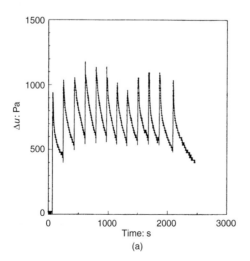

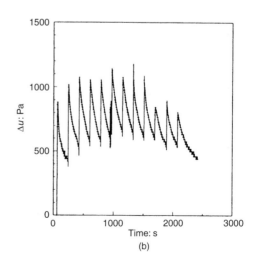

Fig. 11. Induced pore water pressures during dynamic compaction, drop height 60 cm and drop number 12: (a) conical-based pounder, initial relative density 60%; (b) flat-based pounder, initial relative density 63%

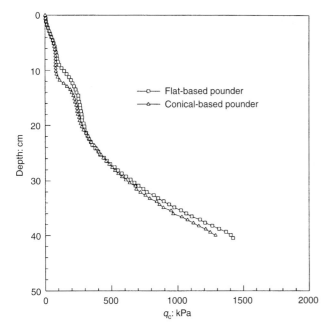

Fig. 13. Measured data of electric cone penetration resistance at location of impact

confined compression in that area, whereas the conical-based pounder induces a lateral flow in that area. A combination of larger penetration and less compression by the conical-based pounder results in a crater almost identical to that induced by the flat-based tamper. Note that the above discussions are based on test results from specimens of 40% relative density and have improved by dropping pounders from a height of 60 cm. Different efficiencies may be obtained from different pounders when these conditions are changed.

CONCLUSIONS

Densification of loose sandy soils by dynamic compaction with flat-based pounders has been found to be successful in many projects. An innovative idea of using a conical-based pounder to improve the efficiency of dynamic compaction is suggested by the experimental findings. Results of the laboratory dynamic compaction tests on both dry Mai-Liao sand specimens of initial densities of 30% and 40% (loose) and wet Mai-Liao sand specimens of initial density of 60% (medium dense) show that the volume of the craters induced by the conical-based pounder is 29–95% larger than that induced by the flat-based pounder. Heave was not found in these tests. The grain size characteristics and contractive response of the Mai-Liao sand specimens could be responsible for the lack of heave around the crater. The results of the electric cone penetration

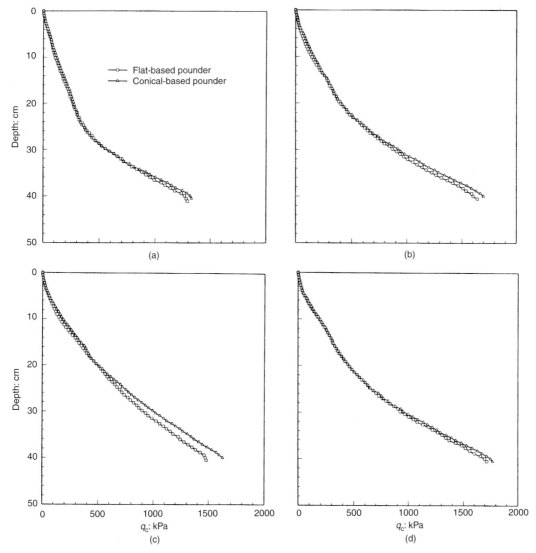

Fig. 14. Measured cone penetration resistance for dry Ottawa sand, 8 cm away from location of impact: (a) location 1; (b) location 4; (c) location 7; (d) location 10

tests on the Mai-Liao sand specimens also show that the conical-based pounder results in a larger influence zone than that produced by the flat-based pounder. The higher efficiency of dynamic compaction using the conical-based pounder can be explained by a higher degree of sand particles shearing. Lateral displacement of sand particles around the conical-based pounder is higher than that produced by the flat-based pounder, thus resulting in a stronger shearing behaviour. However, the results of two laboratory dynamic compaction tests on Ottawa sand show that the crater induced by the conical-based pounder is about identical to that induced by the flat-based pounder. Heave was appearing around the crater in these tests. The grain size characteristics and dilative response of the Ottawa sand specimens could be responsible for the heave. It can be concluded from these test results that the conical-based pounder has an efficiency better than or equal to that of the flat-based pounder in dynamic compaction.

ACKNOWLEDGEMENTS

This research is supported by the National Science Council of Taiwan under grant number NSC 86-2221-E-033-007. The Department of Education of Taiwan provides funds for building the apparatus. The writers are grateful to both organizations for their continuous support.

REFERENCES

Charles, J. A., Burford, D. & Watts, K. S. (1981). Field studies of the effectiveness of dynamic compaction. *Proc. 10th Int. Conf. Soil Mech. Found. Engng, Stockholm* **3**, 617–622.

Cheney, J. (1983). Physical modeling in geotechnical engineering. *Proceedings of workshop on geotechnical centrifuge modeling*, University of California, Davis, July.

Chow, Y. K., Yong, D. M., Yong, K. Y. & Lee, S. L. (1994). Dynamic compaction of loose granular soils: effect of print spacing. *J. Geotech. Engng Div., ASCE* **120**, No. 7, 1115–1133.

Ellis, G. W. (1986). Dynamic consolidation of fly ash. *Proceedings of international symposium on environmental geotechnology*, 564–573.

Gambin, M. P. (1983). The Menard dynamic consolidation method at Nice Airport. *Proc. 8th Eur. Conf. Soil Mech. Found. Engng, Helsinki* **1**, 231–234.

Jessberger, H. L. & Beine, R. A. (1981). Heavy tamping: theoretical and practical concepts. *Proc. 10th Int. Conf. Soil Mech. Found. Engng, Stockholm* **3**, 695–699.

Leonards, G. A., Cutter, W. A. & Holtz, R. D. (1980). Dynamic compaction of granular soils. *J. Geotech. Engng Div., ASCE* **106**, No. 1, 35–43.

Lukas, R. G. (1980). Densification of loose deposits by pounding. *J. Geotech. Engng Div., ASCE* **106**, No. 4, 435–446.

Lukas, R. G. (1986). *Dynamic compaction for highway construction: 1, Design and construction guidelines*, Report FHWA/RD-86/133. US Department of Transportation, FHWA.

Lukas, R. G. (1995). *Dynamic compaction*, Geotechnical Engineering Circular No. 1, Publication No. FHWA-SA-95-037. US Department of Transportation, FHWA.

Mayne, P. W., Jones, J. S. & Dumas, J. C. (1984). Ground response to dynamic compaction. *J. Geotech. Engng Div., ASCE* **110**, No. 6, 757–774.

Menard, L. & Broise, Y. (1975). Theoretical and practical aspects of dynamic consolidation. *Géotechnique* **25**, No. 1, 3–18.

Miller, G. F. & Pursey, H. (1955). On the partition of energy between elastic waves in a semi-infinite solid. *Proc. R. Soc., London* **233**, 55–69.

Orrje, O. & Broms, B. (1970). Strength and deformation properties of soils as determined by a falling weight. *Swedish Geotechnical Institute Proceedings* No. 23, 1–25.

Poran, C. J. & Rodriguez, J. A. (1992). Design of dynamic compaction. *Can. Geotech. J.* **29**, No. 5, 796–802.

Senneset, K. & Nestvold, J. (1992). Deep compaction by vibro wing technique and dynamic compaction. In *Grouting, soil improvement and geosynthetics*, GSP 30, pp. 889–901. American Society of Civil Engineers.

Smoltczyk, U. (1983). Deep compaction. *Proc. 8th Eur. Conf. Soil Mech. Found. Engng, Helsinki* **3**, 1105–1114.

Van Impe, W. F. (1989). *Soil improvement techniques and their evolution*. Rotterdam: Balkema.

Welsh, J. P. (1983). Dynamic deep compaction of sanitary landfill to support superhighway. *Proc. 8th Eur. Conf. Soil Mech. Found. Engng, Helsinki* **1**, 319–321.

Merrifield, C. M. & Davies, M. C. R. (2000). *Géotechnique* **50**, No. 6, 675-681

A study of low-energy dynamic compaction: field trials and centrifuge modelling

C. M. MERRIFIELD* and M. C. R. DAVIES†

This paper describes research into understanding the efficiency of the low-energy dynamic compaction process, and the development of a novel technique of real-time monitoring that can demonstrate soil improvement in quantitative engineering units during the process. The research, undertaken in the field and using a 500 gTonne geotechnical centrifuge, investigates the validity of applying the principles of the WAK (wave-activated stiffness [K]) test analysis to monitor the progress of compaction, allowing the process to be halted once the required degree of improvement has been reached. The analytical procedures underpinning the interpretation of the compaction process data in the time and frequency domain are presented, including the derivation of dynamic stiffness, depth of compaction, and damping factors associated with the combined footing/soil system. The methods of instrumentation and data acquisition are described for both field and centrifuge test programmes, along with the procedures adopted for real-time signal conditioning and data recognition. The results of the field and centrifuge test programmes are discussed, and conclusions are drawn on the effectiveness of the compaction process, confirming the validity of the WAK test analysis in the prediction of improvement in soil stiffness with increasing number of blows.

KEYWORDS: centrifuge modelling; compaction; full-scale tests; ground improvement; model tests; monitoring

Cette étude décrit la recherche menée dans le but de comprendre l'efficacité du processus de compactage dynamique à faible énergie ainsi que le développement d'une technique novatrice de suivi en temps réel capable de quantifier l'amélioration d'un sol en unités industrielles, en cours de processus. La recherche, menée sur le terrain et utilisant une centrifugeuse géotechnique de 500 g Tonne, a pour but de montrer s'il est valide d'appliquer les principes de l'analyse de l'essai WAK (rigidité [K] activée par ondes) pour suivre le progrès du compactage, permettant de stopper le processus une fois que le niveau d'amélioration requis a été atteint. Nous présentons les procédés analytiques soustendant l'interprétation des données du processus de compactage dans le domaine temps et fréquence, notamment la déduction de la rigidité dynamique, de la profondeur de compactage et des facteurs de tassement associés au système combiné assise/sol. Nous décrivons les méthodes d'instrumentation et d'acquisition de données pour les programmes d'essais sur le terrain et en centrifugeuse, avec les procédés adopté pour la préparation des signaux en temps réel et la reconnaissance des données. Nous analysons les résultats des programmes d'essai sur le terrain et en centrifugeuse et nous en tirons des conclusions quant à l'efficacité du processus de compactage, confirmant la validité de l'analyse de l'essai WAK à prévoir l'amélioration de la rigidité du sol avec un nombre de frappes croissant.

INTRODUCTION

The decline of heavy industry in the latter half of the twentieth century saw the growth of large areas of derelict land, comprising both natural and filled ground, resulting from the demolition of buildings and the accumulation of building waste and domestic or commercial refuse. Common in urban areas, and surrounded by existing structures, many of which are sensitive to excess ground vibration, this derelict land has often been considered unsuitable for development owing to its unacceptably high compressibility and variable mechanical properties. Construction on such sites can be achieved only by relatively expensive foundation options (such as piles), by removal of the existing ground and replacement with a material of acceptable engineering properties, or by improvement of the engineering properties of the existing ground.

One of the methods available for the mechanical modification of ground properties is dynamic compaction. While is has been used for many years as a soil improvement process, dynamic compaction was only formalized in a scientific manner in the last 25 years (Menard, 1976; Menard & Broise, 1975). The system conceived by Menard, frequently referred to as *heavy tamping*, involves repeated blows on the ground surface from a mass. Typically, the mass used in this technique is of the order of 10–20 t, falling through a height of up to 20 m, although an application in which a mass of 220 t falling through 24 m has been reported (Gambin, 1979). As would be expected, the depth to which improvement of soil properties may be obtained using this technique has been demonstrated to be a function of the energy imparted to the soil by the falling mass, and improvement at depths of up to 33 m have been proved (Mayne *et al.*, 1984). However, the effectiveness of this technique in improving soil stiffness to a significant depth is countered by the effects of induced ground vibrations on nearby sensitive structures, which are exacerbated as the energy per blow is increased. Because of this environmental restraint, in urban environments other ground modification techniques (such as vibro compaction) are often selected.

It is not always necessary to achieve ground improvement to great depths: for example, when the ground to be modified consists of a layer of loose material with a depth of only a few metres, or when only a small increase in bearing capacity is required without concern about settlement. In an attempt to provide a rapid, economic and efficient ground improvement system, this has led to the development of *low-energy dynamic compaction*, or *repeated compaction*. Involving typically a mass of 7 t falling through 2 m onto a movable steel target 1·5 m in diameter, this process has been effective in compacting ground to depths of 3–4 m (Watts & Charles, 1993; Allen, 1996). Since the energy per blow is less than in conventional dynamic compaction, the consequential risk of damage to the existing infrastructure is potentially reduced, although much lesser than normal dynamic compaction impact energies that would be adopted for comparable depth of treatment.

As with all ground modification techniques there is a requirement for monitoring to ensure that the required degree of improvement has been achieved. Given the rapidity of the low-energy dynamic compaction process, the time required in conventional site investigation techniques to evaluate progress accounts for a significant proportion of the treatment time. The research reported in this paper is focused on the evaluation of the compaction process and the development of a real-time

Manuscript received 31 January 2000; revised manuscript accepted 16 May 2000.
* Manchester School of Engineering, University of Manchester, UK.
† Department of Civil Engineering, University of Dundee, UK.

monitoring procedure to provide immediate data to terminate tamping once the required improvement has been achieved, thus allowing the process to be used optimally.

The results of a field trial in which the compaction process was evaluated are described. Site investigations were conducted prior to and following compaction to assess the effectiveness of the treatment. In addition, during this trial an attempt was made to monitor continuously the improvement of ground stiffness using a geophysical technique based on the principles of the wave-activated stiffness (WAK) test developed to measure the stiffness of ground beneath a footing (Briaud & Lepert, 1990). However, while—as is discussed below—conventional site investigation indicated clearly that the compaction process increased the strengths and stiffness of the ground, the change in stiffness was not detected by the geophysical technique. This was because problems were encountered when using conventional geotechnical instrumentation (geophones) under conditions of impulsive forces with accompanying high accelerations. As a result of this, a series of centrifuge tests was conducted to investigate under highly controlled, repeatable conditions both the process of ground improvement using the low-energy dynamic compaction technique, and the methodology for an immediate monitoring of the process.

APPLICATION OF THE WAK TEST ANALYSIS

The method proposed for assessing ground improvement by monitoring the change in stiffness is based on the wave-activated stiffness (WAK) test (Briaud & Lepert, 1990; Maxwell & Briaud, 1991; Tschirhart & Briaud, 1992). In the original WAK test, soil stiffness beneath a footing is measured by impacting the footing using a sledgehammer instrumented with a force transducer and monitoring the velocity response of the footing using geophones. The stiffness of the soil is obtained by considering the soil to be vibrating as a single degree of freedom (SDOF) system. For a variety of sites Maxwell & Briaud (1991) reported very good comparisons of the stiffness (subgrade reaction modulus), K_s, obtained using this technique with stiffness measured using the conventional plate bearing apparatus. The mass of the footing and vibrating soil is obtained during the analysis to obtain the stiffness value; from this it is possible to make an assessment of the depth of influence of the WAK test. When used to monitor the process of dynamic compaction, this provides a measure of the depth of improvement as outlined below.

The analysis relies on a relationship developed between the dynamic input to the system (impact force, $F(t)$) and the dynamic output (vertical velocity, $v(t)$). The most important feature of this test for the design of foundations is that, although essentially a dynamic test, the excitation and free-body response are of such low frequency that the output may be used to predict the static input–output relationship of the system. This is achieved by the derivation of a transfer function that is representative of this relationship. Note that, since the system is assumed to be linear, the input–output relationship and the transfer function will be independent of the magnitude of the input (Herlufsen, 1984). Although the load–displacement relationship of a natural soil is elasto-plastic over large displacements, the justification for the use of the WAK test is that it is believed that the displacements induced by the low-frequency impact, while initially high as the soil compacts, become smaller and essentially within the linear range of the load–displacement curve as the end of the compaction process is approached.

The analysis is used to predict the stiffness of the soil underlying the footing, K, the internal damping of the system, C, and the theoretical mass of soil and footing, M, which contributes to the behaviour of the system (Tschirhart & Briaud, 1992). A derivation of the analysis is presented in Appendix 1.

The depth of influence of the impact, d, can be estimated by assuming the mass of the vibrating soil beneath the footing to be spherical in shape. Knowing the mass and the density of the bulb, the volume may be determined, and hence the depth from the geometry, using the following equation:

$$\gamma = \frac{(M - m)}{\frac{4}{3}\pi \left(\frac{d}{2}\right)^3} \tag{1}$$

where M = mass of the footing/vibrating soil system, m = mass of the footing, γ = density of the soil, and d = diameter of assumed sphere (i.e. depth of influence of impact).

FIELD TRIALS
Equipment specification

The compactor used in the trial comprised a dropweight system mounted on the rear of a standard crawler crane and lifted by a hydraulic pile-driving hammer. The dropweight was of 7 t mass and fell through a height of 1·2 m. It was contained within an adjustable cage with a vibration damper to eliminate high-frequency vibrations. The energy was transmitted to the ground by way of a drive cap onto a circular anvil block (follower) situated inside a foot (1·5 m in diameter), which was attached to the compactor by chains, allowing it to be moved from one compaction location to the next by the compactor itself.

Instrumentation on the compactor included a specially fabricated dynamic load cell together with two commercially available geophones mounted on the foot (Allen, 1996). The data acquisition system was designed to process the data, develop the transfer function, and produce the soil stiffness and depth of improvement achieved after each impact, based on the WAK test analysis.

Test procedure

Representing a typical development site suitable for treatment by dynamic compaction, a test site was prepared by creating an excavation 10 m × 10 m in plan and 3 m deep in a high-plasticity sandy clay. This was backfilled with a graded fill material, including sand and small fragments of rock and concrete, by end tipping.

Prior to treatment the site was subjected to a site investigation, which included SPT, dynamic probe and zone loading tests. This provided the baseline data for the improvement. The compaction procedure included 50 impacts per location, at the rate of about one per second, on a square grid at 1·5 m spacing in a single pass. The site investigation was repeated after compaction.

Test results

Prior to the compaction process it was noted that the depth of penetration of the dynamic cone was of the order of 100 mm per 2 to 3 blows throughout the depth of the backfilled excavation. The dynamic cone penetration test results in Fig. 1 show large improvements in the strength of the ground extending to a depth of approximately 3 m, being the boundary of the fill and the underlying in-situ clay. The greatest improvement was at a depth of approximately 2 m, immediately above the soil interface. The SPT results are shown in Fig. 2; they confirm the findings of the dynamic cone penetration tests. The evidence is that the improvement extended to a depth of between 2 and 2·5 m.

The zone loading tests, carried out over an area of 2 m × 2 m, showed a fourfold increase in stiffness. The load–displacement curves for the soil state before and after compaction are shown in Fig. 3, and the increased stiffness, as a subgrade reaction modulus, K_s, is given for each case.

The impulse resulting from the impact of the falling weight onto the 1·5 m diameter footing plate initially produced velocities in excess of the range of the geophones: therefore part of the velocity record was lost. Although it was possible to process these data, the results obtained with part of the signal missing produced unreliable transfer function determinations.

The field trials demonstrated the effectiveness of the compactor in the improvement of fill properties. While the concept of the on-board instrumentation to provide real-time monitoring of

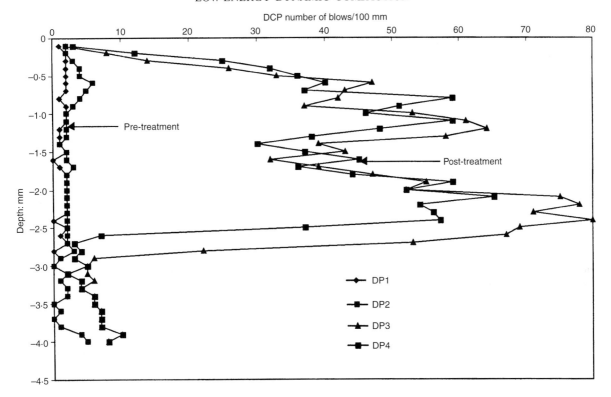

Fig. 1. Field trials: dynamic cone penetrometer test results

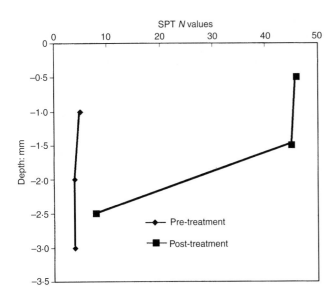

Fig. 2. Field trials: standard penetration test results

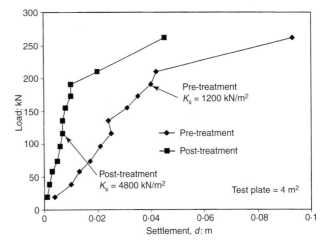

Fig. 3. Field trials: zone loading test results

compaction progress was intact, a need presented itself for further tests in a controlled environment to prove that reliable stiffnesses may be found in this manner. The centrifuge programme was designed specifically with this as an objective.

CENTRIFUGE MODELLING STUDY

Model design

The model compactor was designed to represent as closely as possible the compactor used in the field trials described by Allen (1996); see also Merrifield et al. (1998), Parvizi (2000). The compaction process in the centrifuge tests was simulated by a model dropweight mass of 0·875 kg falling through 100 mm onto a target 100 mm in diameter and 10 mm thick (Fig. 4). The design of the dropweight release mechanism was such that it ensured that the height of fall, and hence the energy trans-

mitted into the soil during each impact, was constant, notwithstanding the increase in the settlement of the target footing with increasing number of impacts. This provided a basis for an investigation of the improvement of soil stiffness with each successive impact, confirmed by a number of tests undertaken at the same g level.

The process in the centrifuge differed from that in the field in one respect. While the impact frequency in the field was 0·5 Hz—a specific characteristic of the equipment that highlights its attraction—this rate of impact was not attempted in the centrifuge. To scale this rate correctly would have involved an extra level of complexity of the apparatus, and therefore a blow rate of approximately one blow per minute was used. However, since the soil bed in the centrifuge tests comprised dry soft sand (typical relative density = 40%), no pore water pressures could develop, and so their dissipation was not considered. Therefore, in both the field trials and model tests, each impact was considered as a separate event: that is, the ground was subjected to repeated loading rather than cyclic loading.

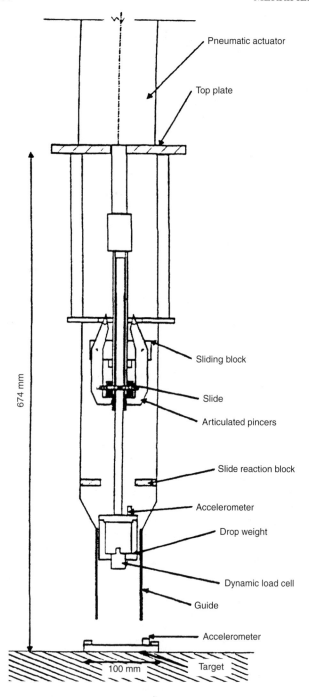

Fig. 4. Section through centrifuge model dropweight mechanism

Table 1. Centrifuge scaling factors for dynamic compaction

Parameter	Model scale	Field scale
Dynamic stiffness, K: (kN/m)	1	N
Subgrade reaction modulus, K_s: (kN/m³)	1	$1/N$
Mass of vibrating system, m: kg	1	$1/N^3$
Damping coefficient, C: kN/m/s	1	N^2
Depth of improvement, d: m	1	N
Time: s	1	N
Frequency, f: Hz	1	$1/N$

Table 2. Summary of Mersey River sand properties

Property of Mersey River sand	Value
Effective grain size, d_{10}: mm	0·15
Uniformity coefficient	1·8
SG of particles	2·655
Peak angle of internal resistance, $\phi'(R_d = 40\%)$	29·5°
Dry density, ρ_{dry}: kg/m³	
max.	1726
min.	1407

achieved by dry pluviating through a fine aluminium mesh, placed on the sand surface and drawn up through the sand in increments to provide a uniform bed of low density. The relative densities thus achieved varied between 37·5% and 41%. The results obtained, and presented below, show that with this deviation of relative density there was very little difference in the magnitudes of stiffness, vibrated mass and depth of influence achieved in each test at the same g level.

In order to measure stiffness using the principles of the WAK test it was necessary that both the impact force from the falling mass, $F(t)$, and the velocity of the footing, $v(t)$, were recorded. Since the experience of the field tests indicated that geophones (to measure velocity directly) were not appropriate for this application, in the model these were replaced by robust solid-state, low-mass miniature accelerometers. Velocity was obtained by integration of the acceleration signal. The dropweight was instrumented with both an accelerometer and a dynamic load cell, while an accelerometer was secured to the target (Fig. 4). Additional miniature accelerometers and dynamic earth pressure cells were placed in the sand bed at strategic locations to monitor the attenuation of the propagating stress wave initiated by the falling mass.

The data acquisition system comprised an on-board 32 channel transient recorder with a maximum synchronous sample rate of 50 k samples per second at 8-bit resolution. The signal processing and data manipulation was based on that proposed by Allen *et al.* (1994). This system, conceived to allow in-process estimation of compaction efficiency, monitors sequential impacts occurring in quick succession, returning values of estimated dynamic stiffness after each impact.

Test procedure

The majority of tests were undertaken at a centrifuge acceleration of $20\,g$. In addition, in order to validate the scaling laws governing the process, tests were also conducted at unit gravity, $10\,g$, $20\,g$ and $30\,g$. The model target dimensions, the mass of the dropweight and height of fall, however, were kept constant for all the tests. The consequence of this was that the equivalent field scale input energy per impact and the equivalent footing area increased with increase in centrifugal acceleration level.

The sand bed model was built at unit gravity, placed in the centrifuge and accelerated to the test g level. The target, located centrally on the sand bed, was then subjected to twelve impacts typically. Outputs from the load cell and accelerometers were recorded for each impact together with the signals from the instrumentation located in the soil. On selected tests, after

A convenient model scale of 1/20 was selected for the majority of the tests, the centrifuge acceleration being measured at the soil surface. Since the centrifuge scaling factors for mass and length are $1/N^3$ and $1/N$ respectively, at this scale the model represented a mass of 7 t falling though 2 m onto a target footing 2 m in diameter. Scaling factors are well established for dynamic events in centrifuge models (e.g. Taylor, 1995). The appropriate generic scaling factors together with scaling factors derived in this study specifically for modelling dynamic compaction are presented in Table 1.

Materials and instrumentation

Mersey River sand was used in the centrifuge model tests. Its properties are summarized in Table 2. Since the purpose of the experiments was to investigate the process of dynamic compaction and monitor its progress, it was essential that the sample should be prepared with a low relative density. This was

Labels on figure: Pneumatic actuator; Top plate; 674 mm; Sliding block; Slide; Articulated pincers; Slide reaction block; Accelerometer; Drop weight; Dynamic load cell; Guide; Accelerometer; 100 mm; Target

completion of the compaction process and still at test g level, a static loading rig was traversed into place above the target and a plate load test conducted.

RESULTS OF CENTRIFUGE MODELLING STUDY
Monitoring of compaction

As discussed above, because in the WAK analysis the system is assumed to be linear, the determination of the dynamic stiffness, K, is independent of the magnitude of the input signal. The method therefore measures the dynamic stiffness of the soil bed in the model irrespective of g level (see also the scaling factors in Table 1). To allow comparisons between the experiments and prototype conditions this value is converted into a subgrade reaction modulus, K_s:

$$K_s = \frac{K}{A} \qquad (2)$$

where K_s = subgrade reaction modulus (kN/m^3), K = WAK stiffness (kN/m), and A = surface area of target (m^2).

Both the subgrade reaction modulus, K_s, and the theoretical depth of compaction influence were determined by the WAK test analysis after every impact. The measurement of improvement can be clearly seen in Fig. 5, where the results of four tests are presented at field scale values. In each case the sand bed was constructed to a relative density of 38% ($\pm$4%) and subjected to the same energy per impact. The calculated depth of improvement, d, throughout each test is shown in Fig. 6. There is good agreement between each test for both K_s and d, with a clear demonstration that there was of the order of 65% increase in K_s while the depth of improvement increased by

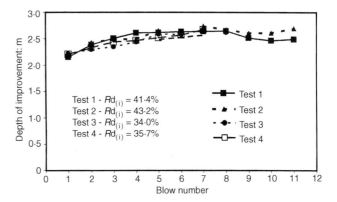

Fig. 6. Summary of centrifuge WAK test results: depths of improvement

30%. This translates to a depth of improvement in the field of 2·75 m and a maximum subgrade reaction modulus achieved after seven impacts of approximately 30–35 MN/m^3, representing an increase in stiffness of the order of 50–75%.

After seven impacts a reduction in both K_s and d is apparent. This may be attributed to a loosening of the sand within the compaction zone due to overcompaction and consequential shearing, volume change and reduction in stiffness (Lukas, 1986). Judicious use of the WAK test analysis makes it possible to predict the onset of this condition, allowing the timely termination of the compaction process.

Evidence of the depth of improvement may be inferred from the responses of the dynamic earth pressure cells located within the sand bed at depths of 100, 200 and 300 mm immediately below the model target on its centreline. These depths are translated into prototype values of 2, 4 and 6 m respectively. Fig. 7 shows the attenuation of the peak pressure from the point of impact in test 1 for impacts 2, 5 and 11 respectively. With increase in the number of impacts the level of stress at each location became greater, owing to increased stiffness of the soil immediately below the target. The stress level in a propagating stress wave is a function of the wave velocity, which in turn is a function of the stiffness of the soil (e.g. Clough & Penzien, 1975). Therefore, as compaction proceeds and the stiffness of the soil is increased (as monitored by the WAK test, Fig. 6), an increase in stress level at a location with number of impacts would be expected. Superimposed on Fig. 7 are the theoretical depths of compaction limits for each impact, computed by the WAK test analysis. These cross the stress attenuation curves over a range of 200–400 kPa. The contact pressure beneath

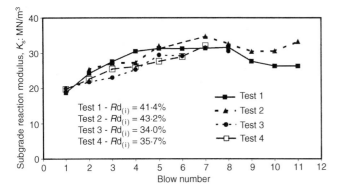

Fig. 5. Summary of centrifuge WAK test results: subgrade reaction moduli

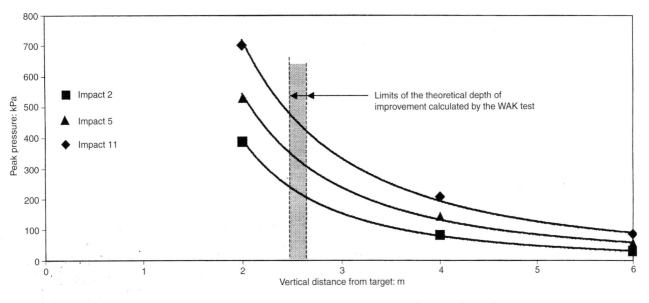

Fig. 7. Attenuation of peak particle pressure induced by impacts 2, 5 and 11 of a typical compaction series

conventional compaction plant is typically 300–400 kPa for smooth-wheeled rollers and 600–700 kPa for pneumatic rubber-tyred rollers (e.g. Das, 1984). Therefore the depth of compaction limits would appear to coincide with a zone beneath which compaction would not be expected.

Assessment of the stiffness monitoring system

The monitoring system was assessed in a series of model tests conducted at different acceleration levels. Values of the subgrade reaction modulus calculated for each test at both model and prototype scales are plotted in Fig. 8. This figure shows an increase in model K_s with increased acceleration. If the soil were linear elastic the value of subgrade reaction modulus would be independent of the acceleration level. However, soil stiffness is highly influenced by stress level (one of the fundamental reasons for geotechnical centrifuge modelling), and hence, as would be expected, the model value of K_s increases with the number of gravities.

It is well established that the subgrade reaction modulus, K_s, varies with the size of the plate used in a plate bearing test (e.g. Terzaghi, 1955). Since the same size of target was used in all the experiments, when the tests are scaled up the prototype size of the plate increases. Hence with increased gravity one would expect there to be a decrease in the prototype value of subgrade reaction modulus. The results of the experiments show this (Fig. 8).

Lysmer & Richart (1966) proposed an analogue to approximate the dynamic response of a *rigid* footing to vertical motion that included a spring constant, k_z, as follows:

$$k_z = \frac{4Gr_0}{1-\nu} \tag{3}$$

where G = shear modulus, ν = Poissons ratio, and r_0 = radius of footing. It can be easily shown that, with $K_s = k_z/A$, and where A is the area of the footing of radius r_0:

$$K_s = \frac{2E}{(1-\nu^2)\pi r_0} \tag{4}$$

where E = Young's modulus: that is, if the soil is assumed to be linear elastic, the subgrade reaction modulus, K_s, is inversely proportion to the radius of the plate, r_0.

In Fig. 9 the normalized prototype values of subgrade reaction modulus obtained from the experiments are plotted, as a percentage of the value of K_s at $1g$, against centrifugal acceleration together with a least-squares-curve fit of the data points. Also shown on the plot is the value of K_s predicted using equation (4), assuming that the soil properties do not vary. Comparison of the two curves shows that the trend in the experimental values follows that of the theory. However, while of the same order, the experimental values of K_s from the tests at elevated gravity are greater than the predicted values. This reflects the increases in the measured values of dynamic stiffness of the soil with increasing gravity. Nevertheless, these results, combined with those presented in Fig. 8, show that the centrifuge model WAK instrumentation successfully detected

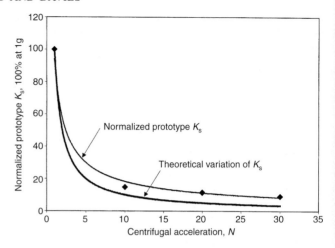

Fig. 9. Variation with centrifugal acceleration of normalized field K_s, and theoretical variation of K_s assuming linear elasticity

variations in soil stiffness with centrifuge acceleration, and that the variation in prototype subgrade reaction modulus with footing size could be modelled correctly.

Plate bearing tests were conducted in two tests undertaken at $20g$ immediately after compaction on the same sample of soil. To eliminate size effects, the plate used for these tests had the same dimensions as the target used in the compaction apparatus. It was located at a distance of 100 mm (that is, one diameter) from the compaction test footing. Because the g level was not changed during the process of positioning the loading device and testing, the soil was not subject to any change in stress that might affect its stiffness. However, stress waves spreading laterally from the point of compaction will probably have caused some modification to the soil properties. To take some account for this the average value of K_s measured over the first five impacts in the same tests is shown in Table 3. Comparison of the values of subgrade reaction modulus measured in the static and dynamic tests is very good quantitatively. Young's modulus, E, for the soil may be obtained by equation (4), using the values for K_s in Table 3. This gives average values of 40·3 MPa and 42·5 MPa for the static and dynamic tests respectively for the full drained condition of $\nu = 0·20$.

CONCLUSIONS

In common with other ground improvement processes, the effectiveness of ground modification using low-energy dynamic compaction may be assessed by conducting pre- and post-treatment site investigations. Whilst these investigation are essential, constraints of cost and time will limit the range of these investigations, and will not provide any information about the effectiveness of the process until treatment is complete. In-process monitoring is therefore most attractive, because it allows a continuous record of the effectiveness of the treatment at each location, which, combined, provide a complete record of the area treated. Additionally, the effectiveness of the technique may be monitored during operation, permitting treatment to be stopped, with associated financial savings, once predetermined values have been reached or continued treatment at one location is no longer effective.

Monitoring of the compaction process in the field trial using a technique based on the principles of the WAK test proved to

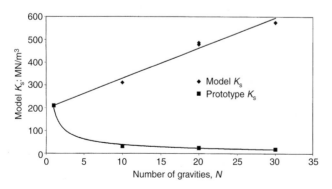

Fig. 8. Model and prototype K_s plotted against centrifuge acceleration, N

Table 3. Comparison between static K_s, measured by a plate load test, and dynamic K_s, measured by the WAK test conducted in centrifuge tests at $20g$ (model values)

Static K_s: MN/m^3	Dynamic (WAK) K_s: MN/m^3
537·9	575·7
524·1	552·9

be unsuccessful. However, the trial demonstrated that the process very effectively improved the engineering properties of a typical backfill material to a depth of about 2·5 m.

A series of centrifuge model tests were conducted to investigate under more highly controlled conditions the process of low-energy dynamic compaction together with its monitoring. The tests demonstrated most conclusively that a monitoring system based on the principles of the WAK tests could measure changes in dynamic stiffness, k_z, of the soil in tests conducted at different centrifuge accelerations. The monitoring system was also able to resolve the increase in model subgrade reaction modulus, K_s, between successive impacts during dynamic compaction. These results indicate that a prototype scale version of the monitoring technique should be able to be used for real time in-process monitoring of dynamic compaction. An additional benefit of the WAK technique is that it returns parameters that may also be measured using conventional site investigation techniques. This permits the values to be compared with those from conventional investigations or used in design.

ACKNOWLEDGEMENTS

The contributions of Dr Sarah Allen and Dr Mansour Parvizi to this investigation are most gratefully acknowledged. In addition, the authors wish to acknowledge the following organizations that contributed to various aspect of this study: BSP, International Foundations AMEC, Civil Engineering and DERA.

APPENDIX 1. DERIVATION OF THE WAVE ACTIVATED STIFFNESS (WAK) TEST ANALYSIS (BRIAUD & LEPERT, 1990)

The differential equation of motion describing the response of the single-degree-of-freedom system to a harmonic force is

$$F(t) = F_0 e^{i\omega t} \tag{5}$$

where F_0 = maximum force amplitude, t = time, and $i = \sqrt{-1}$. That is,

$$Mx''(t) + Cx'(t) + Kx(t) = F_0 e^{i\omega t} \tag{6}$$

where $x(t)$ is the vertical displacement of the footing due to $F(t)$, and ω is the angular frequency of the force function.

For the case of harmonic excitation it can be shown that $x(t)$, the particular integral of equation (3), is (Clough & Penzien, 1975)

$$x(t) = \frac{F_0}{\sqrt{(K - M\omega^2)^2 + C^2\omega^2}} e^{i(\omega t - \beta)} \tag{7}$$

where K = soil/footing stiffness, M = mass of the soil/footing system, C = damping coefficient, and $\tan \beta = C\omega/(K - M\omega^2)$.

Differentiating with respect to t gives the velocity of the system:

$$v(t) = \frac{F_0\omega}{[(K - M\omega^2)^2 + C^2\omega^2]^{1/2}} e^{i(\omega t - \beta + 90^0)} \tag{8}$$

The transfer function is the ratio of the response (velocity, v) to the input (dynamic force, F) and in terms of circular frequency, ω, and is expressed as

$$\left|\frac{v}{F}(\omega)\right| = \frac{\omega}{[(K - M\omega^2)^2 + C^2\omega^2]^{1/2}} \tag{9}$$

Since the maximum value of the transfer function occurs at the undamped natural frequency of the system, ω_n, this value may be found by differentiating equation (9) with respect to ω and setting to zero:

$$\frac{1}{[(K - M\omega^2)^2 + C^2\omega^2]^{1/2}} - \frac{C^2\omega^2 - 2M\omega^2(K - M\omega^2)}{[(K - M\omega^2)^2 + C^2\omega^2]^{3/2}} = 0 \tag{10}$$

Hence

$$\omega = \omega_n = \sqrt{\frac{K}{M}} \tag{11}$$

Substituting equation (7) into equation (5) gives

$$\left|\frac{v}{F}(\omega)\right|_{\omega=\omega_n} = \frac{1}{C} \tag{12}$$

Therefore, from investigation of the maximum value of the transfer function the damping coefficient, C, may be obtained along with a relationship between K and M.

By investigation of another point on the transfer function curve, that is

$$\omega, \left|\frac{v}{F}(\omega)\right|_{\omega=\omega_2}$$

the mass, M, may be calculated by solution of equation (9) and K by subsequent substitution into equation (11).

Verification of these values is obtained by substituting K, M and C into equation (9). This curve may then be compared with the experimental data.

Signal processing

Since the transient signal is aperiodic, it is necessary to represent it by a discrete Fourier transform. In this study it was convenient to find the transfer function by determination of the ratio of the cross-spectral density to the power spectral density:

$$H_{(\omega)} = \frac{S_{xy(\omega)}}{S_{xx(\omega)}} \tag{13}$$

where $S_{xy(\omega)}$ is the cross-spectral density, found by the Fourier transform of the cross-correlation function, and $S_{xx(\omega)}$ is the power spectral density, found by the Fourier transform of the autocorrelation function (Bendat & Piersol, 1980; Brooke & Wynne, 1988).

REFERENCES

Allen, S. (1996). *The low energy dynamic compaction of soil.* PhD thesis, University of Wales, College of Cardiff.

Allen, S., Davies, M. C. R. & Brandon, J. A. (1994). In process assessment of the stiffness of soil footings during dynamic compaction. *Proc. 19th Int. Seminar on model analysis, Leuven, Belgium*, 935–946.

Bendat, J. S. & Piersol, A. G. (1980). *Random data: analysis and measurement procedures*, 2nd edn. London: John Wiley & Sons.

Briaud, J.-L. & Lepert, L. (1990). WAK test to find spread footing stiffness. *ASCE* **116**, No. GE3, 415–432.

Brooke, D. & Wynne, R. J. (1988). *Signal processing: principles and applications.* London: Edward Arnold.

Clough, R. W. & Penzien, J. (1975). *Dynamics of structures.* New York: McGraw-Hill.

Das, B. M. (1984). *Principles of foundation engineering*, Monterey, CA: Brooks/Cole Engineering Division.

Gambin, M. P. (1979). Menard dynamic consolidation. *Proceedings of ASCE seminar on ground reinforcement*, Washington, DC, pp. 27–40.

Herlufsen, (1984). *Technical review: dual channel FFT analysis (Part 1).* Bruel & Kjaer.

Lukas, R. G. (1986). *Dynamic construction for highway construction*, Vol. 1, *Design and construction guidelines.* Report FHWA/RD.86/133. Washington: Federal Highway Administration, US Department of Transport.

Lysmer, J. & Richart, F. E. (1966). Dynamic response of footing to vertical loading. *J. Soil Mech. Found. Engng Div., ASCE* **92**, No. SM1, 65–91.

Maxwell, J. & Briaud, J. S. (1991). *WAK tests on 53 footings*, Department of Civil Engineering Report, Texas A&M University.

Menard, L. (1976). A discussion of dynamic compaction. In *Ground treatment by deep compaction.* London: Institution of Civil Engineers, 106–107.

Menard, L. & Broise, Y. (1975). Theoretical and practical aspects of dynamic consolidation. *Géotechnique* **25**, No. 1, 3–18.

Merrifield, C. M., Cruickshank, M. & Parvizi, M. (1998). Modelling of low energy dynamic compaction. *Proc. Int. Conf. Centrifuge 98, Tokyo*, 819–824.

Parvizi, M. (2000). Centrifuge modelling of low energy dynamic compaction. PhD thesis, University of Manchester.

Taylor, R. N. (1995). *Geotechnical centrifuge technology.* Chapman & Hall, London.

Terzaghi, K. (1955). Evaluation of the coefficient of subgrade reaction. *Géotechnique* **5**, No. 4, 197–326.

Tschirhart, A. R. & Briaud, J.-L. (1992). The WAK test: a method of measuring the static stiffness of a soil. *Users manual*, Texas A&M University.

Watts, K. P. & Charles, J. A. (1993). Initial assessment of a new rapid ground compactor. *Proceedings of conference on engineered fills.* London: Thomas Telford, 399–412.

Mayne, P. W., Jones, J. S. and E'Dinas, D. C. (1984). Ground response to dynamic compaction. *Journal of Geotechnical Engineering, ASCE* **110**, No. GT6, 757–773.

INFORMAL DISCUSSION

Session 2

CHAIRMAN MR BARRY SLOCOMBE

Barry Slocombe, *Chairman*

Before the discussion, a brief contribution from David Nash of Bristol University which follows on from Professor Gens paper on the effectiveness of temporary surcharge in reducing secondary compression.

David Nash, *University of Bristol*

I want to address the issue raised by Professor Gens, about design of pre-compression for reducing secondary settlements in soft clays. We are all familiar with Bjerrum's model of consolidation for soft clay. (Just to remind you, he developed this hypothesis actually from previous work by Taylor and Marchant.)We are probably all familiar with the diagram that shows loading, from in-situ stress, past yield stress to some final stress, and thereafter creep consolidation. The sloping lines indicate creep delay time. Now, when we are considering reducing the stress during creep, we are talking about moving to the left on this diagram and thereby, in some way, gaining some time. The coefficient of secondary consolidation is a very awkward parameter and it is necessary to have some sort of reference time to relate it to. Actually, it is more satisfactory to express the same information in terms of strain rate, as Bjerrum himself recognized.

We have a very similar diagram of the process. The consolidation process starts from a point, A, on the plot of strain or void ratio versus effective stress. We apply a stress increment and get primary consolidation to a point B. During this process both creep and dissipation of excess pore pressure is going on. Thereafter we get creep settlement. If we unload, as is going to be our strategy (as Professor Gens was suggesting), we move left on the diagram and then creep onwards at a reduced effective stress.

The parallel lines on this diagram do not represent time, they represent strain rate. I believe that it is more productive to use a stress-strain strain-rate model for consolidation analysis than to use stress-strain time. I have been playing around with the model suggested by Yin and Graham, and draw your attention to what is observed if strain is plotted against time. The primary consolidation is followed by creep, and if stress is not reduced, creep continues with a linear relationship, giving us a constant C_α with time. If we remove our surcharge, however, then indeed there is a major reduction in the rate of creep and C_α is reduced. What you notice from the stress-strain strain-rate model, and I believe this has been observed in the field, is that C_α, the slope on this line, will increase again with time. In the very long term, even though there may have been a temporary reduction in the coefficient of secondary compression, you may still end up with quite large secondary settlements.

I have used this analysis to explore the reduction of strain rate, or alternatively the reduction or change in slope of C_α, when the surcharge is immediately removed. Interestingly, there is a relationship between the reduction of strain-rate or C_α and overconsolidation ratio, very similar to the one found empirically, first of all by Ladd, and shown again at this symposium in the work of Professor Gens. There is a curve that has exactly the same shape, and the reduction is dependent on overconsolidation ratio, and the ratio of C_α in

the normally consolidated range to C_c. We get a family of curves depending on that ratio $C_\alpha : C_c$. It is not appropriate to go into the details of the other data points here, but I just make the point again that this initial reduction in C_α is not necessarily a guide to what is going to occur in the long term.

Professor Antonio Gens

David Nash brings up a quite valid point. What we have shown in the paper is only part of the story. The paper was of limited length so we chose to focus on what the field measurements were telling us of the real behaviour of soils in place. But Dr Nash is quite right, for the practical application we chose to base the analysis on a C_α sort of method, a well-known approach. It is true that it has to be used carefully as, in some cases, such an approach may lead to inconsistencies.

In fact, we did look at the existing strain-rate based models and we were not fully satisfied by them, indeed we are devising a new one at present. Anyway, what those models indicate is that, at some time, the initial, normally consolidated, secondary compression rate may be recovered, and this certainly has implications. We are quite aware of it and, in fact, we managed to recover this type of behaviour in laboratory tests. Fortunately, when considering practical implications, the difference between the two approaches is not that large. When you again pick up again the higher secondary consolidation rate, actual settlement rates are still quite small since we are working in a logarithmic scale.

In summary, although from the theoretical point of view the conceptual distinction is significant, from a practical point of view for computing long-term settlements, differences are not large.

Dr Angus Skinner, *Skinner & Associates*

It is very interesting that Professor Gens has developed the secondary consolidation model on the one-dimensional basis, where as in fact in reality the design of the structures, being water-retaining structures, are wholly based on differential settlements rather than one-dimensional single settlements. What are the anticipated secondary consolidation differentials anticipated to be on your structures?

Professor Antonio Gens

In fact, you can get an estimate of differential settlements using a basic one-dimensional model by considering the variation of the stresses at different points. This has been done by the people doing the final plant design; they are estimating differential settlements from one-dimensional secondary settlement calculations. However, from a geotechnical engineering point of view I always found it risky to base the design just on differential settlements because, if we are not that good at predicting total settlement, the difference between two uncertain magnitudes becomes even more doubtful. So my personal preference is to limit total settlements and let differential settlements take care of themselves. By limiting the total settlement, you are automatically bounding differential settlements. However, if differential settlements have to be computed, it can indeed be done, even with one-dimensional analysis.

Professor Tao-Wei Feng

I conducted my PhD study ten years ago and surcharging to reduce secondary settlement was one of my research topics. I think that Professor Gens and Professor Nash in the reports that they make both miss some important aspects of surcharge. When we unload and remove the surcharge the soil is going to rebound immediately and initially and the magnitude of rebound and the duration of rebound is also a function of the magnitude of the surcharge removed. The higher OCR you produce, the longer the rebound period before secondary compression reappears should be considered. There is a duration or period they are missing in this design.

Professor Antonio Gens

This was considered in design, and it is included in the last figure of the paper.

David Greenwood, *Geotechnical Consulting Group*

Question to Mike Jefferies: what is the energy release from a single charge? And as a comment, you made a remark about the different void ratios resulting from hydraulic filling and following explosive compaction. Surely, this is simply due to the different modes of deposition? – in the first case, layer by layer for particles without the full self-weight stress on; in the second, with the full weight on, settling rather quickly after explosion resulting in compaction. I suggest that because the strains are different you do not get the same structure in the ground.

Mike Jefferies

Considering the situation as energy from a single charge is a tad misleading. Only about a third, or less, of the explosive's theoretical energy goes into the cavity expansion, which is what does the work and fluidizes the soil. The rest of the energy gets radiated in shock waves – you see the Δp effect, in terms of the measured pore pressures, as the shock wave passes but it then dies off immediately. You then also have the worry about the coupling effect between the explosives, the explosive detonation velocity, and the way the soil responds. We have actually used charges between 1·5 kg and about 18 kg, increasing with depth (as noted in the presentation). In Table 1 in the paper, we have given the trade name of the various explosives we have used. Our experience is that it makes very little difference whether 40% or 80% relative-energy explosive is used.

David Greenwood, *Geotechnical Consulting Group*

Can you quantify the relative energies?

Mike Jefferies

No, I do not have those numbers available. It would be possible, using the information in Table 1, to look it up in the manufacturers' literature.

In terms of the consolidation process, I also think along the lines that different mechanisms are involved. I would also note that when we placed the fills hydraulically, they do go in as layers. They are not 20 m of fill placed in one shot, they are actually placed in typically 0·3 m lifts, so there is not necessarily a huge difference. The reason for raising the issue was to get everyone thinking, given the direction of this conference, and to start getting some science into what we do. Until we solve the problem of how to predict the result of different, for want of a better phrase, flocculation processes, we are not going to be able to predict exactly what we get from, for example, explosive compaction. We

are forced back into relying on precedence, which is what is in the paper.

Simon Wheeler, *delegate*

A question for Mike Jefferies. In your presentation you said that you have to keep the peak particle velocity quite low when close to existing structures. All the videos you showed were in areas where there aren't any buildings close by. I do not know if you are trying to sell this for the UK, but surely it would not work because of the amount of structures that we have. Therefore, how close can you be to an existing structure?

Mike Jefferies

The short answer is 6 m is the closest we have undertaken. The first video shown was actually from the Trans-Ex site, which is in the industrial development area of Vancouver. There were concerns at that site over adjacent structures. Our experience has been that the 2 inches per second rule, which is what we tend to get stuck with, is a little conservative and it is actually possible to go an awful lot higher. What tends to control these jobs is how well you do in informing the neighbours about what is going to go on.

The job with the 6 m distance was actually on a steel oil rig, where the structural engineer was concerned about damage to the rig. It was strain gauged and we went approximately 8 times over the 2 inches per second. There was no damage to the steel, as you would expect, but what did happen is that we sheared off every light on the rig. When you are causing vibrations, everything should be seismically proofed – in this instance it wasn't and we just wiped out all the lighting. The light bulbs just sheared off during the accelerations, but there was no damage to the structure. So you need to think about one or two things, depending on the frequency content.

We have tried other controls as well, such things as putting damping trenches in the way. The Trans-Ex site had a trench cut around the edge to control the propagation of surface waves. So there are things you can do, but the public relations perception is important as well.

Simon Wheeler, *delegate*

When using these compaction methods, all of them, has anybody actually considered the effects it has on the surrounding sites? Presumably, all compaction methods are site specific, but nobody seems to have considered what effect this is having on the surrounding sites that haven't had compaction. For example, blowing the site up – surely this will pushing water under the adjacent site.

Mike Jefferies

We have seen very little lateral transmission of pore pressure. What I have seen is what we had on the video, with vertical break out in pipes. We are not aware of having experienced much densification of adjacent properties due to what we are doing on our sites, with any of the techniques I have used. But having seen explosives working, it seems to be once the ground has been disturbed it will continue going down much more easily than the ground adjacent to it – I am not that surprised about that.

Barry Slocombe, *Chairman*

I can add a similar comment from the dynamic compaction point of view. If you do have problems on adjacent sites, it is vibration, not settlement.

Professor Michael Davies

On a similar subject, with repeated compaction you have to be a little bit careful Although locally it is a discrete event, similar to the heavy dynamic compaction, I have undertaken work where we have measured vibrations approximately 100 m away and because of lateral spreading of stress waves it becomes a cyclic loading, not a discrete loading. You then have to be careful about the frequency of compaction. This is work we have done and which we are about to publish.

Bill Craig, *University of Manchester*

A question for Mike Jefferies: in your presentation you demonstrated that compaction leads to liquefaction. The liquefaction eventually breaks out, you showed, a couple of minutes later, at the surface. That is an instability problem that results from some sort of fingering of the fluidized soil. It comes out through a pipe – while you may densify the site globally, the pipe must surely be very, very loose. Do you do any post treatment to the pipe area or do you just assume the site is uniform?

Mike Jefferies

An interesting question. It is the same problem around the explosive charge, you have a cavity which is then caved in and is also loose. The way we handle it is by undertaking repeated passes, and we tend to scale the charges down on the second or third pass. The reality is, however, that there are pockets of loosish soil in the final passes. They are not particularly loose because we have seen that re-consolidation can actually produce quite reasonable densities. But that is, I think, the explanation for the loosenings we see on the CPT traces and why I am very keen to show the lower bound values on our CPT plots in our paper, rather than just quoting average results.

David Greenwood, *Geotechnical Consulting Group*

Question to Professor Feng: does he not consider when dropping a flat pounder, that a wedge forms in front of it, essentially conical or tetrahedral, rather like a Prandtl wedge in a footing bearing problem, so that you get the same sort of shearing motions anyway?

I remember going with Louis Menard in the early 1970s to see some work at the French nuclear submarine base at Brest. That work was through about 2 m of water into the sea-bed. For that reason, he chose to use a conical pounder. This was to reduce the resistance as it fell through the seawater and, as far as he was concerned, he did not find that it gave him any benefit in practice on shore. He reverted every time on shore to the flat pounder. What would be Professor Feng's views on that?

Professor Tao-Wei Feng

The laboratory tests we conducted we did carefully and the results we obtained I think are most likely true. However, regarding the point you raised that a conical shaped soil wedge is formed below the flat based tamper, I think maybe it is an ideal situation and a theoretical point of view and maybe different from the actual pounder shape that we made manually. Also the conical base pounder impact is simple. The tip starts to penetrate the ground from the ground surface all the way down, so during that process the sand particles in the neighbourhood of the cone are pushed sideways, which will not happen with a flat pounder.

David Greenwood, *Geotechnical Consulting Group*

I think you may have a small problem in practice with a conical pounder, particularly of limited area – it depends on the area – in penetrating too much, then recovering the weight. If the pounder buries then it is not possible to keep the process going. It is important to be careful not to bury the weight.

David Johnson, *Bachy Soletanche*

Question for Professor Feng. There is clearly a difference in efficiency between the two sands you used for your experiments. Do you feel able to make any preliminary conclusions about where you think a shaped tamper would be beneficial as compared with a flat-bottomed tamper. For example, would it be dependent on different grading or different soil types?

Professor Tao-Wei Feng

The two soils I tested are quite different. For the Ottawa sand, there was no apparent difference between the different tampers. The results are about the same. However, for the finer Mai-Liao natural sand the conical pounder is more effective.

David Johnson

Can you make any wider conclusions, for example, for in a given set of conditions perhaps a shaped tamper would be beneficial but in another set of conditions perhaps it would not.

Professor Tao-Wei Feng

I would have to study further to collect more data and test more sand with different grain size distribution and different relative densities before I can make any such conclusions.

Alonso, E. E., Gens, A. & Lloret, A. (2001). *Géotechnique* **51**, No. 9, 822–826

WRITTEN DISCUSSION

Precompression design for secondary settlement reduction

E. E. ALONSO, A. GENS and A. LLORET

D. Nash, *University of Bristol*

The authors have presented a very interesting case study that supplements the meagre collection of such field data in the literature. The writer has recently considered the similar problem of using temporary surcharge to reduce the long-term settlement beneath embankments on soft clay (Nash & Ryde, 2001). While the authors have adopted a conceptual model in which creep strain and void ratio exhibit a linear relationship with time, in using such a model there is frequently difficulty in deciding the time origin, particularly if the time for primary consolidation is long.

As noted by the authors, when considering the effectiveness of precompression for secondary settlement reduction (Bjerrum (1972) proposed the use of a conceptual isotache model similar to that shown in Fig. 17. Here the parallel lines or isotaches connect soil states at which the plastic volumetric strain rate is constant. The solid line ABD in Fig. 17 indicates a path that might be followed by an element of soil loaded by an increment of total stress, $\Delta\sigma$, as it consolidates. As the effective stress increases from the initial state at point A, the creep strain rate increases as the soil state moves to the right, but the effective stress and strain rate are limited by the ability of the pore water to escape. After further consolidation the strain rate decreases until excess pore pressures have dissipated at point B, after which creep continues at decreasing rates towards point D. If the loading is reduced with a consequential reduction of effective stress, say from C to E, there may initially be some elastic swelling followed by a resumption of creep compression, with the creep rate reduced to that of the corresponding isotache at E. Thereafter creep continues at a reduced rate towards point F. Although it is known that in practice a large reduction of stress may result in secondary swelling, in this simple isotache model there is no lower limit to creep. Several such models have been proposed, including that of Yin & Graham (1996), which has been incorporated into a model of consolidation accelerated by vertical drains by Nash & Ryde (2001).

This model may be used to predict the creep behaviour when the effective stress is reduced (say from point C to point E in Fig. 17). This is illustrated in Fig. 18, which shows the creep behaviour predicted for an oedometer test under constant effec-

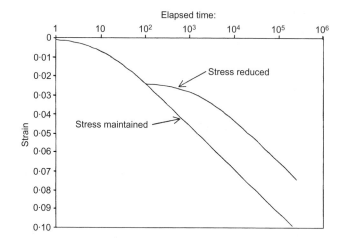

Fig. 18. Creep behaviour predicted using isotache model showing effect of reduction of applied stress

tive stress, and the effect of reducing the effective stress after an elapsed time $t = 100$. It may be seen that reduction of effective stress reduces both the creep rate and the coefficient of secondary compression C_α, and that C_α (but not the creep rate) gradually increases with time, behaviour that has been reported in the literature. It may also be seen that the subsequent settlement at large time is quite a significant proportion of that which would have occurred if there had not been any unloading.

The reduction of creep rate predicted by the model is related to the degree of overconsolidation achieved. Assuming that equally spaced isotaches indicate a logarithmic reduction of strain rate it may be shown from the geometry of Fig. 17 that when the effective stress is reduced from σ_1' to σ_2':

$$\frac{\dot{\varepsilon}_1}{\dot{\varepsilon}_2} = \frac{C_{\alpha_1}}{C_{\alpha_2}} = \left(\frac{\sigma_1'}{\sigma_2'}\right)^{(\lambda-\kappa)/\psi} \tag{4}$$

where the values of creep rate and C_α are those just before and just after the change of stress. Here λ and κ are the slope of the normal consolidation and swell lines (from critical-state soil mechanics), and ψ, which determines the separation of the isotaches, is approximately equal to $C_{\alpha NC}/2\cdot3$, where $C_{\alpha NC}$ is obtained for the normally consolidated soil. The parameters λ, κ and ψ may be obtained for example from incremental load oedometer tests, or the ratio $\psi/(\lambda-\kappa)$ may be approximated by the ratio $(C_\alpha/C_C)_{NC}$, for which typical values are given by Mesri & Godlewski (1977) and Holtz et al. (1991).

The relationship between the immediate reduction of creep rate and C_α and the degree of overconsolidation predicted by equation (4) is shown in Fig. 19 for various values of $\psi/(\lambda-\kappa)$. It is interesting to note that this is similar to that found empirically by Ladd (1971) and the authors in their Fig. 14, although the values predicted by the isotache model are independent of the surcharge time (contrary to Ladd's findings), and are smaller than those found empirically. This difference may arise because in practice determination of C_α after unloading requires measurements over a significant time period, and as noted above C_α is likely to increase with time. The writer is of the opinion that a varying C_α is not a very helpful quantity, and can more usefully be replaced by a single value that defines the spacing of the isotaches.

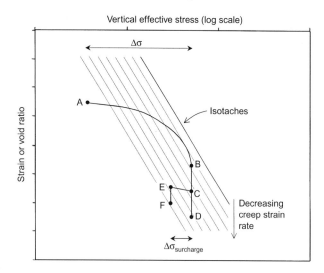

Fig. 17. Isotache model for compression of soft clays

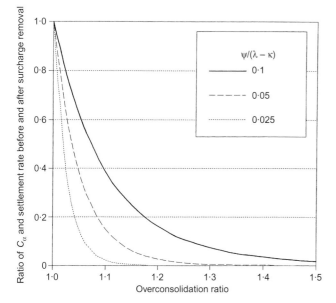

Fig. 19. Reduction of creep rate and $C\alpha$ due to overconsolidation predicted using the isotache model

In practice it may be necessary to reduce the applied stress before all excess pore pressures have dissipated, even if vertical drains are used to accelerate consolidation. The isotache model has been incorporated into a consolidation analysis, and used to predict the effect of precompression on the long-term settlement of a hypothetical test fill placed at the Bothkennar soft clay research site (Nash, 2001). The effects of a permanent loading of 100 kPa with a temporary additional surcharge of 50 kPa were modelled, to explore the effectiveness of the surcharge in reducing the long-term creep settlement. The analyses included vertical drains at approximately 2 m centres, but also considered consolidation without drains (and one case without creep). The settlement–time plots in Fig. 20 show that the long-term total settlement is independent of the time at which the surcharge is removed, but as might be expected the timing strongly influences the residual settlement after surcharge removal. Points A,

B and C in Fig. 20 indicate unloading after surcharge periods of 1, 2 and 3 years respectively, and the figure shows clearly the penalty of removing the surcharge before the long-term settlement has been achieved.

Such isotache models offer a framework for interpreting field and laboratory creep data involving loading in the normally consolidated region followed by a small degree of unloading. The data given by the authors in Fig. 9 appear to be consistent with such a model. It would be interesting to plot the field and laboratory data for each depth in the form logarithm of strain rate against strain (Nash & Ryde, 2001). Such a plot would indicate whether the separation of the isotaches (given by the slope of the lines) is indeed the same before and after removal of the surcharge as expected.

T.-W. Feng. *Chung Yuan Christian University, Taiwan*

The authors have presented a surcharging case history in which, in particular, both in-situ preload test and oedometer surcharging test have been carried out to study the reduction of secondary compression by overconsolidation. The term 'surcharging' denotes a special case of precompression, in that a surcharge load is removed at the end of precompression (Feng, 1991).

Rebound of the ground upon the removal of surcharge is an important aspect in using the surcharging technique to reduce the secondary compression. The deformation behaviour of the ground upon the removal of surcharge load includes primary rebound as a result of dissipation of negative pore water pressure, followed by secondary rebound, which stops when the secondary compression reappears (Feng, 1991; Mesri & Feng, 1991). The time of the end of primary rebound can be predicted by using the Terzaghi theory of one-dimensional consolidation (Mesri *et al.*, 1978). The time of the end of secondary rebound—that is, the time of reappearance of secondary compression—can be predicted from an empirical correlation developed using both field and laboratory data (Feng, 1991). It was found from the empirical data that the time of the end of secondary rebound increases significantly with an increase in the magnitude of the surcharge load. For instance, field experiences (Lea & Brawner, 1963; Holtz & Broms, 1972; Veder & Prinzl, 1983; Samson, 1985; Jorgenson, 1987) show that the time of reappearance of secondary compression after the removal of surcharge is about 10–40 times the time of the end of primary rebound. It is not clear how the rebound behaviour of the ground is analysed

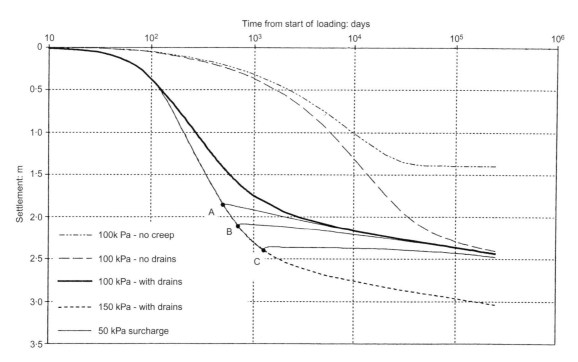

Fig. 20. Predicted time–settlement behaviour of hypothetical test fill at Bothkennar showing the effect of temporary surcharge

or computed by the authors. It seems from Fig. 15 of the authors' paper that only the primary rebound was dealt with.

It is not known, in relation to the *in situ* preload test, why the surcharge load was removed as late as 2 years after the construction of the embankment. The reason for raising this question is that the curves of measured settlement against time as shown in Fig. 9(a) indicate a maximum time of the end of primary consolidation of around 100 days. It is rare to leave the surcharge load in place for such a long time. It is more efficient to reduce the secondary settlement by using a higher surcharge load and a shorter precompression time. It is unfortunate that the preload test was concluded only 8 months after the removal of surcharge load. This means that caution should be taken in using the data presented in Fig. 14. Since the preload test has already been concluded, there is no way to find out how close the predicted post-surcharge secondary settlement is to the measured post-surcharge secondary settlement. It may be mentioned that the writer and co-workers successfully analysed the post-surcharging secondary settlement at the Ska-Edeby test fill site (Feng, 1991; Mesri *et al.*, 1994), at which the ground settlement measurement programme was maintained for 30 years.

There are other suggestions for the attention of the authors. In Fig. 5(a) it can be seen that the surcharge load was removed after 3 days of oedometer consolidation, but information on the time of the end of primary consolidation is absent. The only clue from Fig. 5(a) is that it must be less than 1000 seconds. Thus it can be concluded from this observation that at least two log cycles of secondary compression had occurred before the removal of surcharge load. The contribution of the long-term secondary compression to the increase in OCR should be added to the OCR induced by the removal of surcharge load to obtain an actual OCR value. The actual OCR values should then be used to correlate with the ratio $C_\alpha / C_{\alpha NC}$.

Finally, the definition of the ratio $C_\alpha / C_{\alpha NC}$, as used in both Fig. 5(b) and Fig. 14, is not consistent with the literature. In precompression operations with a surcharge load, the permanent (final) load after the removal of surcharge is the same as the final load without surcharge. Therefore both C_α and $C_{\alpha NC}$ are in fact measured under the same effective stress, and are used conventionally to evaluate the effectiveness of surcharging to reduce secondary compression. Apparently, the authors did not follow this practice.

Authors' reply

The authors appreciate the interest of the writers in the work presented, and welcome the opportunity to discuss further some of the issues raised in the paper.

To D. Nash

One of the main points of the paper is to show how site investigation data and field monitoring records are used in design and prediction. In this context, the conceptual model adopted was kept simple since the practical nature of the problem dominated the entire modelling approach. One of the major concerns was to identify *in situ* the main factors controlling long-term secondary behaviour of the deep soil profile under study.

The authors also maintain an interest in fundamental issues related to long-term deformation of clays. Empirical models such as Bjerrum's (1972) isotache concept are probably useful, especially for low OCR, as the writer notes. Bjerrum's conceptual model predicts a progressive increase of the C_α coefficient after a reduction in effective stress. C_α will eventually recover the NC value, previous to unloading, as Fig. 18 indicates. Although there is some experimental evidence in support of this trend there are also long-term field observations that show that C_α is consistently reduced to a fraction of the NC value. This is shown, for instance, in the data presented by Monoi & Fukuzawa (1998) for the sites 'A' Sapporo and 'B' Chiba. In the second site, for example, the average C_α value, which may be derived 7 years after unloading, seems to be stabilised at a value that is only 8% of the NC C_α value previous to the unloading. Any potential

subsequent increase in the C_α coefficient would correspond to times too long to be of practical concern.

The authors share the writer's opinion that C_α is not a convenient parameter, especially when considering creep phenomena from a fundamental point of view. In a recent work aimed at deriving more accurate models for secondary consolidation strains (Alonso *et al.*, 2001), it was found that secondary strains depend on the reloading stress increment ratio, on the initial strain rate, and on time. The model developed in the work just mentioned is based on the original ideas of Mitchell (1981) (secondary strain rates and time, both in log scale, are linearly related), and on a fairly comprehensive laboratory test programme involving long-term oedometer tests for several combinations of preconsolidation, unloading and reloading (final) stresses.

This recent model, as well as Bjerrum's (1972) isotache concept, remain largely empirical and have significant limitations. One is mentioned in the writer's discussion and refers to the intensity of the unloading step (or OCR). This aspect has been recently investigated by means of a series of long-term oedometer tests. In Fig. 21 plots of vertical strain against time are given for three tests on remoulded samples of the low-plasticity clay described in the paper. The three tests maintain a common value for OCR of 2 (the initial stress, $\sigma_i' = 0.4$ MPa, was reduced to $\sigma_f' = 0.2$ MPa), but the loading time under $\sigma_i' = 0.4$ MPa varied in the three tests (10 min, 100 h and 500 h respectively). Once the transient primary swelling is essentially finished, a secondary behaviour may be observed in the three tests. This secondary behaviour is better observed in Fig. 22, where strains after unloading are plotted with a common time origin (100 s after unloading). The three samples exhibit a swelling secondary behaviour. In two of them ($t = 10$ min; $t = 100$ h) the swelling strain rate decreases with time and may even reverse sign (this is the case for the specimen loaded for $t = 10$ min), but in the specimen loaded for 500 h prior to unloading the swelling strain rate is approximately constant during the test period. This pattern of behaviour is not well predicted by the types of framework mentioned before.

It should be noted, also, that the initial slope of the plots of strain against log t of Fig. 22 (or initial C_α values) depend on the surcharge time, unlike equation (4). Even for small overconsolidation ratios, a changing initial C_α value with the surcharge time is also recorded, as the additional data plotted in Fig. 22 for the same soil and for OCR = 1.05 show. These experimental results are in agreement with Ladd's (1971) findings reported in the paper (Figures 5(b) and Fig. 14).

Therefore models based on the isotache framework also have difficulties in reproducing with some generality the observed one-dimensional creep behaviour of clays. Even more sophisticated empirical expressions, rooted in direct phenomenological observations of creep behaviour under a variety of loading sequences (Alonso *et al.*, 2001), are incapable of modelling accurately a number of settlement-swelling creep transitions.

The authors believe that a more fundamental approach, which links creep behaviour to local fluid transfer phenomena at a microstructural level, may provide more consistent results. This idea has also received attention by the authors (Alonso *et al.*, 1991), and in a recent work (Navarro & Alonso, 2001) a model for secondary compression of clays has been proposed in which relatively simple, yet comprehensive, models for primary/secondary behaviour of clays are presented. This approach incorporates in a natural way the transient compression-swelling secondary behaviour, when the OCR or the preloading times take arbitrary values.

To T.-W. Feng

The writer bases the main part of the discussion on a description of the deformation behaviour of soils on unloading in terms of a primary rebound, a secondary rebound and a final secondary compression. This framework is probably adequate for some ranges of OCR and preloading times. However, it is hardly a general description of the unloading behaviour of soils.

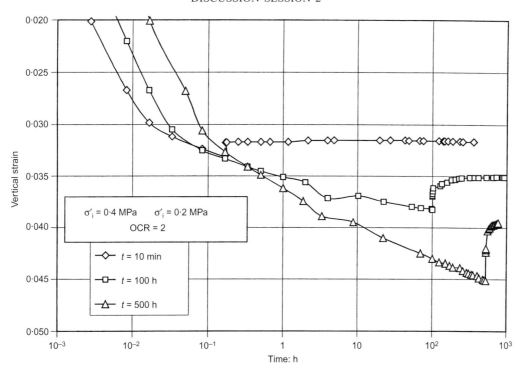

Fig. 21. Vertical strains measured in the oedometer loading–unloading tests on remoulded samples of Barcelona low-plasticity deltaic clays. Initially applied vertical stress, $\sigma'_i = 0.4$ MPa; final vertical stress, $\sigma'_f = 0.2$ MPa

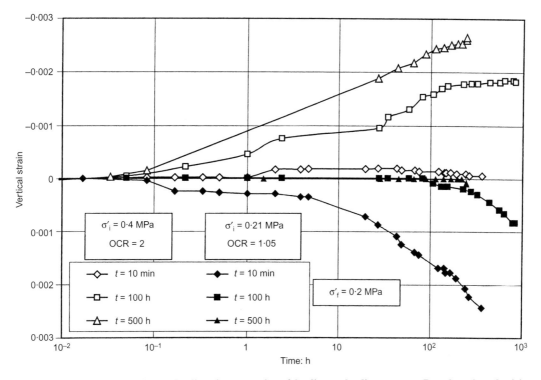

Fig. 22. Strains measured after unloading, in two series of loading–unloading tests on Barcelona low-plasticity deltaic clay for OCR = 2 and OCR = 1·05. In all cases a constant creep stress, $\sigma'_f = 0.2$ MPa, is applied

For instance, when OCR increases, only long-term swelling is recorded, as shown in Figs 21 and Fig. 22. The empirical law mentioned by the writer, which defines the initial time of the secondary compression as 10–40 times the final time of the primary rebound, does not fit—even in an approximate manner—the experimental data recorded by the authors. In Fig. 21, the test for OCR = 2 and $t = 10$ min follows a pattern of behaviour compatible with the framework mentioned above. The thickness of samples in those experiments was reduced on purpose to 8 mm to minimise the period of primary deforma-

tion. Accurate records of the early stages of deformation immediately after loading or unloading show that the time for primary consolidation was about 100 s The time at which secondary compression resumes in the test (OCR = 2, $t = 10$ min) may be located at $t = 10^5$ s, which implies a ratio $10^5/10^2 = 10^3$ (against the 10–40 range suggested by Feng). This ratio is much higher for the test (OCR = 2, $t = 100$ h) plotted in the same Fig. 21, and has no meaning for the test (OCR = 2, $t = 500$ h) that only exhibits secondary expansion. As pointed out above, the authors believe that the development of a general framework for secondary consolidation of clays may require the incorporation of the physical phenomena underlying this type of behaviour.

Concerning the period over which the preload test was carried out, it must be pointed out that the test embankment was left in place for the maximum possible time compatible with the schedule of the project, in order to define as precisely as possible *in situ* primary and secondary soil deformation characteristics. Of course, the loading time of the preload test does not presuppose the time adopted for the preload design, where the usual strategy of using a higher surcharge load applied over a limited time (6 months, as indicated in Fig. 16) has been used. The unloaded trial embankment offered reliable results during eight additional months. Afterwards, construction of the designed preload embankments for the water treatment plant began to affect the test area, and measurements could no longer be used to interpret unloading effects. Naturally, this means that the effects of unloading could not be fully observed, but they will be checked in the future from the monitoring data of the water treatment plant. In summary, the preload test was a useful investigation tool, the purpose of which was to define *in situ* soil behaviour, and in no way was intended to follow exactly the preload design steps.

Regarding the last two points raised by the writer, the authors have plotted the data in a manner that is considered to be more relevant for the purposes of the problem considered. For instance, the question of which OCR should be used to analyse changes in C_α as a result of unloading is still an open one. In the authors' experience, the secondary deformation history cannot be interpreted on the basis of a unique relationship between rate of secondary deformation and updated OCR. An additional function of time is required to describe secondary behaviour, as the recent empirical model reported in Alonso *et al.* (2001) suggests. The nominal OCR, which does not require additional assumptions for its definition, was therefore used in Fig. 5(a) to plot the data. With respect to the last remark of the writer, the C_α values compared in Figures 5(a) and Fig. 14 of the paper correspond in fact to different confining stresses: the consolidation load and the load remaining after unloading. But this is precisely what is required in practice when the effects of partial unloading are investigated. Available field creep deformation records provide information for the trial embankment load before and after unloading. If the same practice is followed in laboratory tests, oedometer and field data can in principle be represented together in the same plot, as in Fig. 14. As a final comment, the practical approach followed by the authors, which relies on a simple C_α model for secondary soil behaviour, is probably suited only for small OCR values. Under these circumstances, the confining effective stresses before and after unloading are relatively close, and their effect on the creep coefficient C_α should be of minor relevance.

REFERENCES

Alonso, E. E., Gens, A. & Lloret, A., (1991). Double structure model for the prediction of long-term movements in expansive materials. In *Computer methods and advances in geomechanics* (eds G. Beer, J. R. Booker and J. P. Carter), Vol. 1, pp. 541–548. Rotterdam: Balkema.

Alonso, E. E., Lloret, A., Gens, A. & Salvadó, M. (2001). Overconsolidation effects on secondary compression rates. *Proc. 15th Int. Conf. Soil Mech. Geotech. Engng*, Istanbul (in press).

Bjerrum, L. (1972). Embankments on soft ground. State of the art report. *Proceedings ASCE specialty conference on performance of earth and earth-supported structures*, Purdue University, Vol. 1, pp. 1–54.

Feng, T. W. (1991). *Compressibility and permeability of natural soft clays, and surcharging to reduce settlements*. PhD thesis, University of Illinois at Urbana-Champaign.

Holtz, R. D. & Broms, B. (1972). 'Long-term loading tests at Ska-Edeby, Sweden.' *Proceedings ASCE specialty conference on performance of earth and earth-supported structures*, Purdue University, Vol. 1, pp. 435–464.

Holtz, R. D., Jamiolkowski, M., Lancellotta, R. & Pedroni, R. (1991). *Prefabricated vertical drains: design and performance*, CIRIA Ground Engineering Report: Ground Improvement. Oxford: Butterworth-Heinemann.

Jorgenson, M. B. (1987). Secondary settlements of four Danish road embankments on soft soils. *Proc. 9th Eur. Conf. Soil Mech. Found. Engng.* **2**, 557–560.

Ladd, C. C. (1971). *Settlement analysis of cohesive soils*, Research Report R71-2. Cambridge, MA: MIT.

Lea, N. D. & Brawner, C. O. (1963). Highway design and construction over peat deposits in lower British Columbia. *Highway Res. Record*, No. 7, 1–32.

Mesri, G. & Feng, T. W. (1991). Surcharging to reduce secondary settlement. *Proceedings of the international conference on geotechnical engineering for coastal development: theory to practice*, Yokohama, Vol. 1, pp. 359–364.

Mesri, G. & Godlewski, P. M. (1977). Time- and stress-compressibility interrelationship. *J. Geotech. Engng. ASCE* **103**, No. GT5, 417–429.

Mesri, G., Ullrich, C. R. & Choi, Y. K. (1978). The rate of swelling of overconsolidated clays subjected to unloading. *Géotechnique* **28**, No. 3, 281–307.

Mesri, G., Lo, D. O. K. & Feng, T. W. (1994). Settlement of embankments on soft clays. *Proc. Settlement 94*, ASCE Geotechnical Specialty Conference, Special Publication No. 40, Vol. 1, pp. 8–76.

Mitchell, J. M. (1981). Soil improvement. State of the art report. *Proc. 10th Int. Conf. Soil Mech. Found. Engng. Stockholm* **4**, 509–565.

Monoi, Y. & Fuzukawa, E. (1998). Measurement of long-term settlement of highly organic soil improved by preloading. In *Problematic soils* (eds E. Yanagisawa, N. Moroto and T. Mitachi), Vol. 1, pp. 83–87. Rotterdam: Balkema.

Nash, D. F. T. (2001). Modelling the effects of surcharge to reduce long-term settlement of reclamations over soft clays: a numerical case study. *Soils Found.* (in press).

Nash, D. F. T. & Ryde, S. J. (2001). Modelling consolidation accelerated by vertical drains in soils subject to creep. *Géotechnique*, **51**, No. 3, 257–278.

Navarro, V. & Alonso, E. E. (2001). Secondary compression of clays as a local hydration process. *Géotechnique* (in press).

Samson, L. (1985). Postconstruction settlement of an expressway built on peat by precompression. *Can. Geotech. J.* **22**, No. 2, 308–312.

Veder, Ch. & Prinzl, F. (1983). The avoidance of secondary settlements through overconsolidation. *Proc. 8th Eur. Conf. Soil Mech. Found. Engng.* **2**, 697–706.

Yin, J.-H. & Graham, J. (1996). Elastic visco-plastic modelling of one-dimensional consolidation. *Géotechnique* **46**, No. 3, 515–527.

Gohl, W. B., Jefferies, M. G., Howie, J. A. & Diggle, D. (2002). *Géotechnique* **52**, No. 8, 622–623

WRITTEN DISCUSSION

Explosive compaction: design, implementation and effectiveness

W. B. GOHL, M. G. JEFFERIES, J. A. HOWIE and D. DIGGLE

W. H. Craig, *University of Manchester*

The authors have demonstrated with their video clips that a picture can indeed be worth a thousand words. Following the detonation of subsurface explosions a volume of the ground liquefies and the pore fluid seeks a route to the surface. The process of pushing an instability, through some fingering mechanism, to the surface can clearly take some considerable time—of the order of 2 mins. The subsequent outrush of soil and liquid from a pipe in an unpredictable location can be quite violent, and can also last for some time. While the overall result may be densification and consolidation of the ground, the pipe outlet itself may remain as a very loose zone. Is there any evidence of this from post-detonation probings?

Authors' reply

The discusser has identified a potential source of loose zones within the densified mass. As noted, loose zones may be created by the flow of water displaced by reduction in void ratio due to densification. However, the largest volume changes occur during the first pass of explosive compaction, and the site illustrated in the video (Kelowna) was extremely loose (apparent relative density of about $D_r \sim 25\%$), which emphasised this aspect of the ground response.

The videos were shown to make the point that explosive compaction is really an induced consolidation, at least initially, with density change following explosive detonation on a very different and independent timescale. Attempts to simply correlate the achieved final density with the explosive weighting miss an important mechanism. However, the video perhaps overemphasised the prevalence of piping.

Large-scale liquefaction is commonly associated only with the first few shots of the many used in compacting a site, and even then only with the very loose site. Typically we detonate only 9–12 holes in a single shot, and the first pass will correspondingly involve many shots to cover the entire area of the site. These later shots induce excess pore pressures, but settlements are reduced and the compaction of the site becomes more uniform. Wide-scale liquefaction is simply not seen then—indeed this might even be taken as proof of adequacy of treatment if liquefaction was the primary concern. Settlements may be induced some distance from the shot area once the soil in the area settling has been disturbed by its own shot. Typically after two or three passes of blasting the zone of settlement extends at approximately a 45° line from the deepest depth of blasting.

Compaction uniformity is important because many compaction projects are related to liquefaction concerns, and liquefaction (and cyclic mobility) is controlled by the looser pockets in the soil. This control by the loosest end of the spectrum appears to have been first reported on the basis of centrifuge studies of the Osterschelde closure caissons (Rowe and Craig 1976). Subsequent numerical studies have reinforced the conclusion in the case of cyclic mobility in hydraulically placed sand (Popescu *et al.*, 1997) and static liquefaction of hydraulically placed sand. It appears that the controlling behaviour (*characteristic property* in limit states jargon) is dominated by the loosest 10–20% of the deposit. Being aware of the importance of loose pockets, we have

been more concerned with the collapse of material into the expansion cavity caused by the explosive charge, primarily because there is always a regular grid of such features, and these features are always there even in the last stages of compaction. Fig. 6 of the paper showed the approximate 97·5th and 2·5th percentile limits (95th percentile range) of penetration resistance of compacted soils because of this concern for the variation in soil properties. Fig. 8 adds to this plot by presenting the variation penetration resistance encountered when testing a fill using a regular grid. As can be seen, there are substantial fluctuations in penetration resistance on a scale of a few metres.

Figure 9 shows penetration resistance for CPT soundings targeting the explosive zone, with the results of soundings 0·4 m, 1·4 m and 3·0 m from the blast borehole being shown. The depths of the explosive charges in the borehole are also indicated. Comparing Fig. 9 with Fig. 8, there is some support for a conclusion that there may be a looser zone within about a 1 m radius of the detonated borehole. However, the data at 9 m depth are less clear on this point (although at this depth we have less certainty in actual distance because of CPT deviation in the ground). Also, there is not a clear correlation with proximity to the charge because the loosest zone at 0·4 m radius appears about midway between the two upper charges. On the other hand, the fluctuation between the soundings at 1·4 m and 3 m radius are entirely consistent with the general pattern of fluctuations seen as a whole as plotted on Fig. 8.

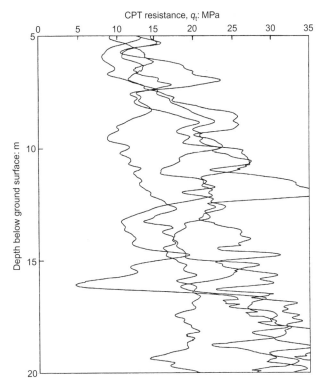

Fig. 8. Illustration of soil variability when tested by CPT on a regular grid (densification zone started at 5 m depth)

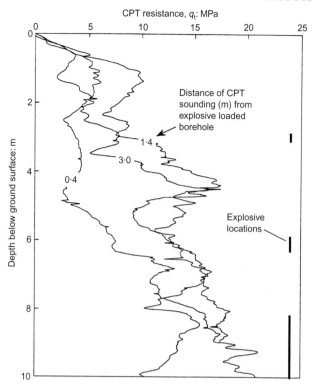

Fig. 9. CPT resistance adjacent to blast hole

Recent numerical simulations suggest that the important feature is the degree to which loose zones are aligned along the direction of possible shear zones; vertically inclined looser columns are unlikely to be a concern in this regard. Nevertheless, we have always used at least two passes in production explosive compaction to compact loose zones remaining after the larger settlements of the first pass and to improve the uniformity of the densified mass.

In conclusion, while piping in the early stages of explosive compaction is often dramatic, its occurrence is a clue to the nature of explosive compaction rather than a cause for concern for the results of properly executed work. Loose columns do appear to arise around the explosive charges themselves, and this is one reason to limit charge sizes in favour of more passes. Too much explosive is not a good thing as far as effective compaction is concerned. Our work has used the criterion that about 85% of the soil should be denser than the desired target as a way of reasonably accommodating the concerns about property variation.

REFERENCES

Popescu, R., Prevost, J. H. & Deodatis, G (1997). Effects of spatial variability on soil liquefaction: some design recommendations. *Géotechnique* **47**, No. 5, 1019–1036.

Rowe, P. W. & Craig, W. H. (1976). Studies of offshore caissons founded on Oosterschelde sand. In *Design and construction of offshore structures*, Institution of Civil Engineers, London, 49–55.

Session 3

Bell, A. L. (2004). *Ground and soil improvement* 103–111

The development and importance of construction technique in deep vibratory ground improvement*

A. L. BELL†

Deep vibratory methods for improving the ground for foundation engineering and other geotechnical applications are now used extensively on a worldwide basis. Originating in Germany in the 1930s, the methods have developed from densification of sands to stone column reinforcement in a range of natural soils and filled ground including soft clays. The paper commences by indicating the process nature of deep vibratory ground improvement, and identifies the construction technique employed as an integral element, which is too often given poor consideration. The development of the two main methods currently employed—vibrocompaction and vibrated stone columns—is described, together with key applications and limitations. Examples of how attention to construction technique can transform the performance are given for each of the main methods. Possible future trends are discussed.

KEYWORDS: bearing capacity; case history; clays; compaction; liquefaction; sands; silts; vibration

BACKGROUND

Deep vibratory ground improvement is best understood as a process rather than a product. It can be applied most effectively if all the elements of the process are understood in relation to each other, and if each is given proper attention at all stages.

The key elements are as follows:

(a) site evaluation
(b) ground investigation
(c) development of concept
(d) design
(e) construction technique
(f) process evaluation
(g) commissioning and maintenance.

This sequence is apparently chronological, although this may not always be the case. This paper aims to illustrate the development and importance of *construction technique* as an often neglected element in the application of deep ground treatment. A brief description of these elements follows. For a fuller treatment the reader is directed to ICE (1987a,b), Slocombe & Moseley (1991), NHBC (1992), Moseley & Priebe (1993), and BRE (2000).

Construction technique is the effective deployment of site resources to meet the design and other project objectives in an optimum manner. It normally includes: the employment of trained and experienced supervisors and operatives; purpose-designed plant and equipment; materials that are carefully specified, stored and used in correct quantities; and the application of the most appropriate construction method for the site and ground conditions.

Site evaluation includes all the practical considerations needed to ensure a safe, timely and trouble-free operation on site, as well as a desk study to determine previous uses of the site and any other pertinent geological or practical information. The issues to be addressed include: access to the working area; watertable and drainage considerations; overhead and underground services; adjacent structures or buildings such as railways, roads or residences; and topography, including proposed changes such as the introduction of fill or the proximity of excavations or embankments.

Ground investigation (BSI, 1999) for ground treatment needs to identify the properties of the ground to be treated. Too often the underlying better-bearing strata are identified and characterised but the softer or looser layers are ignored, thereby jeopardising the possible use of ground improvement techniques, as appraisal is limited by lack of data. Investigation and test methods should, in addition to identifying geotechnical properties, also be aimed at revealing the variability of the ground. They should be selected with the post-treatment testing in mind to ensure comparability. Shallow layers can effectively be examined with trial pits, which enable detailed section descriptions to be obtained, and are of particular benefit in characterising made ground. Deeper layers will require boreholes, and these are usefully supplemented with in-situ tests such as SPT, CPT or DPT soundings.

The *concept* or guiding principles on which the ground improvement scheme is to be based needs to be developed in the light of the other key elements in the process. It is also important to keep the overall project objectives in mind, including non-technical issues such as: programme, specific quality or safety objectives; particular constraints, such as limitations in working hours, water supply or spoil disposal; and how the ground improvement work interacts with subsequent construction in practical and technical terms. A clear view of the concept will usually lead to the optimum selection of the method to be employed.

The *design* should include a review of all the factors likely to be influenced by the construction technique as well as the performance requirements. These need to be clearly defined and communicated to all parties to the contract. Several design assumptions, such as the level of improvement available at the site, can often be verified only during or after construction, and this needs to be reflected in the approach to the work.

Except where extensive experience is available, formal *evaluation* of the success of the treatment is essential. This must as a minimum include a review of the site treatment records indicating the energy input, the depths reached for each penetration, the materials employed, and details of the actual equipment and method employed on site. Post-treatment in-situ testing for comparison with similar pre-treatment testing is sometimes appropriate. Full-scale load testing is expensive but represents a useful means of proving the efficacy of the improvement achieved. Evaluation is further discussed in Moseley & Priebe (1993).

Finally, due consideration must be given to *handover and*

Manuscript received 12 June 2002; revised manuscript accepted 29 August 2002.
* Keynote Lecture presented to *Géotechnique* Symposium in Print on Ground and Soil Improvement, 6 February 2001.
† Keller Ground Engineering, Coventry, UK.

maintenance. Adequate protection must be given. For example, if stone columns in cohesive soil are to remain open to the elements for a long period, temporary cover may be needed; the close proximity of following operations such as excavation may need planning to avoid undermining the earlier work.

Deep vibratory ground improvement can be achieved with a range of different tools. This paper is concerned with the most widespread methods on a worldwide basis, and these rely on the use of the *depth vibrator* (also sometimes known as a *vibroflot*). A depth vibrator (Fig. 1) can be suspended from a jib or hung off the leaders of crawler-mounted base machines. It is used in a vertical orientation to penetrate to the required depth, sometimes with prebore assistance, and applies lateral vibratory energy to the ground by means of an internal rotating eccentric weight assembly. The casing houses an electric or hydraulic motor to drive the eccentric, and has a reinforced nose cone to reduce wear maintenance. Passages for air and/or water, depending on the method employed, are attached. Fins are connected to prevent twist when the vibrator is suspended. Significant depths of treatment, of 50 m or more, can be attained by connecting a series of follower tubes to the vibrator with flexible couplings.

Depth vibrators were originally developed in Germany in

the 1930s to densify clean granular sands and other cohesionless soils in-situ by a method known as *vibrocompaction.* This can be used to densify to depths of well over 50 m, and is still in wide use today.

The other main method of deep vibratory treatment was developed in the late 1950s as a derivative of the vibrocompaction method, in order to extend the application to cohesive soils, which do not respond effectively to the vibratory effect of depth vibrators. It employs vibrators to form dense vertical columns of crushed rock or gravel to reinforce the ground, and in some cases to provide enhanced drainage. This new method, known as *vibrated stone columns,* makes possible treatment of a range of cohesive soils and fills, and more recently has even been found to enhance performance in cohesionless soils.

The key geotechnical effects of these two main methods are summarised in Table 1.

DEVELOPMENT OF VIBROCOMPACTION METHODS
The basic process is known as *wet vibrocompaction,* as it requires the use of water. Equipment (Fig. 2) consists of a depth vibrator (Fig. 1(a)), together with follower tubes suspended from a crane jib, an electric or hydraulic power pack, and a water supply of 12 000–20 000 l/h. Vibrators

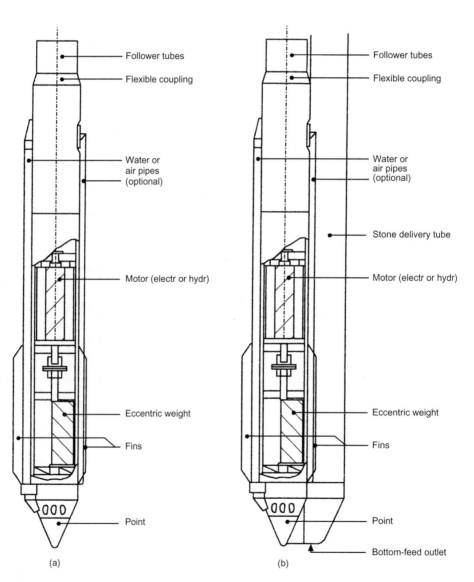

Fig. 1. Depth vibrators (after Slocombe et al., 2000a): (a) depth vibrator; (b) bottom-feed depth vibrator

Fig. 2. Wet vibrocompaction plant ready to commence

developed for compaction are usually designed to be more efficient in densification than in penetration. Accordingly they have high amplitudes of 20 mm and more, and a large side-force of the order of 200–300 kN; they operate at relatively low frequencies, in the range 1200–2000 rev/min; they have high power outputs, and tend to be heavy, with large barrel diameters. The sequence of operation during one cycle of compaction is shown in Fig. 3. During penetration to the required depth the vibrator is assisted by its weight,

its lateral vibratory action and the water flow. The key requirement of the water is to ensure that an annulus is maintained between the vibrator and the borehole, and flow rate is therefore important. Considerable amounts of wet, silty spoil are generated by flushing out the fines in the annulus to the surface, and must be properly channelled and disposed of, sometimes after being placed in lagoons to separate the solid and fluid components. During the compaction phase the water flow and pressure are reduced (Fig. 4). Infill is introduced, and, by repenetration over a short depth, and lateral vibratory action, the infill and surrounding ground are densified. This is repeated until the surface is attained to complete the compaction cycle. The densification effect is reduced with radial distance from the compaction centre. It is necessary to repeat the compactions on a triangular or square grid over the treatment area, at spacings ranging from 1·5 m to 4·0 m depending on the nature of the ground, the densification required, the equipment specification, and the construction technique employed. If the backfill is provided from overlying deposits, the site surface level will be reduced by the ground densification, and this needs to be allowed for or avoided by the use of imported backfill.

Recent development has shown that stone backfill enhances the level of densification achieved. This is thought to be partly because of the coarser-grained stone addition, which densifies better than sands, but also because of improved transmission of the vibratory effects. It has been noted (Slocombe et al., 2000b) and in the author's experience, for example, that the radius of improvement appears to be enhanced using stone backfill. A further advantage applies when fast drainage is important to the application, as the stone columns normally provide considerable drainage capacity, even in predominantly granular soils. These advantages now often lead to a preference for stone backfill in compaction, rather than sand alone, although the likely enhancement in treatment must be set against the increased

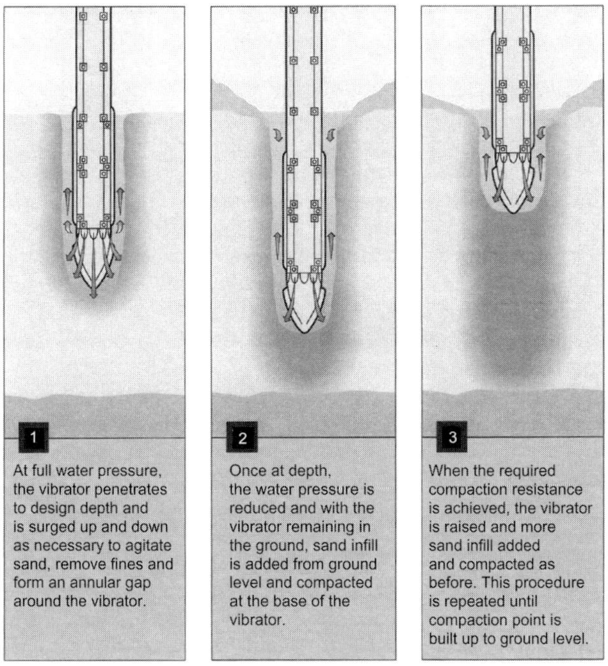

1	2	3
At full water pressure, the vibrator penetrates to design depth and is surged up and down as necessary to agitate sand, remove fines and form an annular gap around the vibrator.	Once at depth, the water pressure is reduced and with the vibrator remaining in the ground, sand infill is added from ground level and compacted at the base of the vibrator.	When the required compaction resistance is achieved, the vibrator is raised and more sand infill added and compacted as before. This procedure is repeated until compaction point is built up to ground level.

Fig. 3. Construction sequence: vibrocompaction

Fig. 4. Wet vibrocompaction during the compaction phase

Table 1. Potential geotechnical effects of deep vibratory ground improvement

Method	Sands, gravels and granular fills	Cohesive soils and mixed fills
Vibrocompaction	Densification	–
Vibrated stone columns	Densification, reinforcement and drainage	Reinforcement and drainage

cost of stone. This can be justified in a range of applications, for example those in which there are very heavy foundation loads, or where earthquake liquefaction limitation is particularly demanding.

Another significant development has been the introduction of *bottom-feed systems* (Fig. 5). The systems are described below in connection with vibrated stone columns, but have more recently also been applied very successfully in vibrocompaction to avoid the need for water supply and spoil disposal (see Slocombe *et al.*, 2000a).

APPLICATIONS, LIMITATIONS AND CONSTRUCTION TECHNIQUE IN VIBROCOMPACTION

Applications for vibrocompaction are wide-ranging (e.g. Greenwood & Kirsch, 1983). Originally the main application was for foundations on cohesionless soils, and this remains a major area of application. Structural loads up to about $500 \, kN/m^2$, and embankments and dams of the order of 40 m in height, have been successfully founded on ground improved by vibrocompaction.

More recently a significant new area of application has arisen from the need to reduce the potential for soil liquefaction arising from earthquakes and the minimisation of extensive damage and risk to human life (Kirsch & Chambosse, 1981). Extensive case histories are now found in the literature, including positive performance observed during actual earthquakes. Vibrocompaction is also often considered suitable in seismic retrofit programmes for existing structures and dams (e.g. Dobson, 1987; Baez & Martin, 1992; Mitchell *et al.*, 1995; Raison *et al.*, 1995).

Other applications include: pre- and post-densification of filled ground; reduction of lateral loads behind retaining structures; reduction of permeability by densification; and compression of landfills to provide enhanced cell capacity (e.g. Bell *et al.*, 2000; Sidak, 2000).

It is vital with vibrocompaction to appreciate the soil types and conditions that require care to achieve the treatment objectives. In certain cases they may render the method unsuitable, even with the best techniques available. It is widely understood that, when the content of the fines (silt particles less than 60 µm) increases beyond 10–15% by weight of the soil, or 2–3% clay, densification is significantly reduced. Although some argue that little or no densification occurs beyond 20% fines, recent evidence indicates that some densification can be achieved at much more than this figure by using stone backfill and working the ground hard (Slocombe *et al.*, 2000a). Occasionally it has been observed in experience known to the author that naturally occurring cohesionless soils in apparently loose states contain bonding between the particles. Such bonds can be broken down by the action of the vibrators, but not at sufficient radius from the centres of compaction to make the treatment viable. Indeed all that may be achieved is a weakening of the ground. Very coarse dense soils, typically with particle sizes in excess of 150–200 mm, may prove difficult or impossible to densify. In other cases it is not the soil grading or state that is the issue so much as the soil profile. Where, for example, there are relatively thin layers (less than the vibrator barrel diameter) of loose soil sandwiched between cohesive layers, or where there are thin cemented layers between loose zones, local vibration dampening may occur, and the approach must identify and address this.

The importance of construction technique often gets lost in the midst of all the other effects considered in vibrocompaction, and is seldom covered in the literature. Slocombe *et al.* (2000b) report an example. Two 18·4 m diameter sheet-piled cofferdams were to be constructed in 15 m of water, as temporary protection for dry dock works at Devonport Docks in the UK. The cofferdam fill was hydraulically placed crushed limestone aggregate, with nominal 6 mm to 3 mm grading. The placed fill required further compaction to achieve a minimum 80% relative density, in order to provide adequate weight to generate sufficient base sliding resistance and to avoid liquefaction in the face of the design earthquake. Vibrocompaction was the selected method to achieve the improvement required. Cone penetration testing was employed to provide a measure of relative density together with a pre-agreed means of interpreting the data.

An initial stage of wet vibrocompaction was conducted by the general contractor using a hired 90 kW depth vibrator together with an experienced crew. This failed to meet the density requirements, and was followed by a second stage that left some zones untreated (Fig. 6). The results were especially poor at depth, and refusal to penetrate below about 12 m was common. Consequently the author's company was invited to complete the vibrocompaction. The results are shown in Fig. 6 for comparison, and indicate a very different level of densification. In all CPT profiles final-stage densification was well beyond the minimum required, including the full depth. Average relative densities of 90% for the west cell and 87% for the east cell were calculated. The difference in performance of the later vibrocompaction was due solely to a different approach to the construction technique employed. The final-stage compaction employed a 125 kW vibrator, higher pressure, flow water pumps to assist penetration of the previously treated ground and to ensure that the required depths could be attained, and close monitoring by experienced staff. They ensured that the ground was worked hard—that is, repeated penetrations and review of the energy demand by means of the ammeter reading As can be seen from Fig. 6, the higher power of the vibrator, although an important factor, does not in itself explain the significantly improved performance. The application of experience in working the soil has been the key factor in dealing with a difficult problem.

Fig. 5. Dry bottom-feed vibrocompaction using stone backfill

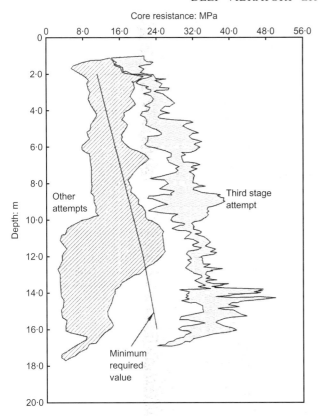

Fig. 6. Densification measured from cone penetration testing in Devonport cofferdam fills (after Slocombe et al., 2000b)

THE DEVELOPMENT OF VIBRATED STONE COLUMN METHODS

Vibrators for forming stone columns originally were identical to those needed for vibrocompaction—heavy, powerful, low-frequency units with water-flushing jets. These are still used for stone columns where very large-diameter columns are needed, as in liquefaction prevention work for example. More commonly, stone column vibrators are often higher frequency than sand compactors, typically up to 3000 rev/min, to assist in penetration, and with lower power, side force and amplitude—enough to compact the stone in situ, but not sufficient for large-radius vibratory influence.

The original method for forming stone columns employed a wet process derived from the vibrocompaction method. Following penetration to the required depth using water flush, the vibrator remains in the hole, and—as with vibrocompaction—the borehole is maintained open by the action of continued water flushing at reduced pressure. A charge of stone is introduced from the surface, by means of a movable stone skip, into the annulus between the vibrator and the borehole side. The charge is compacted by the action of repenetration of the vibrator and the effects of vibration. By repeating this cycle a column of stone is built up. It is important that each charge is properly compacted so that the column of stone is able not only to enhance its own shearing resistance by densification, but also to optimise the support of the surrounding soil, by increased local lateral pressures.

It was readily found that in some cohesive soils, and some mixed soils above the water table, the vibrator could be removed from the bore without loss of stability. Consequently a dry process with no water flushing was developed for such soils (Fig. 7). Compressed air is employed to help keep the vibrator running free in the bore, but this must be used with care to avoid damaging the surrounding soil, as it provides the column support. Once the vibrator forms the borehole, it can be removed completely, unlike the wet process. A charge of stone is then fed into the open borehole, and the vibrator is re-introduced in order to penetrate and compact the stone. It is again removed from the borehole and the cycle is repeated until the column is built up and compacted to the surface. As the stone can be fed into open bore, coarser gradings than for the wet system can be employed.

The drawbacks with the use of water and spoil generation in the wet process gave considerable impetus to the search for an alternative method of construction. This arrived via the *bottom feed* system (Moseley & Priebe, 1993), which was a patented method of feeding stone to the base of the borehole by means of a delivery tube attached to the vibrator (Fig. 1(b)). In this way the need to feed stone via the annulus between the vibrator and the bore wall was avoided. The full development of the system included a new base machine (known as a *vibrocat*), to which the special vibrator could be attached by a leader system (Fig. 8). The stone is fed into the poker by means of a stone skip, which can travel up and down the leaders to the top of the vibrator where the entry chamber is located. Compressed air is used to assist the stone entry and flow. The stone feed into the skip is mainly independent of the stone column construction cycle. Other benefits include the prevention of unwanted inclusions of soil in the column, and assistance in penetration by additional downforce of up to 150 kN by the vibrocat winch. The new method therefore provides productivity gains as well as a clean, water-free approach. As a consequence the advent of the bottom feed methods has been the most significant step forward with stone column treatment to date. As well as eventual application to vibrocompaction, as described above, the bottom feed method opened the door to better monitoring of the method, as described below.

The selection of the stone is a further key issue in forming vibrated stone columns correctly. The stone needs to be free of fines, and should be hard, inert material having due regard for the ground in which it is to be placed, consisting typically of natural gravels, crushed rock, hardcore or slag or well-burnt shale. It is vital to ensure that the stone has sufficient hardness and resistance to wear to prevent degradation from the construction process forming sand- and silt-sized particles that would fail to generate the design angles of friction, and which might further degrade with time. Certain weak limestones, sandstones and chalks have this propensity. The grading of the stone employed is partly governed by the method selected. The dry method typically uses 75 mm down to 40 mm, often relatively single-sized. The wet and bottom feed methods respectively need finer stone to improve flow in the annulus or stone tube, typically in the 20–40 mm range. Further guidance on stone specifications may be obtained in ICE (1987a) and BRE (2000).

Correct construction of stone columns requires an adequate charge of suitable stone to be placed into the base of the bore. It then needs to be compacted and forced into the surrounding soil in a controlled way, requiring repeated penetrations with appropriate energy input assessed from the vibrator output, for example via an ammeter in the case of electric vibrators. Successive charges need to be placed so as to form a continuous column without compromising the quality of work already done by contamination with unwanted soil, loosening the stone, or weakening the surrounding ground. Consequently, good practice requires the monitoring and recording of the penetration rates, repenetrations, stone consumption and energy input. The development of the leader-mounted bottom feed vibrators meant that the above parameters could be measured and recorded automatically. Fig. 9 provides an example of a printout from a

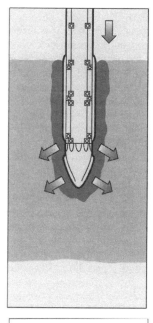
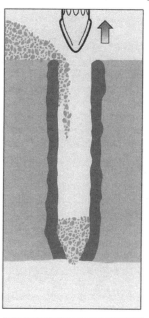
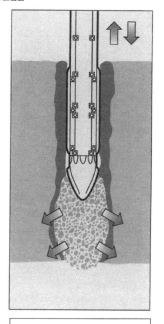
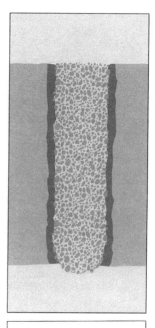

1. Penetration to required depth by action of weight, vibration and compressed air

2. Vibrator is removed and stone backfill placed in the base of borehole

3. Vibrator is placed into stone, which is compacted and forced laterally into the surrounding soil

4. The cycle is repeated until a continuous compacted column of soil is formed

Fig. 7. Construction sequence: vibrated stone columns by dry method

Fig. 8. Purpose-designed plant for bottom-feed vibrated stone column construction

vibrocat illustrating the construction of several stone columns. As supervision of construction is often left solely to the specialist contractor, considerable variation in quality of stone columns is possible, and automatic construction recording is a large step forward in quality assurance and control. This type of information is now routinely produced by the leading specialists, and enables them to achieve and demonstrate the consistency of approach to build confidence. Such information can readily be made available to clients direct from the site. More recently this type of information has been used to provide a further improvement in quality control—automatic control of column construction independent of the machine operator, whose function changes to set-up at each position and maintenance. The latest vibrocats operate with this facility.

Other developments in machine design relate to scale and reliability. Smaller lighter 'minicats' specifically designed to work in relatively low-headroom sites, with tight access conditions and less than ideal working surfaces, have recently been introduced. These operate full-scale vibrators, and are fully instrumented with winch pulldown: as a result the effects of scale on technical performance have been found to irrelevant. An example of the mini machine is shown in Fig. 10.

APPLICATIONS, LIMITATIONS AND CONSTRUCTION TECHNIQUE FOR VIBRATED STONE COLUMNS

Applications for vibrated stone columns are widespread (eg Chambosse, 1983; Greenwood & Kirsch, 1983; Mitchell & Huber, 1983; Bell et al., 1986; Kirsch et al., 1986; Davie et al., 1991; Sondermann, 1996).The technique is well suited for lightly loaded structures up to 100 kN/m^2, such as strip foundations for housing, pads for steel and concrete-framed industrial structures, or area treatment for sewage treatment works or similar in relatively weak soils including soft clays and silts. Even in soft soils, stone columns are commonly used for heavier, less settlement-sensitive structures such as road and railway embankments, tanks for oil, gas or water, and storage or large process areas. Heavier structures such as framed warehouses, industrial units and racking systems can be used in a narrower range of soils, including (with care) firm cohesive clays, silts and fills. Indeed, in these types of soil the method has been applied very successfully for heavy foundations—up to about 250 kN/m^2—and in mixed fills and granular soils loads up to 450 kN/m^2 have been carried for industrial structures, medium-rise commercial and residential buildings, silos and similar. The method has also seen application in slope stabilisation and remediation works.

This base of experience over many years provides con-

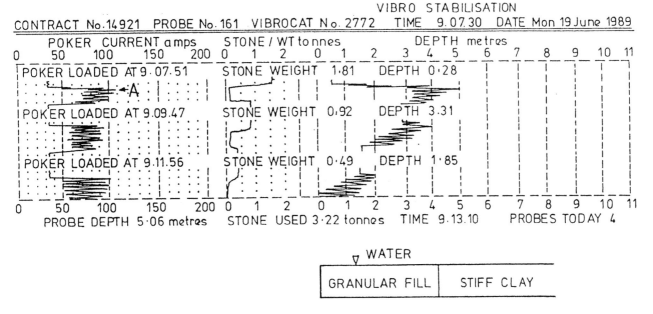

Fig. 9. Example of printout from automatic recording of the construction of vibrated stone columns (Slocombe & Moseley, 1991)

Fig. 10. Purpose-designed small-scale plant for vibrated stone column construction

fidence in the vibrated stone column systems. However, the processes do need identification and consideration of the limitations on the methods to avoid poor application (BRE, 2000). Care in design is needed where the variability in the ground can lead to potential large differential settlements even following treatment—for example construction over a buried valley or quarry edge. Deep unsaturated fills, especially non-engineered fills, may be subject to large settlements unrelated to loading conditions, as a result of inundation from rainfall, floods, or other means (Charles & Watts, 1996). Contaminated soils need special consideration aside from toxicity issues, because by their nature vibrated stone columns need the support of the surrounding ground. In many cases contaminated soils contain materials that may degrade or lead to degradation with time, or for which very limited long-term experience is available. Very low strength cohesive soils, of less than about 15 kN/m², and very sensitive clays and silts may provide profound construction difficulties in addition to obvious design issues. Construction in such deposits may be possible if the layers are thin, or if construction time is taken to build up the columns carefully and to sufficient diameter without unnecessarily disturbing the surrounding ground. The wet method is sometimes useful in achieving these ends. Peats and highly organic clays

similarly cause formidable difficulties in construction and design. Where they occur in discrete layers, about one or two vibrator diameters in thickness, it is usually possible to construct locally large columns through them. Alternatively it may be possible to locally strengthen the stone column with concrete or grout to bridge the very soft, creep-prone zones. Filled ground may also give rise to difficulties owing to the presence of refuse, voids or other degradable materials. These require particular care as the degradable elements may be quite widespread through the deposits and may not be detectable during construction—the ground investigation with such materials needs to be adequate. Finally, problems in penetrating the ground at all may be caused by obstructions or hard or very dense deposits overlying the layers to be treated. Preboring or the employment of a different or more powerful approach may be able to address this.

An example of the range of quality possible arising from insufficient attention to construction technique was provided by what should have been a routine foundation project in mixed soils in the UK, which the author was asked to comment on. A typical soil profile is shown in Fig. 11. The vibrated stone column treatment was intended to provide adequate bearing capacity for a range of foundations. It would have been expected that the columns would have been formed continuously to varying depths of penetration through the firm materials into the better underlying granular material below. Following in-situ load tests that caused concern, an investigation was put in hand. This was progressed by in-situ testing and by exposing, excavating and logging several columns along their axis. This indicated that many columns had been badly constructed. In the worst cases, site records suggested that treatment had been performed to depths of up to 4 m. While there is no way of telling whether the vibrator had penetrated to such depths, column A in Fig. 11 would appear not to have been constructed properly beyond 2·5 m in depth. Even within this depth it was apparent that the stone column was discontinuous. Column B was a better column, but the nominal stone diameter was reducing with depth. In the lower reaches it is less than the nominal vibrator diameter— again the result of lack of building up the column in discrete lifts, each of which is compacted to predetermined limits such as a target ammeter reading. Accurate records of the

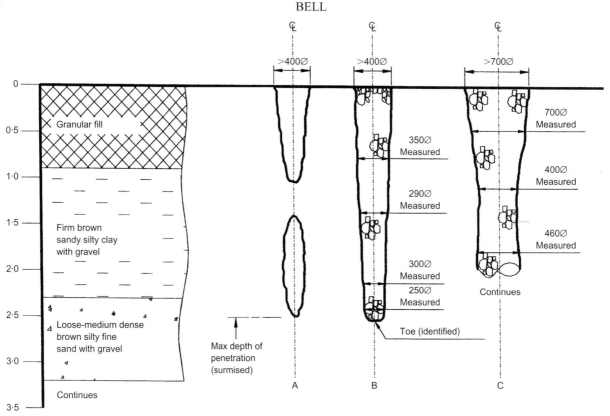

Fig. 11. UK site: poor stone column quality

amount of stone consumed by each column would also have revealed the deficiencies of the construction.

A complete retreatment was eventually instructed, and an example column C from this work is also shown in Fig. 11. A different standard of construction technique was clearly employed, although the work was conducted with identical equipment to the original. The column is continuous, is constructed fully to the correct depth, and has a minimum diameter of about 400 mm, which was adequate for the design. Some variation in diameter is to be expected for properly compacted columns, as strata or layers with differing lateral resistances are encountered. The key is to compact each stone charge to predetermined limits prior to placing and compacting the next charge.

FUTURE DEVELOPMENT IN DEEP VIBRATORY GROUND IMPROVEMENT TECHNIQUES

It is unwise to speculate too far on the direction or the significance of future development with the vibro methods. However, several trends appear relevant in current practice, and the direction these may continue to take is noted.

Improvements in productivity and efficiency on site have been a key trend during the last 25 years, and it would be surprising if this did not continue. Mechanical improvements in the reliability, power and efficiency of vibrators should continue. At present the most powerful sand compaction vibrators output up to as much as 180 kW, and centrifugal force of up to 300 kN. Improvements in bearing wear resistance, electric and hydraulic motor efficiency and dynamics will continue to push these parameters.

Variable-frequency vibrators represent a possible step forward, holding out the promise of matching machine characteristics with resonant frequency of the immediate surrounding ground. This type of vibrator has, to date, been expensive to develop and operate, and any efficiency improvements noted have been as inconsistent as the ground.

However, it would be surprising if practical variable-frequency vibrators suited for a limited range of cohesionless soil types were not to emerge in the next few years.

Information technology has been a useful vehicle for recent innovation, and further developments in communication will undoubtedly be seen. Automatic monitoring to ensure quality control will continue to become more sophisticated and to include more parameters, such as variations in backfill and power consumption with depth for each compaction or stone column. These improvements will be more easily communicated between all parties to the treatment to improve technical control. This will lead to increased automation, just recently introduced, aimed at better control of the processes, with the objective of eliminating the human factor from construction technique, and leaving this to robotics.

Environmental pressures are rightly being applied to all site processes, in order to minimise waste, increase sustainability and reduce pollution to water, land and air. The more immediate effects of noise and vibration are also relevant issues. It would be expected, therefore, that more use of recycled or other low environmental impact materials will be employed in construction. There has been limited use to date, for example, of crushed concrete or brick as a substitute for natural materials in vibrated stone columns, and this needs to be developed after appropriate examination. The major processes will need to aim at reductions in noise and emissions from plant, and the issues surrounding water supply and effluent disposal will become increasingly large hurdles for the wet processes.

CONCLUSIONS

Deep vibratory methods for ground improvement are now widespread worldwide, and are arguably the leading means of mechanical improvement of soils for geotechnical purposes. Vibrocompaction methods densify cohesionless soils, and development has led to more efficient sand vibrators, the benefits of using stone backfill, and the advent of cleaner, dry methods. Vibrated stone column methods have been developed from vibrocompaction and can be applied to reinforce a wide range of soil types including soft cohesive soils. Wet, dry and dry bottom feed methods are all in current use, with the latter opening the door to significant environmental improvements, productivity gains, and quality control development. Both vibrocompaction and vibrated stone column methods remain subject to considerable variation in performance arising from construction technique, which requires close attention to ensure successful application.

ACKNOWLEDGEMENTS

The author gratefully acknowledges the experience, insight and observations from many engineering and site colleagues in the ground improvement industry, and especially within the Keller Group.

REFERENCES

Baez, J. I. & Martin, G. R. (1992). Liquefaction observations during installation of stone columns using the vibro-replacement technique. Geotech. News Mag., September, 41–44.

Bell, A. L., Kirkland, D. A. & Sinclair, A. (1987). Vibro replacement ground improvement at General Terminus Quay. *Proceedings of the conference on marginal and derelict land*, Glasgow, pp. 697–712.

BRE (2000). *Specifying vibro stone columns*. Garston: Building Research Establishment.

BSI (1999). *Code of practice for site investigations*, BS 5930, pp. 85–100. Milton Keynes: British Standards Institution.

Chambosse, G. (1983). Liquefaction problems in the Fraser Delta and protection of a LNG tank. *Beitrage zu Standammbau und Bodenmechanik*, 105–112.

Charles, J. A. & Watts, K. S. (1996). The assessment of collapse potential of fills and its significance for building on fill. *Proc. Instn Civ. Engrs Geotech. Engng* 119, 15–28.

Davie, J. R., Young, L. W., Lewis, M. R. & Swekosky, F. J. (1991). Use of stone columns to improve the structural performance of coal waste deposits. In *Deep foundation improvements: Design, construction and testing* (eds M. I. Esrig and R. C. Bachus), STP 1089, pp. 116–130. Philadelphia: American Society for Testing and Materials.

Dobson, T. (1987). Case histories of the vibro systems to minimise the risk of liquefaction. In *Soil improvement: A ten year update* (ed. J. Welch), ASCE Geotechnical Special Publication No. 12, pp. 167–183. New York: American Society of Civil Engineers.

Greenwood, D. A. & Kirsch, K. (1983). State of the art report on specialist ground treatment and vibratory and dynamic methods. *Proceedings of the conference on piling and ground treatment*, London.

Institution of Civil Engineers (1987a). *Specification for ground treatment*. London: Thomas Telford.

Institution of Civil Engineers (1987b). *Specification for ground treatment: Notes for guidance*. London: Thomas Telford.

Kirsch, K. & Chambosse, G. (1981). Deep vibratory compaction provides foundations for two major overseas projects. *Ground Engng* 11.

Kirsch, K., Heere, D. & Schumacher, N. (1986). Settlement performance of storage tanks and similar large structures on soft alluvial soils improved by stone columns. *Proceedings of the conference on marginal and derelict land*, Glasgow, pp. 651–663. London: Thomas Telford.

Mitchell, J. K. & Huber, T. R. (1985). Performance of a stone column foundation. *J. Geotech. Engng* 111, No. 2, 205–223.

Mitchell, J. K., Baxter, D. C. P. & Munson, T. C. (1995). *Performance of improved ground during earthquakes*, ASCE Geotechnical Special Publication No. 49, pp. 1–36. New York: American Society of Civil Engineers.

Moseley, M. & Priebe, H. (1993). Vibro techniques. In *Ground improvement* (ed. M. P. Moseley). Glasgow: Blackie Academic.

NHBC Standards (1992). Vibratory ground improvement techniques.

Raison, C. A., Slocombe, B. C., Bell, A. L. & Baez, J. I. (1995). North Morecambe Terminal, Barrow, ground stabilisation and pile foundations. *Proc. 3rd Int. Conf. Recent Adv. Geotech. Earthquake Engng Soil Dynam., St Louis* 1, 187–192.

Sidak, N. (2000). Soil improvement by vibrocompaction in sandy gravel for ground water reduction. *Proceedings of the international workshop on compaction of soils, granulates and powders* (eds Kolymos and Fellin), pp. 23–32. Rotterdam: Balkema.

Slocombe, B. C. & Moseley, M. P. (1991). The testing and instrumentation of stone columns. In *Deep foundation improvements: Design, construction and testing* (eds M. I. Esrig and R. C. Bachus), STP 1089, pp. 85–100. Philadelphia: American Society for Testing and Materials.

Slocombe, B. C., Bell, A. L. & Baez, J. I. (2000a). The densification of granular soils using vibro methods. *Géotechnique* 50, No. 6, 715–725.

Slocombe, B. C., Bell, A. L. & May, R. E. (2000b). The in-situ densification of granular infill within two cofferdams for seismic resistance. *Proceedings of the international workshop on compaction of soils, granulates and powders* (eds Kolymos and Fellin). Rotterdam: Balkema.

Sondermann, W. (1996). Ruttelstopverdicthung zur Baugrundverbesserung für die Feste Fahrbahn in Schnell bahnbau. *Proc. 3rd Darmstädt Geotech. Colloq., Darmstädt.*

Kovacevic, N., Potts, D. M. & Vaughan, P. R. (2000). *Géotechnique* **50**, No. 6, 683–688

The effect of the development of undrained pore pressure on the efficiency of compaction grouting

N. KOVACEVIC*, D. M. POTTS† and P. R. VAUGHAN†

Compaction grouting involves the injection of stiff grout that does not penetrate the ground. Bulbs of grout are formed. It is applied to free-draining granular soils. The aim is to increase the density of the soil being treated by applying high local pressure. The design of compaction grouting is based largely on experience and empiricism. If the soil is not sufficiently permeable for consolidation to occur as it is treated, excess pore pressures may be generated, which will dissipate after treatment. The potential effect of such excess pore pressures on the compaction achieved is considered here. It is found that the efficiency of treatment may be reduced substantially.

KEYWORDS: compaction; consolidation; ground improvement; grouting; numerical modelling and analysis.

La cimentation de compactage passe par l'injection d'un ciment rigide qui ne pénètre pas dans le sol. Des bulbes de ciment se forment. Cette méthode est pratiquée sur des sols granulaires à drainage libre dans le but d'augmenter la densité du sol à traiter en appliquant une pression locale élevée. La solution de cimentation compactante est basée largement sur l'expérience et l'empirisme. Si le sol n'est pas assez perméable pour que la consolidation ait lieu pendant le traitement, il peut se produire des pressions de pores excessives qui se dissiperont après le traitement. Nous étudions ici l'effet potentiel de ces pressions excessives sur le compactage obtenu. Nous constatons que l'efficacité du traitement peut en être réduite de manière significative.

INTRODUCTION

Compaction grouting is a technique for densifying soils and lifting structures by pumping very stiff grout into the ground down a pipe of relatively large (50–100 mm) diameter at pressures up to 10 MPa. This forms an inclusion of an approximately spherical shape. Pumping continues as the tube is either driven or withdrawn after driving, and a cylinder of grout is formed. The grout acts as a high-viscosity non-penetrating fluid. The soil around the grout inclusion must fail in order to yield and displace to the relatively large diameters achieved, and large pressures can be exerted equivalent to the limit pressure in the pressuremeter test. Hole spacing is of the order of 2 m. The method is an attractive one, particularly for increasing the liquefaction resistance of layers of loose granular soils at depth. There is a minimum of ground and site disturbance compared with other methods of treatment. Ground treatment by compaction grouting can involve the development of excess undrained pore pressures. The possible effect of this on the efficiency of treatment is considered here.

COMPACTION GROUTING

Compaction grouting is normally used in granular soils, and quite large compressions have been recorded. Generally heave of the ground surface is small, much smaller than the volume of grout injected. Fig. 1 (from Boulanger & Hayden, 1995) shows the increases in cone penetration resistence recorded in several trials of compaction grouting, as a function of the average replacement ratio, which is the ratio of the volume of grout injected (corrected for ground heave) to the volume of ground treated. The estimated average final radius of the treatment holes and their spacing is also indicated. Boulanger & Hayden (1995) report that there is some loss of cone penetration resistance with time after treatment. This cannot be due to an increase in porosity, and it possibly indicates a loss of lateral stress as the soil creeps.

If the material being treated is free draining then it will consolidate as the concrete is pumped, and the process will be a

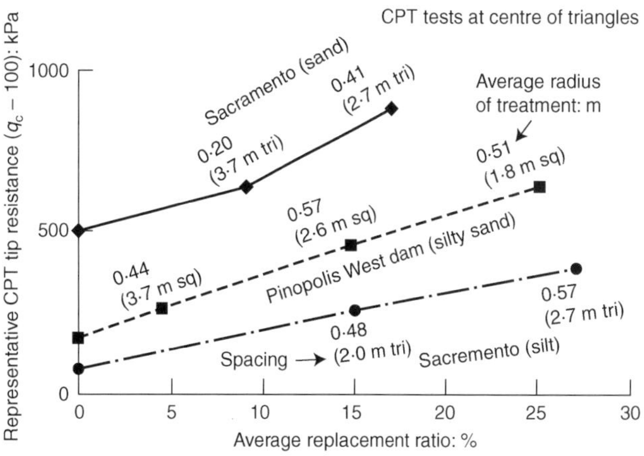

Fig. 1. Effect of compaction grouting on CPT tip resistance in granular soils (after Boulanger & Hayden, 1995)

drained one. If the permeability is lower the soil may not consolidate as the concrete is pumped. It will be displaced undrained, and this will be followed by consolidation while the grout column remains of fixed diameter.

A SIMPLE ONE-DIMENSIONAL MODEL OF DRAINED AND UNDRAINED COMPACTION GROUTING

Figure 2 shows an idealized model in which a soil sample of a specific thickness *l* is contained in a cylinder and loaded by a piston and a spring. The piston and spring represent the confinement of the surrounding soil. The uniaxial stress is σ. Δ is the compression of the soil, which is assumed to be saturated with a linear stiffness $m = \Delta/(l.\delta\sigma')$. ω is the displacement of the piston. The spring stiffness ($\delta\sigma/\sigma x$) has two values, E_1 for first loading and E_2 for unloading/reloading. A layer of 'grout' of thickness *g* is injected between the piston and the soil sample.

Drained injection

If the soil consolidates as the injection of thickness *g* is made, the soil compresses by an amount Δ_d and the spring compresses by an amount ω_d. The increase in total stress in the soil, $\delta\sigma$, is equal to the increase in effective stress, $\delta\sigma'$, and is given by the compression of the spring as

Manuscript received 2 February 2000; revised manuscript accepted 7 September 2000.
* Geotechnical Consulting Group.
† Imperial College, London, UK.

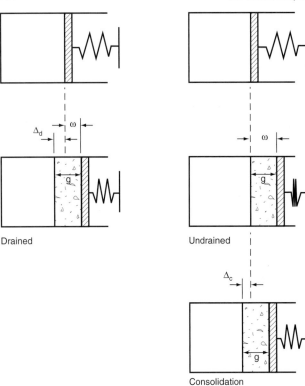

Fig. 2. One-dimensional analogy for drained compaction grouting and for undrained grouting followed by consolidation

$$\delta\sigma_d = \delta\sigma_d' = E_1.\omega_d \tag{1}$$

From the compression of the soil:

$$\delta\sigma_d = \delta\sigma_d' = \Delta_d/lm \tag{2}$$

and

$$g = \omega_d + \Delta_d \tag{3}$$

Solving:

$$\delta\sigma_d' = \frac{E_1 g}{1 + E_1 lm} \tag{4}$$

and

$$\Delta_d = \frac{E_1 g lm}{1 + E_1 lm} \tag{5}$$

Undrained injection followed by consolidation

If the soil sample is saturated then, during undrained loading, $\delta u_u = \delta\sigma_u$ and $\delta\sigma_u' = \delta\sigma_u - \delta u_u = 0$. δu_u and $\delta\sigma_u$ are the increases in pore pressure and total stress respectively. There is no compression or densification of the soil. The increase in total stress is given by the compression of the spring due to the thickness of grout injected, g, and:

$$\delta\sigma_u = E_1 g \tag{6}$$

During consolidation the soil sample compresses, and the spring expands by Δ_c. There is a change in pore pressure and a change in total stress, $-\delta\sigma_c$, and, noting that Δ_c is negative:

$$\delta\sigma_c = -E_2\Delta_c \tag{7}$$

Since the final pore pressure is equal to the initial one, the change in effective stress due to undrained loading and consolidation, $\Delta\sigma_{uc}'$, is the same as the change in total stress:

$$\delta\sigma_{uc}' = \delta\sigma_u + \delta\sigma_c = gE_1 - \Delta_c E_2 \tag{8}$$

The strain in the soil is Δ_c/l, and the change in effective stress is

$$\delta\sigma_{uc}' = \Delta_c/lm \tag{9}$$

Solving equations (8) and (9) gives

$$\Delta_c = \frac{gE_1 ml}{1 + E_2 ml} \tag{10}$$

and from equation (9):

$$\delta\sigma_{uc}' = \frac{gE_1}{1 + E_2 ml} \tag{11}$$

Comparison between the two drainage cases

The drained solution and that with undrained injection followed by consolidation compare as follows:

$$\frac{\Delta_{uc}}{\Delta_d} = \frac{1 + E_1 lm}{1 + E_2 lm} = \frac{\delta\sigma_{uc}'}{\delta\sigma_d'} \tag{12}$$

There is a monotonic increase in effective stress in the grouted soil in both drained loading and undrained loading followed by consolidation. Since there is no reduction in stress, non-reversible stiffness of the soil being grouted does not come into play. The stiffness m is for first loading in both cases. The spring is analogous to the mass of soil surrounding and confining the soil element being grouted. In the drained case the load in the spring also increases monotonically. Thus E_1 equivalent to first loading always operates. In the undrained case the spring load increases during injection (first loading) but decreases during consolidation (unloading) when E_2 applies. It would be expected that $E_2 > E_1$. Thus consolidation after the grout is injected undrained produces less densification than drained grouting using the same amount of grout. This hypothesis is now examined by finite element analysis.

TWO-DIMENSIONAL SOLUTION FOR COMPACTION GROUTING BY FINITE ELEMENT ANALYSIS

The analysis

The total stresses applied by compaction grouting decay radially away from the injection borehole, and are not one-dimensional as presumed in the simple model. To simulate this situation a finite element analysis has been performed, using the geotechnical program ICFEP developed at Imperial College. The analysis was performed in two dimensions using a large strain formulation and radial coordinates. The computational soil model used was modified Cam Clay (Gens & Potts, 1988). The parameters adopted are summarized in Table 1. Typical isotropic volume changes in first loading and unload-reloading are shown in Fig. 3. This behaviour is based on that of a sandy

Table 1. Parameters defining the modified Cam Clay soil model used in the analysis

Parameter		Description	Value adopted
v_1	$1 + e_1$	Specific volume at unit pressure	1·788 at 1 kPa
λ	$\Delta v/\Delta\log p'$	Slope of virgin consolidation line	0·066
κ	$\Delta v/\Delta\log p'$	Slope of swelling line	0·0077
G	Δ_T/Δ_γ	Shear modulus	$100 p'$
ψ_{cv}		Angle of shearing resistance (critical state)	30°

A Mohr–Coulomb hexagon and a circle have been used for the shapes of the yield and plastic potential surfaces in the deviatoric plane respectively.

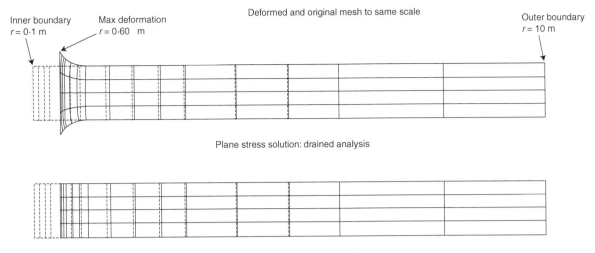

Fig. 3. Soil compressibility assumed in the finite element analysis

till that occurs in an old pre-Devensian buried glacial channel associated with the River Daugava in Latvia.

The mesh used is shown in Fig. 4. It consists of four layers of radially symmetric elements of total thickness 1 m. Results are plotted against an absolute radius, r. When, say, radial displacement is plotted against radius, r, the displacement when the boundary has been displaced from $r = 0.1$ m to $r = 0.6$ m is plotted at $r = 0.6$ m. The outer boundary was at $r = 10$ m. The inner boundary was initially at $r = 0.1$ m. Initial effective stresses were $\sigma'_z = \sigma'_\theta = \sigma'_r = 200$ kPa with $K_0 = 1.0$. This gives the initial specific volume of the soil, which is assumed to be normally consolidated. The initial pore pressure was $u_0 = 200$ kPa. The external boundary was stress controlled and kept at $\sigma_r = 400$ kPa. The inner boundary was frictionless, and was loaded by 500 incremental displacements of 1 mm, giving a final displacement to $r = 0.6$ m. When consolidation was modelled the inner boundary was fixed. Consolidation after undrained loading was modelled by reducing the excess pore pressures throughout the mesh by 100 increments, each of 1% of the initial value. Coupled consolidation was not modelled. The soil was assumed to be weightless, which ensured that the solutions were symmetric about the central plane.

Two alternative options were modelled for the upper and lower boundaries: plane stress and plane strain. In the first the boundaries were frictionless and at constant normal stress. In the second, the boundaries were frictionless but fixed. The deformed shapes of the mesh for the drained solution after a lateral expansion from $r = 0.1$ m to $r = 0.6$ m are shown in Fig. 4 for the plane stress and the plane strain solution. It can be seen that there is extension in the vertical direction in the plane stress solution. In reality, while the grouting process produces a column of grout, grout injection is more like the expansion of a sphere, with strains in three directions. There were differences between the two solutions. The plane stress solution, which has strains in the third direction, is thought to be the more realistic. Both solutions are symmetric about the centre plane of the four layers. Data are plotted for this plane.

Results obtained

Figure 5 shows the variation with radius r of the lateral displacement, the volumetric strain and the stress level, S, for different stages of loading for the plane stress case. The three stages are expansion to $r = 0.12$ m, expansion to $r = 0.2$ m, and the final expansion to 0.6 m. Note that plots represent the current state. Strains and displacements from zero are correct, and are plotted at the radius where the soil is after loading to the appropriate stage. The difference between strains plotted at the same radius at two different stages is not the strain, as the strains are for two different elements of soil.

S is the proportion of the shear strength at the current normal effective stress that is mobilized. Results are plotted for individual Gauss points within the mesh, and the waviness of the plots is a function of the solution. The curves have not been smoothed.

Lateral deformations were larger further from the injection boundary for the drained case than for the undrained case. The stress level (local shear stress) is also higher further from the injection boundary for the drained case, although there is a larger zone of failure next to the injection boundary for the undrained case. Consolidation reduces the stress level markedly. There is no decrease in volume during undrained loading. The decrease that occurs during consolidation after undrained loading is much smaller than the amount that occurs if loading is drained. Integrating the volumetric strain over the area that is influenced gives a total volume change of a layer of thickness 1 m of 0.694 m^3 for the drained case and 0.053 m^3 for the undrained case. The ratio is 0.07. The difference is as marked, as the simple one-dimensional analysis suggests. The volumetric strains at 1 m radius are 10 times greater for the drained case after expansion to $r = 0.6$ m than for the undrained-consolidated case with the same displacement of the inner boundary.

Figure 6 shows the volumetric strains predicted for the drained and undrained consolidated cases from both the plane

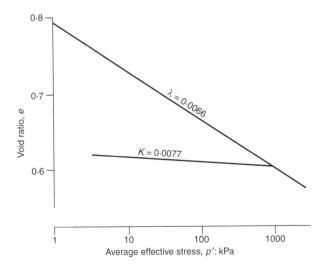

Fig. 4. Finite element analysis of drained expansion, plane stress and plane strain: mesh showing deformed shape after maximum expansion from an initial radius of $r = 0.1$ m to $r = 0.6$ m

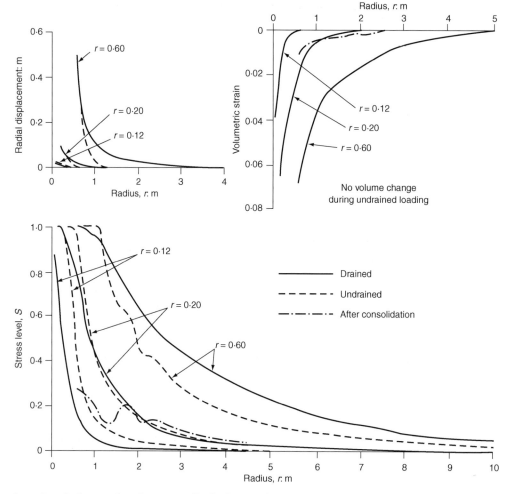

Fig. 5. Finite element analysis assuming plane stress. Drained expansion and undrained expansion plus consolidation. Radial displacement, volumetric strain and stress level plotted against radius for expansion from $r = 0.1$ m to $r = 0.12$ m, $r = 0.2$ m and $r = 0.6$ m

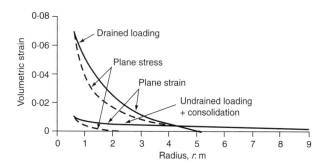

Fig. 6. Finite element analysis assuming plane stress and plane strain. Volumetric strains following drained expansion and following undrained expansion and consolidation. Expansion from $r = 0.1$ m to $r = 0.6$ m

stress and the plane strain analyses. The total volume changes predicted from the plane strain analysis for a 1 m thick layer are 0.95 m³ for the drained analysis and 0.60 m³ for the undrained plus consolidation analysis. In contrast to the result for the plane stress analysis, these are similar. The similarity between the total volume changes predicted for the two drainage cases arises because, in the undrained plane strain analysis, small strains are predicted a long way from the grout boundary. This would be unlikely in reality because ot the three-dimensional effects not allowed in plain strain but which can occur in reality and in the plane stress analysis. In contrast, the volumetric strains predicted at radius $r = 1$ m in the plane strain analysis are 0.05 and 0.007 respectively, more like the results predicted by the plane stress analysis. The strains at 1 m radius

from both types of analysis are similar, and are thought to be a better guide to the effectiveness of the grouting in achieving densification than the overall volumetric strains. If this is so, then the plane stress and plane strain analyses give essentially the same prediction, which is that drained grouting is much more effective at densification than undrained grouting followed by consolidation.

The horizontal total stresses predicted on the surface of the grout column are summarized in Table 2. These are equivalent to grouting pressures. Undrained expansion requires less pressure than drained expansion. Thus the pressure applied to the soil is less, and less compaction is produced. Much of the stress after undrained expansion is lost as consolidation occurs, showing the coupled nature of the consolidation.

Figure 7 shows the pore pressure generated after undrained loading according to both the plane stress and the plane strain analyses, and the mean effective stress after undrained loading and after consolidation. Much more pore pressure is generated in the plane strain analysis, both in magnitude and in the distance from the grout boundary that is affected. This might be expected from the greater constraint of the soil in the plane strain analysis.

The cause of the relative ineffectiveness of the undrained grouting is shown. In both the plane stress and the plane strain analyses the undrained shearing increases pore pressure and reduces mean effective stress. Although it has remained at constant volume the soil has become, in effect, over-consolidated. Most of the compression accompanying consolidation is on the re-load line of Fig. 3, and volume changes are small. Thus, while there is an overall increase in total stress during undrained expansion followed by consolidation, the mean effective stress follows an unload/reload loop, as is assumed in the

Table 2. Normal horizontal total stress on the boundary of the grout inclusion at different stages of the analyses

Type of analysis	Maximum horizontal total stress on boundary of grout injection: kPa			
	Initially	After drained expansion from 0·1 m to 0·6 m	After undrained expansion from 0·1 m to 0·6 m	After undrained expansion from 0·1 m to 0·6 m and consolidation
Plane stress	400	883	561	475
Plane strain	400	894	596	483

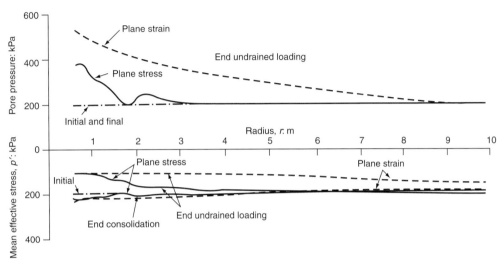

Fig. 7. Finite element analysis assuming plane stress and plane strain. Predicted pore pressure and mean effective stress changes in undrained expansion from $r = 0.1$ m to $r = 0.6$ m, followed by consolidation

one-dimensional analysis presented previously. In the drained analysis the mean effective stress increases monotonically, and compression is on the line for first loading.

GENERAL DISCUSSION

Mejia & Boulanger (1995) indicate that treatment of a single hole may proceed at a rate of approximately 3 m/h. If the grout column is 0·5 m radius then each equivalent spherical expansion might take about 20 min. For treatment to be fully effective in densification, it would be reasonable and conservative to assume that the 95% consolidation time for pore pressures set up at the boundary of the grout by treatment should be about 0·2 h. Soderberg (1962) presents an analysis for consolidation of the excess pore pressure set up around an expanded cylindrical cavity, using the Terzaghi–Rendulic equation. He found that 80% equilibration at the boundary occurred after a time $t = 10 \times r^2/c$ and 95% after $t = 30 \times r^2/c$, where r is the radius of the cavity and c is the coefficient of consolidation. These results are likely to be conservative compared with actual grouting, which will involve three-dimensional coupled consoli-

dation. The zone of excess pore pressure around the point of injection is strongly constrained, and analysis incorporating coupled consolidation would give a different result, probably with faster consolidation. To reflect this a consolidation time of 0·5 h is used in the subsequent calculations. The permeabilities required to give reasonable consolidation times are summarized in Table 3. A coefficient of volumetric compression, $m = 0.1$ m²/MN is assumed in deriving the permeability, k, from $k = cm\gamma_w$, where γ_w is the unit weight of water. The results indicate that undrained pore pressure effects might become noticeable if the bulk permeability of the ground is less than about $k = 10^{-6}$ m/s. Boulanger & Hayden (1995) suggest that compaction grouting may be less effective when applied to silts than when applied to sands, owing to a failure to consolidate during treatment.

Figure 8 shows a pore pressure measurement presented by Mejia & Boulanger (1995). It was made between injection columns in a silt with sand lenses, underlain by sands and silty sands. The test results reported are those shown for Sacramento in Fig 1. The piezometer was in the trial on silt with the 2 m triangular spacing. The piezometer responds to the treatment in

Table 3. Coefficient of consolidation and coefficient of permeability required for substantial pore pressure equilibration during compaction grouting taking 0·5 h

	Coeff. of consolidation, c: m²/s × 10^{-5} $c = Ar^2/t$			Permeability required, k: m/s × 10^{-7} $k = cm\gamma_w$		
	Radius of cylinder: m			Radius of cylinder: m		
	0·2	0·3	0·5	0·2	0·3	0·5
80% equilibration, $A = 10$	22·2	50	139	2	5	14
95% equilibration, $A = 30$	66·7	150	417	7	15	41

Time for consolidation, $t = 0.5$ h. Coefficient of volume compressibility, $m = 0.1$ m²/MN

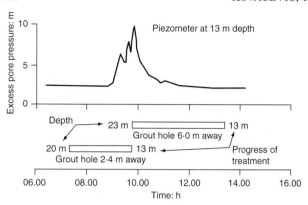

Fig. 8. Pore pressure increases measured by a vibrating wire piezometer in the vicinity of compaction grouting (after Mejia & Boulanger, 1995). Grouting was from the base of the treatment hole upwards

the hole 2·4 m away but not to the treatment in the hole 6 m away. This is in broad agreement with the predictions of Fig. 7. Consolidation is occurring rapidly when treatment stops. This shows that it is also occurring during treatment. In these circumstances reasonably effective treatment could be expected, as was observed. This would be consistent with the predictions made in this paper.

CONCLUSIONS

If a soil being treated by compaction grouting consolidates fully during treatment then mean effective stress increases monotonically. The soil compresses according to its compression curve for first loading. If the permeability is lower, then undrained pore pressures may develop around the grout column as it is formed. If fully undrained, the soil will not compress during treatment. In a contractive soil the finite element analysis suggests that mean effective stress will decrease during treatment. It will increase again as consolidation occurs but the soil will follow an unload/reload compression curve until the original mean effective stress is reached. This will reduce the amount of compression compared with that achieved in the same soil by the same amount of grout if it is loaded drained.

REFERENCES
Boulanger, R. W. & Hayden, R. R. (1995). Aspects of compaction grouting of liquefiable soil. *J. Geotech. Engng Div., ASCE* **121**, No. 12, 844–855.
Gens, A. and Potts, D. M. (1988). Critical state models in computational geomechanics. *Engineering and Computers* **15**, September.
Mejia, L. H. & Boulanger, R. W. (1995). A long term test of compaction grouting for liquefaction mitigation. *Proceedings of session on earthquake induced movements and seismic remediation of existing foundations and abutments, ASCE Convention, San Diego.* (ed. K. Kramer & R. Siddharthan), ASCE Geotechnical Special Publication No. 55, 94–109.
Soderberg, L. O. (1962). Consolidation theory applied to foundation pile time effects. *Géotechnique*, **12**, No. 3, 217.

Muir Wood, D., Hu, W. & Nash, D. F. T. (2000). *Géotechnique* **50**, No. 6, 689–698

Group effects in stone column foundations: model tests

D. MUIR WOOD*, W. HU† and D. F. T. NASH*

Model tests have been performed to determine the mechanisms of response for beds of clay reinforced with stone columns subjected to surface footing loads. An exhumation technique has been used to discover the deformed shapes of the model stone columns and thence to deduce the way in which the columns have transferred load to the surrounding clay. Tests have explored the effect of varying the diameter, length and spacing of the model stone columns. These parameters control whether the columns act as somewhat rigid inclusions transferring load to their tips and eventually deforming either by bulging or by the formation of a failure plane, or whether they are able to compress axially or even, if sufficiently slender, to 'bend' and undergo significant lateral deformation. Miniature pressure transducers have been used to reveal the distribution of contact pressure between columns and clay at various stages during the loading of the footings. It appears that the columns at mid-radius of the footing are typically the most heavily loaded. Results from numerical analysis are used to provide qualitative support for some of the findings from the physical model tests.

KEYWORDS: bearing capacity; clays; model tests; sands; stiffness.

Des essais sur modèle réduit ont été réalisés pour déterminer les mécanismes de réponse de lits d'argile renforcés de colonnes de pierre à des charges appliquées par une fondation de surface.

Une technique d'exhumation a été réalisée afin d'étudier la déformation de ces colonnes et en déduire la façon dont elles transmettent les efforts à l'argile environnante.

Les essais ont servis à étudier les effets des variations de diamètre, de longueur et d'espacement des colonnes. En fonction de ces paramètres, les colonnes agissent comme des inclusions quelque peu rigides transférant la charge à leur pointe pour finir par se déformer en tonneau ou en formant un plan de rupture; ou encore, elles sont capables de se comprimer dans l'axe, et même, si elles sont suffisamment minces, de s'infléchir et de subir une déformation latérale significative.

Des mini-capteurs de pression ont été utilisés pour indiquer la distribution de la pression de contact entre les colonnes et l'argile à divers stades de l'application de la charge sur les fondations. Il en ressort que les colonnes à mi-rayon de la fondation sont les plus sollicités.

Des résultats d'analyses numériques sont utilisés pour appuyer de manière qualitative certaines des conclusions des essais expérimentaux.

INTRODUCTION

There are various techniques available for improvement of the mechanical behaviour of soft clay foundation soils. One of these is the replacement of some of the clay with crushed rock or gravel to form an array of stone columns under the foundation. There are two beneficial effects that result from the presence of the columns of granular material. First, the granular material is stiffer and has higher frictional strength than the soft clay, so that the columns act as piles transmitting the foundation loads to greater depth with load transfer occurring by a combination of shaft resistance and end bearing. Second, the granular material has a high permeability by comparison with the clay so that the columns act as drains reducing the path length for consolidation of the soft clay under the foundation and hence speeding up the strengthening of this material. The work described in this paper is primarily concerned with the first of these effects: the sharing of load between the stone columns and the clay.

The behaviour of single stone columns was studied in classic tests by Hughes and Withers (1974). They used laboratory radiography to observe the deformations occurring in and around a column of sand loaded within a cylindrical chamber containing clay. Their conclusion that the bulging deformation of the column is contained within a length from the surface of about four column diameters has entered the geotechnical canon, and the capacity of individual stone columns has been deduced from an analogy to cylindrical cavity expansion—the bulging stone column expanding radially into the surrounding clay.

However, such an analogy assumes that the stone columns are acting entirely independently, whereas they are in fact often used in groups to provide general improvement of foundation soils under, for example, road embankments, storage tanks or pad foundations. Where columns are closely spaced it is expected that the attempts of one column to bulge will be resisted by the similar attempts of the neighbouring columns and that, in the centre of the group, the load will be transmitted to greater depth. Little information is available to guide the design of groups of stone columns, and the experiments described here have been performed in an attempt to indicate the effect of variation of column diameter, spacing and length on the mechanisms of load transfer within groups of stone columns beneath a rigid foundation.

The model tests have not been performed with any particular prototype in mind but can be seen more generically as a study of the behaviour of regularly inhomogeneous soils that takes its inspiration from prototype stone column reinforcement. The model tests have been performed at single gravity, and there are naturally questions concerning the scaling of the observed effects to prototype dimensions and stress levels. However, on the one hand, the modelling techniques that have been developed will be relevant to other model tests on stone column foundations—for example in a geotechnical centrifuge—and on the other hand the results themselves convey qualitatively messages that are expected to be relevant at all scales.

MODEL TESTS

Full details of the tests and the procedures that were used are given by Hu (1995). Properties of the soils used are given in Table 1. Model tests have been carried out in a loading tank of diameter 300 mm in which kaolin clay was consolidated from slurry to a maximum stress $\sigma_v = 120$ kPa and then allowed to swell back under a stress of 30 kPa leaving a layer thickness of about 300 mm. Details of the model tests are summarized in Table 2. This table shows that in a few tests a higher consolidation pressure was used. It also notes that one test (TL02) was performed in a larger tank of diameter 760 mm. Logistical constraints prevented either further testing at this larger size or

Manuscript received 31 January 2000; revised manuscript accepted 6 June 2000.
* Department of Civil Engineering, University of Bristol, UK.
† W. S. Atkins Consultants, Epsom, UK.

Table 1. Properties of soils used in model tests

Kaolin clay	
Liquid limit	0·63
Plastic limit	0·36
Proportion of clay size particles $< 2\ \mu$m	60%
Drained angle of shearing resistance	23°
Loch Aline fine sand (Belkheir, 1993)	
Dominant mineral	Quartz
Mean particle diameter d_{50}	0·21 mm
Uniformity coefficient d_{50}/d_{10}	1·2
Maximum void ratio	0·84
Minimum void ratio	0·51
Critical state angle of shearing resistance	~30°

systematic investigation of effects of tank size and influence of boundary proximity. A laboratory shear vane was used to give a general indication of the average undrained strength, c_u, within the block of clay at the start of the test: these values are given in Table 2.

Once the required history of consolidation was complete, the clay surface was exposed and model stone columns were formed by a replacement technique: an auger was used to prepare a lined hole in the clay, and sand was then poured into this hole as the lining was removed. (Although a *displacement* technique was used for some tests, in which a closed-end mandrel was inserted into the clay bed to create the hole for the sand column, it was found that the disturbance and ground heave produced were inevitably large and very dependent on the sequence of installation of columns: it was thus difficult to obtain reliably repeatable test specimens. These tests are not reported here.) While the sand column thus formed was only of medium density, the procedure was found to give uniform repeatable columns of good overall quality. The sand chosen for the columns in the present model tests was a fine quartz sand from Loch Aline, on the west coast of Scotland. The properties of the sand are given in Table 1. The relative density of the sand poured into the columns was around 0·5.

The ratio of a typical dimension D of a geotechnical construction to a typical particle size d of the granular material influences many aspects of the behaviour of such constructions. It would be desirable for the sand to be used in model tests to be chosen to give a ratio D/d similar to that found in the prototype structures that are being modelled. In practice, stone columns are formed at typical diameters $D = 0.6$–1 m of crushed rock or gravel with typical particle size $d = 25$–50 mm, so that the ratio D/d lies typically between 12 and 40. The column diameters used in the model tests were 11 and 17·5 mm (column radii r_c 5·5 and 8·75 mm, Table 2); formation of slimmer columns was found not to be practicable. The ratio D/d in the model tests therefore had values of 52 and 83. Although these values are higher than those typical in practice,

the concern in choosing the model sand was to ensure that the sand was sufficiently fine (a) for ease of placement and (b) to allow shear bands up to 20 particles thick (Roscoe, 1970) to form in the columns without undue restraint. Stone columns are used to improve the behaviour of soft soils; the kinematic constraint provided by the soft material surrounding the columns is not great. The present model tests have revealed shear bands forming obliquely across some of the columns; provided the columns are sufficiently long the different ratios of column and particle diameter in model and typical prototype should not then be significant.

The sand columns were placed on a square grid extending just beyond the edge of the area to be loaded (as shown schematically in Fig. 1(a)). (In two tests an area 25% larger was treated, as noted in Table 2; the effect on the observed load–settlement response was small.) Sand columns were formed with different lengths in different tests, but the columns were always 'floating' in that they ended within the clay and were not end bearing. In some tests miniature pressure transducers were placed on the surface of individual columns or on the surface of the clay between the columns and beneath the footing so that some assessment could be made about the distribution of load between the columns and the clay. Loading of the reinforced clay bed was achieved using a rigid circular footing of diameter 100 mm, which was pushed steadily into the clay by an inverted 10 kN triaxial test loading frame. Most tests were stopped when a displacement of 30 mm had been reached, the limit of travel of this apparatus. Test TS20 was stopped at 15 mm in order to obtain an indication of the progress of the mechanism of deformation. Average footing pressure was deduced from the overall footing load measured using a single load cell. The size of the footing ensured that in the 300 mm diameter tank that was used for almost all the tests there was a distance to the boundary equal to the diameter of the footing all round the footing.

After the footing had been unloaded and removed, the damp sand in the columns was sucked out, a wire armature was inserted in each column, and plaster of Paris was poured in and allowed to take up the shape of the deformed columns and set. Subsequent excavation of the clay and exhumation of the columns then produced a very clear plaster model of the deformed state of the columns (Fig. 2), enabling features of the load transfer mechanism to be identified. This exhumation technique was used for tests TS08 onwards, as noted in Table 2.

The parameters that can be varied to control the geometry of the columns are the radius r_o of the footing; the length L of the columns; the radius r_c of the individual columns; and the spacing s of the columns on a square grid (Table 2). An area ratio A_s can then be calculated as the ratio of cross-sectional area of columns to cross-sectional area of foundation soils:

$$A_s = \pi \left(\frac{r_c}{s} \right)^2 \tag{1}$$

Table 2. Summary of model tests

Test	σ_v: kPa	c_u: kPa	r_c: mm	L: mm	s: mm	L/r_c	A_s: %	Note
TS02	180	23	5·5	100	25.3	18·2	15	Trial
TS03	110	5	5·5	100	30·8	18·2	10	Pressure bag leak
TS04	160	16·5	5·5	150	19·8	27·2	24	
TS05	120	10·5	5·5	100	17·6	18·2	30	Extra 25% columns beyond footing
TS07	120	8	5·5	150	30·8	27·2	10	Extra 25% columns beyond footing
TS08	120	15	5·5	100	19·8	18·2	24	Exhumation from this test onwards
TS09	120	11·5	8·75	160	31·5	18·2	24	
TS10	120	11·5	8·75	100	28	11·4	30	
TS16	120	11·5	5·5	100	30·8	18·2	10	
TS17	120	14	5·5	160	19·8	29	24	
TS19	120	10	5·5	160	19·8	18·2	24	No central column
TS21	120	10	5·5	100	19·8	18·2	24	
TL02	160	17	8·75	160	31·5	18·2	24	Large test tank
TS20	120	14	5·5	100	19·8	18·2	24	Load to 15 mm settlement
TS11	120	14	—	—	—	—	—	Plain clay (no sand columns)
TS20/2	120	14	—	—	—	—	—	Plain clay (no sand columns)

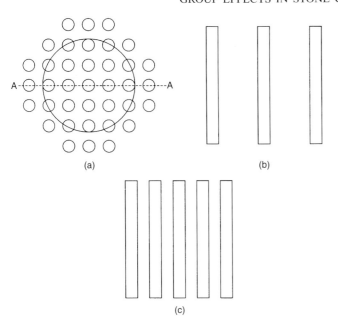

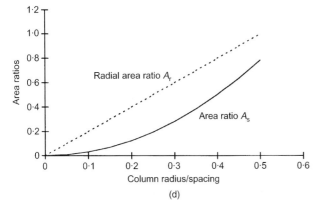

(a)

(b)

(c)

(d)

Fig. 1. Comparison of area ratios: (a) array of stone columns with area ratio $A_s = 24\%$; (b) plane strain arrangement of stone 'trenches' with area ratio $A_r = 24\%$; (c) plane strain arrangement of stone 'trenches' matching area ratio along diameter A–A of footing; (d) numerical comparison of area ratios A_s (equation (1)) and A_r (equation (7)).

Values of L/r_c and A_s are also given in Table 2.

A footing penetration rate of 0.061 mm/min was used for all tests so that loading to 30 mm penetration took about 8·2 h (492 min). An assessment of the likely degree of consolidation associated with this rate of penetration can be obtained in two ways. The standard procedure for predicting consolidation around vertical drains can be used to deduce the time after application of a surface load for any chosen degree of consolidation. Alternatively, a simple one-dimensional analysis can be used to estimate the degree of consolidation in a cylindrical soil region around a vertical drain as the soil is compressed at a steady rate.

Barron's formula (Barron, 1947, quoted by Leroueil et al. 1985) indicates that the degree of consolidation U in a cylindrical block of clay of equivalent external radius R surrounding vertical drains of radius r_c is

$$U = 1 - \exp\left(-\frac{2T_r}{F}\right) \tag{2}$$

where F is a function of drain geometry $\chi = R/r_c$ given by

$$F = \frac{\ln \chi}{1 - \chi^2} - \frac{3 - \chi^{-2}}{4} \tag{3}$$

and T_r is a time factor

$$T_r = \frac{c_h t}{R^2} \tag{4}$$

The coefficient of radial consolidation $c_h \approx 2 \times 10^{-7}$ m²/s for kaolin at the approximate vertical effective stress of these model tests (Al-Tabbaa, 1987). For drains placed on a square array with spacing s, $R \approx 0.564 s$. For the dimensions of the arrays of columns (drains) used in the present tests the times after application of a load for 95% consolidation are given in Table 3. The slowest situation is obtained with columns of radius 5·5 mm and spacing 30·8 mm. The value of t_{95} is then about 21 min and the duration of the loading test is about 24 t_{95}.

Alternatively, the radial variation in pore pressure in a cylindrical block of clay with an impermeable outer boundary of radius R and inner drainage boundary of radius r_c subjected to constant and radially uniform axial strain rate $\dot{\varepsilon}$ is found from solution of the equation of radial consolidation to be

$$u = \frac{\gamma_w \dot{\varepsilon}}{4 k_h} \left[2R^2 \ln\left(\frac{r}{r_c}\right) - (r^2 - r_c^2) \right] \tag{5}$$

assuming purely one-dimensional deformation (i.e. neglecting radial movements) w is the unit weight of water and k_h is the horizontal (radial) permeability of the clay. Integration of this expression over the radius can be used to estimate an average pore pressure and hence an average degree of consolidation by comparison with the effective axial stress generated at any time t (assuming a constant coefficient of consolidation):

$$U = 1 - \frac{F}{2T_r} \tag{6}$$

Equation (6) is directly comparable to the expression introduced by Bishop and Henkel (1962) to assist in selection of strain rates for drained triaxial tests. The times to failure required to give 95% consolidation at failure for the geometries of columns used in these tests are given in Table 3. The slowest time is estimated to be $t_{f95} = 139$ min. The duration of the present tests is about 3·5 times this worst figure. It seems likely then that the tests can be considered to be essentially drained except possibly for these lowest area ratios of sand columns.

NUMERICAL ANALYSIS

The emphasis of this paper is firmly on the physical modelling. However, in order to provide some support for and ability to extrapolate from the results from these physical model tests some numerical analyses have been performed using the finite difference code FLAC (Itasca, 1996). These numerical analyses have been used to confirm the principal mechanisms of soil deformation and to indicate the distribution of contact stresses between clay and sand.

In these fully drained analyses a simple rectangular grid (Fig. 3) has been used to represent the foundation soil, modelled in plane strain. Different patterns of stone columns (different distributions and different lengths) have been created by switching the properties of appropriate zones of the mesh. The footing loading has been treated as an entirely drained process with no allowance for development of excess pore pressures. The footing has been driven down at a steady speed. The clay has been modelled as a strain hardening frictional material with angle of friction varying with plastic shear strain according to the multilinear relationship shown in Fig. 4. Volumetric hardening has not been included. Dilatancy is always taken as the difference between current friction and the ultimate critical state angle of 25°, so for this soil the dilatancy is negative and the plastic volumetric strain is always compressive. The sand has been modelled as a strain softening material with angle of friction varying with plastic shear strain as shown in Fig. 4, again with dilatancy equal to the difference between the current friction angle and the critical state value of 30°: this soil expands as it shears. The stiffness properties of the clay and the sand used in the numerical analyses are summarized in Table 4: the actual values are not particularly important. The ratio of shear modulus to bulk modulus controls Poisson's ratio, here 2/7 for both

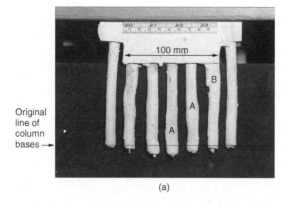

(a)

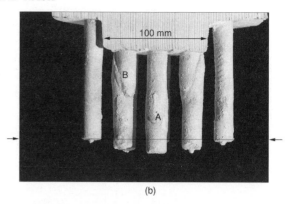

(b)

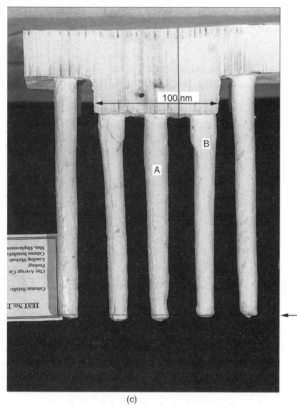

(c)

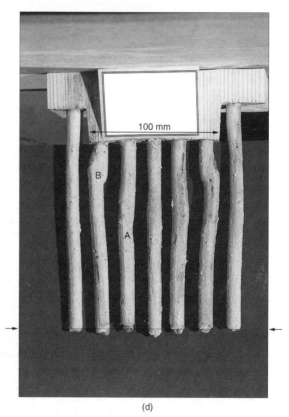

(d)

Fig. 2. Photographs of deformed sand columns exhumed at the end of footing penetration (arrows indicate original level of column bases): (a) $L/r_0 = 2$, $A_s = 24\%$, $r_c = 5.5$ mm; (b) $L/r_0 = 2$, $A_s = 30\%$, $r_c = 8.75$ mm; (c) $L/r_0 = 3.4$, $A_s = 24\%$, $r_c = 8.75$ mm; (d) $L/r_0 = 3.2$, $A_s = 24\%$, $r_c = 5.5$ mm; (e) Sketches of deformation modes

materials. The low initial angle of shearing resistance for the clay ensures that plastic strains will in fact occur at an early stage of the analysis but provides an initial elastic response perhaps equivalent to some overconsolidation. The initial ratio of horizontal to vertical effective stresses has been set at 0·5 in the analyses reported (though varying this ratio had little effect on the mechanisms that were deduced).

There is inevitably a problem in moving between the physical model tests, conducted with an array of individual columns and a circular footing (Fig. 1(a)), and the numerical analyses conducted in plane strain with the columns replaced by equivalent sand filled 'trenches' (Fig. 1(b)). The results are intended to be illustrative. However, it is found that in order to obtain the degree of interaction observed experimentally, the plane strain area ratio needs to be much higher than the area ratio of individual columns. Along a radius such as A–A in Fig. 1(a) intersecting a series of columns the ratio of sand to (sand + clay) is

$$A_r = 2\left(\frac{r_c}{s}\right) \qquad (7)$$

but of course there will be other radii that do not intersect any columns, or which intersect many fewer columns, so that the 'radial area ratio' will be much less than A_r. A comparison of A_r and A_s (equation (1)) is shown in Fig. 1(d). It has been found that the area ratios that need to be used in the plane strain numerical analysis have to be nearer to A_r than A_s, as shown in Fig. 1(c), in order to be able to obtain comparable effects. This suggests that constraint between columns along radii of the footing is particularly significant. It also suggests that although the stone columns do not actually coalesce to form concentric cylinders in the physical model they do seem to experience significant restraint in this circumferential direction.

RESULTS

Modes of deformation

Different modes of deformation can be observed depending on the geometric configuration of the sand columns. These are illustrated in Fig. 2(e) with photographs of exhumed sand

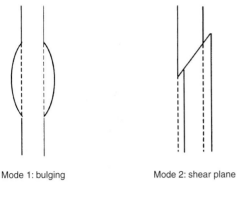

Mode 1: bulging   Mode 2: shear plane

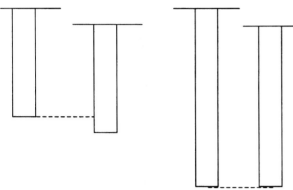

Mode 3: short column penetrates clay; long
column absorbs deformation along its length

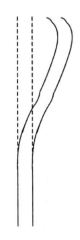

Mode 4: slender column acts as laterally loaded pile

(e)

Fig. 2. (*continued*)

Table 3. Estimates of consolidation times t_{95} (from equation (2)) and t_{f95} (from equation (6))

r_c: (mm	s: mm	t_{95}: s	t_{f95}: s
11	17·6	132	881
11	19·8	229	1531
11	25·3	620	4142
11	30·8	1250	8348
17·5	28·0	334	2229
17·5	31·5	581	3877

columns together with sketches of the modes that are being described.

1. If a column is loaded, and not prevented from expanding radially by closely adjacent columns, then the mean stress increases in the column, and column bulging takes place (A in Fig. 2). Bulging indicates stable ductile deformation.
2. If a column is subjected to high stress ratios with little lateral restraint, and hence little opportunity for increase in mean stress, then a diagonal shear plane may form through the column (B in Fig. 2). Such a shear plane may either indicate global concentration of deformation caused by an overall failure mechanism or suggest that the stress state in the column is such as to encourage bifurcation of material response and occurrence of localization.
3. If a column is sufficiently short for significant load to be transmitted to the base of the column then it will penetrate the underlying clay. As the column length increases the penetration reduces because less load passes to the base of the columns (compare lines of column bases in Figs 2(a) and (b), where the columns under the footings have clearly penetrated the underlying clay, with those in Fig. 2(c), where the penetration of the footing has been absorbed along the length of the columns). This mode can be directly compared with the behaviour of the sand column as a pile.
4. A sand column is intended to be strong under axial load but is not expected to have any significant strength against lateral loading. The displaced line of the centre of each column gives an indication of lateral movements occurring in the clay beneath the footing (this is most obvious in Fig. 2(d)). The sand columns are behaving as laterally loaded piles. For the most slender columns there is probably little flow of the clay past the columns, and they are behaving rather like model inclinometers.

The observed occurrence of modes 1 and 2 of deformation is summarized in Fig. 5(a). Shear planes through the columns are seen towards the edge of the footing, where the displacement discontinuity imposed by the footing is felt most strongly. Bulging is seen at depth, but the depth of the bulging increases as the area ratio A_s increases. This pattern of column deformation can be seen not only as illustrating the effect of local stress state in the columns on the failure mode that occurs but also as an indication of the expected mode of deformation beneath a rough rigid footing (Fig. 6(a)) in which a cone of somewhat undeforming soil is pushed down with the footing. In fact, the edge of this cone is clearly defined near the edge of the footing—the footing itself imposes a severe displacement discontinuity at the ground surface—but it is less well defined under the centre of the footing, where the boundary of the undeforming zone—deduced from the location of the bulging in the columns—becomes somewhat more diffused.

The geometry of deformation as revealed through the observation of mode 3 is summarized in Fig. 5(b), where the penetration at the base of the sand columns is plotted against radius. Penetration is greatest when the columns are short and the area ratio is high. For long columns, particularly when the area ratio is low, no load reaches the base of the columns and the footing penetration is absorbed by compression down the length of the columns.

A combination of the observations made concerning the location of the formation of shear planes or bulging in the columns can be used to estimate the zone of influence of the footing on the underlying reinforced ground. This zone is roughly conical with a lateral angle β (Fig. 6(a)), which increases as A_s increases. If the difference between plane strain and axisymmetric modes of deformation is set aside then this tip angle can be linked with the average mobilized angle of friction $\bar{\phi}$ in the foundation soil:

$$\beta = \frac{\pi}{4} + \frac{\bar{\phi}}{2} \qquad (8)$$

The experimental results are somewhat scattered, but there is some indication that the resulting deduced values of $\bar{\phi}$ fall from around 41° for area ratio $A_s = 30\%$, which is appropriate for an angle of friction of the pure sand at the stress level of the test, to around 23° for area ratios A_s of 24% and 10%, which is

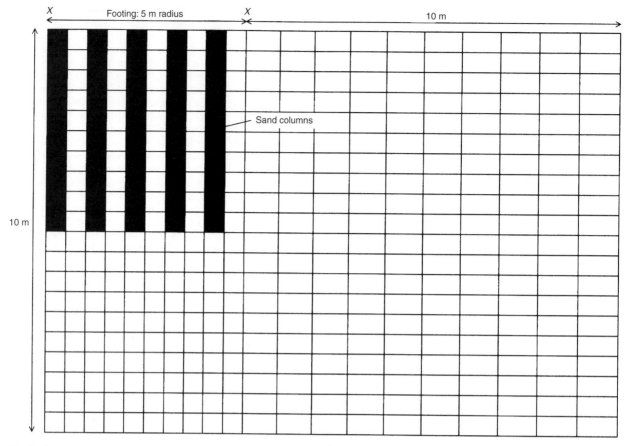

Fig. 3. Grid used for numerical analysis of stone column reinforced foundation. Shaded elements represent stone columns. Rigid footing covers region XX, with left boundary a line of symmetry

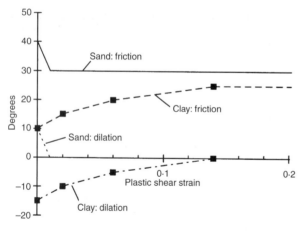

Fig. 4. Assumed variation of friction and dilation with plastic shear strain for clay and sand

Table 4. Properties of soils used for numerical analysis

Clay	
Bulk modulus	10 MPa
Shear modulus	5 MPa
Sand	
Bulk modulus	100 MPa
Shear modulus	50 MPa

7(c)) most of the footing displacement is absorbed in compression of the columns.

The key result to be drawn from the physical observations is that the geometry of the footing interacts with the geometry of the individual columns in producing an eventual mechanism of deformation in the columns. The depth at which the dominant strain is occurring in individual columns is primarily controlled by the diameter of the footing itself, not by the diameter of the individual columns. This can be interpreted both as an interaction between adjacent columns and as the response of the array of columns within an overall mechanism that is driven by the footing. It may be noted, however, that the results in Fig. 6(b) suggest that the geometry of this mechanism is itself influenced somewhat by the presence of the columns.

Typical design procedures for bearing capacity of stone column reinforced soil (e.g. Cooper and Rose, 1999) use average strengths calculated from the strengths of the individual components weighted according to their area ratio. Such a procedure requires that the reinforced area be taken beyond the loaded area so that the complete region of soil that will be affected by a bearing capacity mechanism has been uniformly reinforced. Centrifuge model tests on a rigid footing penetrating soft clay reinforced with 'sand compaction piles' are reported by Kimura et al. (1983). These show a remarkable increase in bearing capacity as the extent of the improved region of clay is

close to the drained angle of friction of the pure clay (Fig. 6(b)).

The numerical parallel to these physical observations can be found in illustrations of typical profiles of vertical displacement through groups of stone columns of different lengths in the numerical analyses (Fig. 7). The gradient of these profiles gives an indication of the vertical strain that is occurring. Near the edge of the footing (column A) the strain is extremely high near the ground surface: the boundary of the penetrating wedge is felt strongly. Towards the centre of the footing (columns D, E) the point of maximum strain becomes lower and the magnitude of the maximum strain decreases, reflecting the more diffuse boundary to the penetrating wedge. With long columns (Fig.

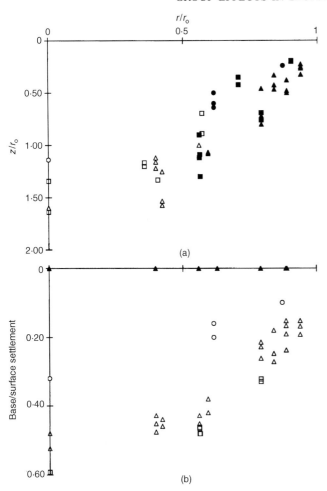

Fig. 5. Geometric effects of footing penetration. (a) Identified bulging (open symbols) and shear planes (solid symbols): $L/r_0 = 2$; $A_s = 10\%$ (○), 24% (△), 30% (□); (b) Penetration of base of columns as proportion of footing settlement: $L/r_0 = 2$ (open symbols), $L/r_0 = 3.2$ (solid symbols); $A_s = 10\%$ (○), 24% (△), 30% (□)

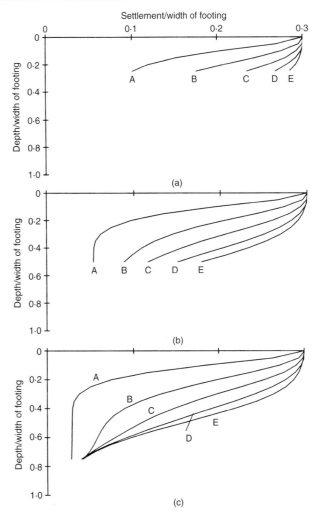

Fig. 7. Settlement profiles of columns in numerical analysis. Column length/footing width: (a) 0·25; (b) 0·5; (c) 0·75. Columns located at distances from centre/footing width: A, 0·425; B. 0·325; C, 0·225; D, 0·125; E, centre

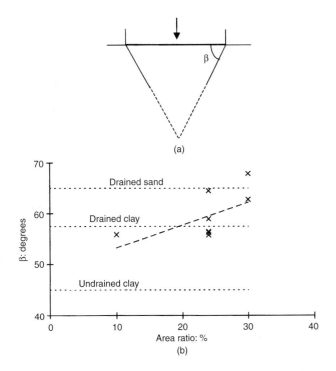

Fig. 6. (a) 'Rigid' cone beneath footing; (b) variation of angle β with area ratio

increased. The external columns have two beneficial effects: they provide a volume of material with higher frictional strength, thus increasing the average shear strength of the ground; and they help to speed the drainage and hence consolidation strength increase of the clay.

Distribution of stresses

Typical results for the variation of average footing pressure, normalized with initial undrained strength, with footing settlement, normalized with footing diameter, are shown in Fig. 8. The results are gathered together in groups showing the effect of area ratio A_s for short and long columns, and the effect of column length. Area ratio has the expected effect: stiffness and strength broadly increase as area ratio increases (Figs 8(a), (b)), though there is experimental scatter and there are some initial bedding effects at the start of some tests. As expected from the observations of the mechanism of deformation described in the previous section, column length is relevant only up to a certain point (Fig. 8(c)); beyond that point, increasing the length of the columns confers no further advantage (Fig. 8(d)). Insufficient data are available to indicate the detail of the link between this critical useful column length and area ratio. However, the tentative dependence of the overall deformation mechanism on area ratio suggested in Fig. 6 indicates an expectation that the critical length should increase as the area ratio increases because the mechanism is being pushed to a greater depth below the footing.

Typical results for the radial distribution of vertical stresses

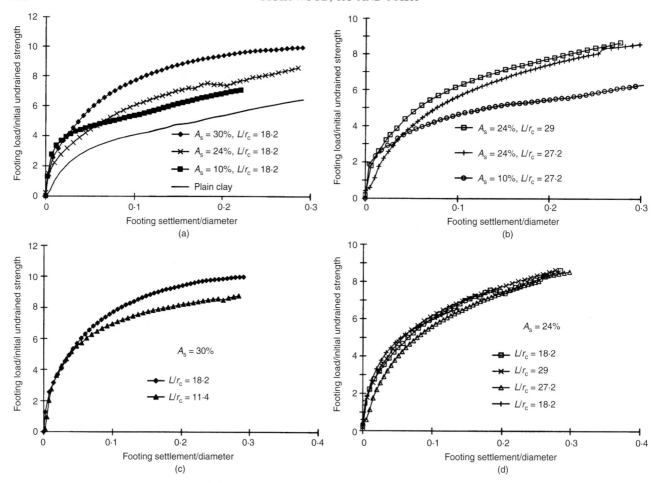

Fig. 8. Normalized load–settlement results for model footings: (a) variation of area ratio (short columns); (b) variation of area ratio (long columns); (c) variation of column length (short columns); (d) variation of column length (long columns)

as measured with the miniature pressure transducers are shown in Fig. 9 for area ratio A_s of 24%. The points indicate the measured contact stresses; the lines join points corresponding to the same stage of loading and do not in any way indicate the supposed radial distribution of normal stress. The high stresses represent the stresses at the top of sand columns; the low stresses represent the stresses at the contact with the surrounding clay. Though there is some initial adjustment as the footing penetrates the soil the relative magnitudes of stresses do not change much thereafter. Allowing for experimental scatter it seems to be fairly consistently observed that the stress on the column under the centre of the footing is lower than that on the column at mid-radius. The stress on the column at the edge of the footing is again lower. Presumably at the edge the sand column is no longer supported by neighbours and consequently

has a lower limiting load. The reason for the lower central stress is less clear.

Numerical analyses on different arrangements of columns show a similar result (Fig. 10). The assumed plastic hardening relationships for the two soils (Fig. 3) lead to a gentle radial redistribution of stress as footing penetration increases: the edge columns tend to weaken a little and the clay tends to strengthen. These numerical analyses show consistently that the central column is less heavily loaded than the adjacent off-centre columns.

A comparison can be made between the physical model observations and some field observations of stress concentration made by Greenwood (1991) near the Humber Bridge (Fig. 11).

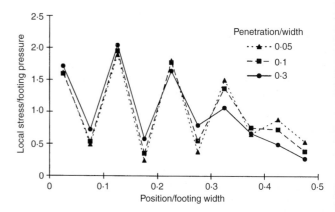

Fig. 9. Contact stresses beneath model footing ($A_s = 24\%$, $L/r_c = 18·2$): points represent measured values; lines join points corresponding to same stage of loading, and do not indicate spatial variation of stress

Fig. 10. Variation of contact stress from numerical analysis: points represent computed values; lines join points corresponding to same stage of loading, and do not indicate spatial variation of stress

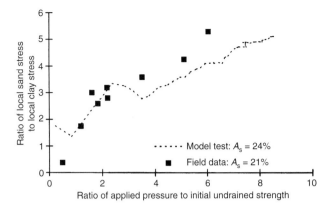

Fig. 11. Stress concentration ratio in model test and field test (data from Humber Bridge trial loading (Greenwood, 1991))

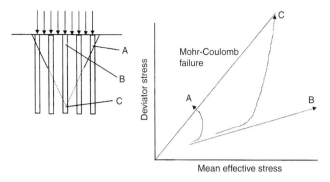

Fig. 12. Schematic stress paths for elements in sand columns

Data from model tests with area ratios of 24% are compared with field data for an area ratio of about 21%. The field data come from a large loaded area and relate to stone columns in the middle of the loaded area. The model data come from the column at mid-radius compared with the stress on adjacent clay. The stress ratio is plotted as a function of the footing average stress or average applied field stress normalized with the initial undrained strength of the soil obtained from field or laboratory vane tests. (The consolidation that occurs during the loading process produces a steady, and desirable, change in undrained strength from the initial value.) The magnitudes of the stress concentration are rather similar. However, it should be remembered that the field data relate to a flexible surface loading whereas the present model data come from tests with a rigid footing.

DISCUSSION AND CONCLUSIONS

The laboratory model tests have shown that there is significant interaction between a footing and the individual stone columns within a group. As a consequence, the load–settlement relationship for neighbouring columns in different locations under a footing will be very different, and it will be pessimistic in design of stone column reinforced foundations to neglect increasing stiffness towards the centre of the group. The kinematic constraints that the base of a rough footing impose push the load to greater depths towards the centre of the footing. At greater depths the clay that is being reinforced will normally have greater strength so that even if the capacity of the individual stone columns is to be calculated from a cavity expansion analogy the depth of the bulging failure should be taken into account in choosing the limiting radial stress. It may be concluded that an analysis that treats each stone column with its surrounding soil as an essentially identical unit cell will lose this actual variation with radial location.

The interactions between the individual stone columns, the loaded area and the surrounding soil can be understood as the behaviour of 'piles' with non-linear, sand-like axial stiffness properties. Axially, like compressible piles (Fleming *et al.*, 1985), beyond a certain length no further load can reach the base of the pile or stone column because it has all been shed through shaft friction: this has been observed in the physical models. The difference from a simple pile analogy is that the radial restraint from the surrounding soil is important in its influence on the mobilization of shear strength within the pile/column. Considering for simplicity only the central column, possessing axial symmetry, the upper part of the column, within the rigid cone beneath the footing, must have a stress path like B in Fig. 12 (plotted in terms of deviator stress $q = (\sigma_a - \sigma_r)$ and mean effective stress $p = (\sigma_a + 2\sigma_r)/3$, where σ_a and σ_r are axial and radial effective stresses), corresponding somewhat to one-dimensional compression. It is only towards the base of the rigid cone that the lateral stress fails to provide the

necessary restraint and the column reaches failure (stress path C in Fig. 12). A similar argument can be used for columns at other radii, though of course the stress state is no longer axially symmetric within those columns. Certainly the edge column has rather limited lateral restraint even near the surface, and early failure is entirely expected (e.g. A in Fig. 12).

This qualitative description of stress paths does not attempt to include the influence of the geometry of the loaded area and of the array of stone columns, which, as seen in the model tests, will certainly affect the depth of influence of the footing. The route towards a failure condition in any column will also depend on the initial stress state in the column and in the surrounding soil, and this will depend very strongly on the installation process used to create the stone columns. In the present tests only a replacement installation process has been used. Displacement installation will clearly produce a very different initial state. While theoretical attempts can be made to assess the effect of simultaneous installation of an array of displacement stone columns (see for example Akagi, 1977) it has not been possible in the present project to devise an appropriate experimental displacement technique for repeatable creation of an array of model columns.

The model tests have been performed at single gravity and at correspondingly low stress level. Insofar as the tests are fully drained, the frictional nature of the soil response should scale directly to prototype dimensions. It cannot be suggested that the load–settlement responses should be quantitatively extrapolated. However, the observations of mechanisms of response of the sand column reinforced foundations—which have been the prime emphasis of the project and this paper—are believed to be relevant. An assessment of the geometry of a failure mechanism (Fig. 6) can be made, making appropriate assumptions about the drainage conditions (and the densities and stress levels) and the corresponding average angles of friction. This can then drive an assessment of the constraint to be felt by each column and hence its likely individual load–settlement response. The general kinematics of the footing–stone column interaction are not expected to be significantly affected by the stress level and dimensions.

The numerical modelling described here is intended to be purely illustrative. As made clear, detailed numerical understanding can only come from fully three-dimensional modelling with properly developed constitutive models for the two soil constituents (e.g. Wehr, 1999), and that is beyond the scope of the present paper. However, the physical model tests have provided insight into the mechanisms of deformation that occur in stone column reinforced soils. Qualitative effects that have been seen here have relevance at other scales. The exhumation technique that was developed for discovering the deformed shapes of the columns may also have wider application.

ACKNOWLEDGEMENTS

The experimental research reported here was supported at Glasgow University from the UK Science and Engineering Research Council (grant reference GR/H15950). Additional financial support was obtained from Glasgow University and from Geotechnical Consulting Group.

REFERENCES

Akagi, T. (1977). *Effect of displacement type sand drains on strength and compressibility of soft clays*. Department of Civil Engineering, Toyo University, Japan.

Al-Tabbaa, A. (1987). *Permeability and stress–strain response of speswhite kaolin*. PhD thesis, University of Cambridge.

Barron, R. A. (1947). Consolidation of fine-grained soils by drain wells. *J. Soil Mech. Found. Div., ASCE* **73**, No. SM6, 811–835.

Belkheir, K. (1993). *Yielding and stress–strain behaviour of sand*. MSc thesis, University of Glasgow.

Bishop, A. W. and Henkel, D. J. (1962). *The measurement of soil properties in the triaxial test*, 2nd edn. Edward Arnold.

Cooper, M. R. and Rose, A. N. (1999). Stone column support for an embankment on deep alluvial soils. *Proc. ICE, Geotech. Engng* **137**, No. 1, 15–25.

Fleming, W. G. K, Weltman, A. J., Randolph, M. F. and Elson, W. K. (1985). *Piling engineering*. Surrey University Press.

Greenwood, D. A. (1991). Load tests on stone columns. In *Deep foundation improvements: design, construction and testing*, ASTM STP1089, pp. 148–171. Philadelphia: American Society for Testing and Materials.

Hu, W. (1995). *Physical modelling of group behaviour of stone column foundations*. PhD thesis, University of Glasgow.

Hughes, J. M. O. and Withers, N. J. (1974). Reinforcing of soft soils with stone columns. *Ground Engineering* May, 42–49.

Hughes, J. M. O., Withers, N. J. and Greenwood, D. A. (1975). A field trial of the reinforcing effect of a stone column in soil. *Géotechnique* **25**, No. 1, 31–44.

Itasca (1996). *FLAC: Fast Lagrangian Analysis of Continua. Version 3.3. User's manual*. Itasca Consulting Group.

Kimura, T., Nakase, A., Saitoh, K. and Takemura, J. (1983). Centrifuge tests on sand compaction piles. *Proc. 7th Asian Regional Conf. Soil Mech. Found. Engng*, Haifa, **1**, 255–260.

Leroueil, S., Magnan, J. P. and Tavenas, F. (1985). *Remblais sur argiles molles*. Technique et Documentation Lavoisier.

Roscoe, K. H. (1970). The influence of strains in soil mechanics: 10th Rankine Lecture. *Géotechnique* **20**, No. 2, 129–170.

Wehr, W. (1999). Schottersäulen—das Verhalten von einzelnen Säulen und Säulengruppen. *Geotechnik* **22**, No. 1, 40–47.

Watts, K. S., Johnson, D., Wood, L. A. & Saadi, A. (2000). *Géotechnique* **50**, No. 6, 699–708

An instrumented trial of vibro ground treatment supporting strip foundations in a variable fill

K. S. WATTS*, D. JOHNSON†, L. A. WOOD‡ and A. SAADI§

Full-scale instrumented load tests have been carried out to study the installation and performance of vibro stone columns supporting a strip foundation in a variable fill and the performance of a similar strip foundation on untreated ground. The results from site investigation and laboratory testing are presented, and the vibro design principles—in particular the use of a modification to a standard formula that reflects the radial densification of soil around the stone column—are outlined. The treated foundation performance is compared with predicted behaviour; calculated settlements are in reasonable agreement with measured values, but stress measurements suggest that current design assumptions significantly overestimate the degree of stress concentration on the stone columns. The degree of radial densification of fill around the stone columns varied according to installation technique and soil type. The implications of the trial observations for existing design methods are considered.

KEYWORDS: design; field instrumentation; full-scale tests; ground improvement; settlement; soil structure interaction.

Des essais de charge entièrement instrumentés ont été effectués pour étudier l'installation et la performance des colonnes de pierre vibro soutenant des fondations bande dans un remblai variable et la performance de fondations bande similaires sur un sol non traité. Nous présentons les résultats de l'enquête sur le site et des essais en laboratoire et nous décrivons les principes de conception vibro, en particulier, l'utilisation d'une variante de la formule standard qui reflète la densification radiale du sol autour de la colonne de pierre. La performance des fondations traitées est comparée au comportement prévu; les tassements calculés présentent une assez bonne concordance avec les valeurs mesurées mais les valeurs de contrainte suggèrent que les suppositions conceptuelles actuelles surestiment considérablement le degré de concentration de la contrainte sur les colonnes de pierre. Le degré de densification radiale du remblai autour des colonnes de pierre varie en fonction de la technique d'installation et du type de sol. Nous considérons les implications des observations expérimentales pour les méthodes conceptuelles existantes.

INTRODUCTION

The ground treatment techniques most commonly adopted in the UK are the various deep vibro processes collectively described as vibro, which involve the construction of stone columns in the ground using a powerful torpedo-shaped vibrating poker (St John *et al.*, 1989). Much of this work is carried out to treat shallow variable fills, and the introduction of stone columns is designed to stiffen the existing fill and create more consistent foundation conditions, thus reducing total and differential settlement.

It is difficult to predict the behaviour of foundations on variable fills in which vibrated stone columns have been installed. The interaction of the modified fill, stone columns and the foundation is complex, and considerable reliance is placed on experience of similar applications. Current design procedures adopt a relatively simplified view of this complex interactive system, and the purpose of this paper is to compare a typical design with detailed field observations. The distribution of stress between stone columns and intervening ground beneath the foundation is critical for the prediction of foundation performance, and is calculated from the relationship between the stiffness of the stone column and the surrounding soil. In practice this is difficult to assess with confidence. A field trial to study the installation and subsequent performance of vibro stone columns in a variable fill was carried out by Bauer Foundations Ltd., the Building Research Establishment Ltd (BRE) and South Bank University to substantiate the findings of a study carried out by Wood (1990a) utilizing finite element methods to model strip foundation behaviour on vibrated stone columns. A paper outlining the field trial was published by Watts *et al.* (1992). The objective of the current paper is to

examine the design principles adopted, in particular the use of a modification to a standard formula that reflects the radial densification of soil around the stone column, and to present observations from the first phase of the trial to load 250 mm thick foundations. The design and behaviour of the strip foundations after stiffening and reloading have been described and analysed by Wood *et al.* (1996). Site investigation activities, instrumentation, ground treatment, foundation construction, loading and monitoring were carried out during the period March 1990 to June 1991. Further site investigations were carried out between January and May 1995. The layout of the trial foundations, site investigation activities and instrumentation is shown in Figs 1 and 2.

SITE INVESTIGATION

An initial site investigation comprising five trial pits and two boreholes identified ash overlying cohesive fill to depths between 3 m and 5 m. There was stiff glacial till below the fill. Further boreholes drilled for instrumentation included SPTs and material sampling. Dynamic probing (DP), described in DIN 4094 (1974) and BS 1377: Part 9 (1990), and geophysics were carried out during ground treatment and the foundation construction phases of the trial. The site investigation data from the immediate vicinity of the treated and untreated test foundations are summarized in Figs 3 and 4 respectively, and illustrate the variability of the fill.

DP tests carried out at several positions on line X–X indicated 3–4 m depth of fill close to Trial Pit 4, and a survey along Y–Y indicated similar ground conditions. DP tests along the line Z–Z suggested variable fill between 5 m and 6 m in depth. Geophysical testing using continuous surface shear Rayleigh waves (Abbiss, 1981) was carried out at two locations along line X–X and one location on Y–Y, and enabled dynamic soil properties to be calculated as a function of depth. There is broad agreement between borehole logs and DP tests regarding the depth of the fill/natural clay interface, and geophysics also helped in identifying soil strata boundaries. Subsequent vibro poker penetration provided a further indicator of depth to the firm clay. A laboratory testing programme was carried out at

Manuscript received 24 May 2000; revised manuscript accepted 7 September 2000.
* Building Research Establishment, UK.
† Roger Bullivant Ltd.
‡ South Bank University.
§ Formerly South Bank University.

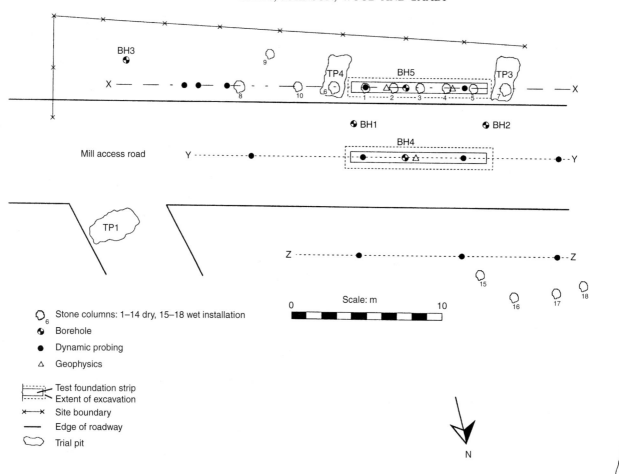

Fig. 1. Plan of trial site

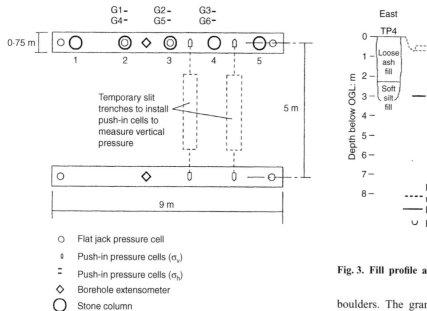

Fig. 2. Instrument layout

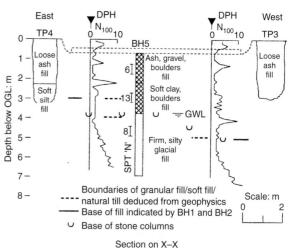

Section on X–X

Fig. 3. Fill profile along line of treated foundation

South Bank University to categorize the fill and underlying strata. Engineering parameters for the fill and natural soil were obtained to confirm design assumptions and predict foundation behaviour. The granular fill material contained a high proportion of black ash, some small pieces of sandstone and limestone, slate, burnt slag, clinker, brick and concrete fragments, sand and gravel and broken sandstone, and sandstone cobbles and boulders. The granular fill also contained pockets of soft silty clay. The lower cohesive fill comprised silty clay with dispersed granular fragments. Particle size analyses were carried out on most of the samples, and the results are plotted in Fig. 5 together with the ranges of soil particle size generally accepted as suitable for treatment by vibro methods (NHBC, 1995). The organic content of the fill ranged from 1·4% to 4%. Classification tests on the cohesive fill indicated inorganic clay of medium plasticity. The soil properties obtained from the field and laboratory investigations are listed in Table 1 with additional properties used to back-analyse treatment performance shown in Fig. 6.

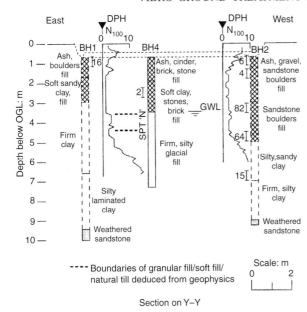

Section on Y–Y

Fig. 4. Fill profile for the untreated foundation

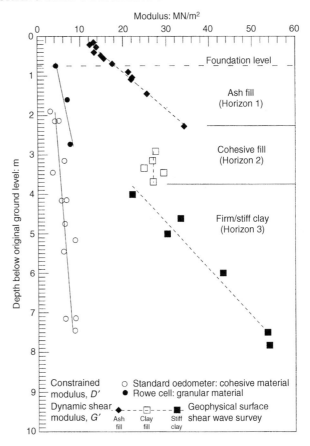

Fig. 6. Soil properties used to analyse treatment performance

TREATMENT DESIGN

A preliminary design, of the kind representative of current practice, was based on the type of limited site investigation information that might ordinarily be available when preparing a tender for a small-scale project. Only data from borehole 1 and the trial pits excavated as part of the original investigation of the site were used. The design consisted of:

(a) an assessment of the ultimate capacity of an individual stone column, and hence the factor of safety against bulging failure of that stone column
(b) an assessment of column spacing
(c) a prediction of the settlement of the loaded composite stone column/soil system.

Stone columns are designed on the assumption that they are acted upon by a triaxial stress system and are assumed to be in a state of shear yield at a critical depth h. Many authors have proposed approximations for the stress system. However, the approach given by Hughes & Withers (1974) for cohesive soils is often adopted. Their operating equation can be rewritten as

$$\sigma'_{vc} = K_{pc}(K_0[\gamma_s h - u_{so}] + u_{so} - u_s + 4c_u) \tag{1}$$

At the trial site the water table was situated at the base of the fill. The total pressure on the top of the column can then be written as

$$P_c = K_{pc}(K_0\gamma_s h + 4c_u) - \gamma_c.h \tag{2}$$

In predominantly granular soils, such as the ash fill at horizon 1 on this site, the current vibro design practice assumes c_u equal to zero, and hence

$$P_c = K_{pc}(K_0\gamma_s h) - \gamma_c.h \tag{3}$$

In granular soils, the assumption that horizontal stresses are governed by K_0 substantially underestimates the strength of the column. If the soil is free draining then the resistance to radial

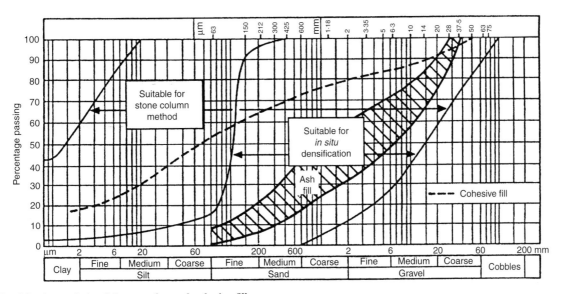

Fig. 5. Particle size analysis of the granular and cohesive fill

expansion will be increased by column formation, and in granular soils there will be a zone of improved ground around each stone column. Thus the use of K_0 is inappropriate in the situation after column construction and prior to loading the stone column. This is recognised by Baumann & Bauer (1974) with the use of a coefficient K_s that has a value between the at-rest and passive coefficients.

In equation 3, K_{pc} is calculated using the average angle of internal friction for the stone column material and improved zone or annulus around the stone column, and is then denoted by $K_{p(ave)}$. If D_c = stone column diameter, D_i = diameter of the improved zone, then $D_i = n \times D_c$ where n = diameter ratio. A typical value for n of 1·5 has been used in this design. Once ultimate stone column capacity has been calculated, the procedure developed by Baumann & Bauer is the commonly adopted method used to determine the distribution of stress between the stone columns and the surrounding soil. In this design the possible benefit of a zone of improved ground around the stone columns in granular soils has been considered. Accordingly modifications have been introduced into the original Baumann and Bauer formula such that it is now

$$\frac{P_i}{P_s} = \frac{1 + 2\left(\frac{E_s}{E_i}\right)(K_s)\ln\left(\frac{a}{r}\right)}{2\left(\frac{E_s}{E_i}\right)(K_i)\ln\left(\frac{a}{r}\right)} \tag{4}$$

In Fig. 7 the model and parameters for the soil and stone column adopted in the preliminary design process are shown. Based on these values, σ'_{vc} for a column is 340 kN/m². The Baumann & Bauer analysis gives P_i and P_s as 248 kN/m² and 12 kN/m² respectively and thus a ratio of 21. The factor of safety against bulging failure, σ'_{vc}/P_i, is 1·4 at the maximum

applied foundation load of 123 kN/m² for column spacing of 1·8 m. The calculated settlement profile for the untreated strip was based on a Boussinesq stress distribution for a strip foundation. In clay soils settlement was calculated using the formula $\sigma_v m_v H$, where σ_v is the average stress in the layer, m_v is the coefficient of volume compressibility for the soil (Table 1), and H is the layer thickness, while in the granular layer (horizon 1 in Fig. 7) the formula $(\sigma_v H)/E_s$ was used. Using the stiffness and compressibility parameters shown in Fig. 7 this indicates up to 40 mm settlement under these loading conditions. The same Boussinesq stress distribution was used to estimate settlement beneath the treated strip. In order to calculate the post-treatment settlement in the granular fill an appropriate average modulus of deformation, E_{Ave}, for the stone column/improved annulus/unmodified soil system in the granular fill was calculated from

$$E_{Ave} \times A_0 = (E_i \times A_i) + (E_s \times A_s) \tag{5}$$

Settlement within the granular layer was then calculated using the formula $(\sigma_v H)/E_{Ave}$. Within the treated clay layer (Horizon 2) settlement without treatment was calculated as described above using the parameters shown in Fig. 7, and a settlement reduction factor after Priebe (1995) was applied to calculate post-treatment settlements. Settlement beneath the treated depth (Horizon 3) is the same as for the untreated strip. In Fig. 7 the predicted settlement for the treated strip is based on immediate settlement for the granular fill layer and immediate and primary consolidation settlement in the underlying clay horizons.

GROUND TREATMENT

Initial treatment was carried out using the dry top-feed 'vibrodisplacement' technique using a hydraulically operated

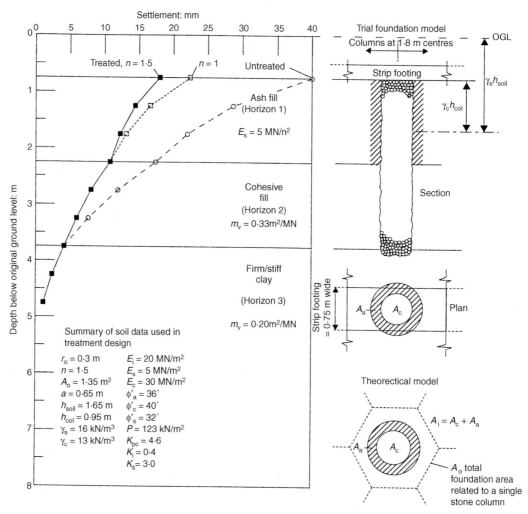

Fig. 7. Soil model and predicted settlements

Table 1. Principal soil parameters and test procedures

Parameter	Soil type				Test method
	Ash fill (0–2·25 m)	Cohesive fill (2·25–3·75 m)	Natural clay (Till)	Stone column	
Bulk unit weight: kN/m³	16·0	20·6	21·2	13·0	Cohesive samples extruded from U100 tubes. Ash fill sample density range estimated from recompacting disturbed material for shear box tests (see below)
Water content: %	Ave = 18·0	22·6	21·6	0	
Dry unit weight: kN/m³	13·1	16·8	17·4	13·0	
Plastic limit: %	–	20	19	–	BS 1377: Part 2: 1990, Test 5.3
Liquid limit: %	–	37	38	–	BS 1377: Part 2 : 1990, Test 4.3
PI: %	–	17	19	–	
Organic content: %	4·1	1·4	–	–	BS 1377: Part 3: 1990, Test 3
Undrained shear strength: kN/m²	–	18–78	75–80		Laboratory vane tests on U100 samples
		(Ave = 40)			Hand vane testing in-situ below treated foundation
Effective cohesion, C': kN/m²		$C' = 5$ $\phi' = 26$	$C' = 0$ $\phi' = 28$		Multi-stage drained triaxial
Shear strength, ϕ': degrees	Loose: $\phi' = 36$ Dense: $\phi' = 46$			Loose: $\phi' = 45$ Dense: $\phi' = 50$	300 mm drained shear box tests
Constrained Modulus, D: MPa	4·3 at 0·0 m 7·2 at 1·5 m		Secant modulus at depths stated, based on trend lines plotted through results from 5 tests on ash fill (horizon 1) and 13 tests on clay fill (horizon 2) and natural clay (horizon 3) in Fig. 7		Hydraulic (Rowe) cell BS 1377: Part 6: 1990, Test 3
		4·2 at 1·6 m 5·4 at 2·8 m			75 mm samples from U100 tubes tested in one-dimensional consolidation apparatus BS 1377: Part 5: 1990, Test 3
			5·4 at 2·9 m 6·9 at 5·0 m		

vibrating poker with a maximum power output of 86 kW and a horizontal force of 25 t. Compressed air was used as the jetting medium. In accordance with normal practice (BRE, 2000), the vibrator penetrated to the design depth and was then fully withdrawn to allow a charge of stone to be placed in the hole. The poker was replaced in the hole and the charge of stone compacted to a dense state. This process was repeated with stone fed in small charges and the column was compacted in stages back to the ground surface. Initially, three isolated columns (8, 9 and 10 in Fig. 1) were installed to establish an appropriate treatment procedure for the trial footings. The degree to which lateral densification of soil takes place during penetration of the poker and compaction of stone is dependent on particle grading, initial soil density, the mechanical specification of the poker and workmanship. It is usually not desirable to operate vibro equipment to the limit of its capacity. In granular soils this can produce a column with a higher load-carrying capability than required, while in cohesive soils overworking may reduce soil strength. For the purposes of this trial, column 8 was installed to the maximum capability of the poker at each stage of the column construction. Columns 9 and 10 were installed in accordance with generally adopted practice for these ground conditions. Five columns (1–5) were similarly installed at 1·8 m centres along the centreline of the treated foundation strip as shown. Columns 6 and 7 were installed to densify fill replaced in trial pits that formed part of the original site investigation. In January 1992 four additional columns (15–18) were installed using the wet top-feed method. The wet 'vibroreplacement' technique uses water jets to maintain a circulating flow in the cylindrical void to remove loose material and maintain stability while stone is compacted to form the column. This technique minimizes disturbance of the in-situ soils. The purpose of these additional columns was to compare the effect of installing stone columns using the wet and dry methods in this fill.

INSTRUMENTATION, FOUNDATION CONSTRUCTION AND LOADING

Details of foundation construction and instrumentation have been described by Watts et al. (1992), and only a brief summary follows. Precise levelling, as described by Cheney (1973), was used to obtain a record of total foundation movement. Six push-in spade-shaped pressure cells were installed from small-diameter boreholes to measure changes in lateral (radial) total earth pressure due to column installation and foundation loading, and four more were installed after foundation construction by jacking horizontally from a slit trench between the trial foundations to measure vertical stress immediately beneath the strips in the fill between columns. Magnet extensometers were installed into the underlying stiff clay in boreholes 4 and 5 to measure settlement with depth midway between columns 2 and 3 of the treated foundation strip and in a similar position under the untreated strip. Pneumatically operated 0·3 m diameter flat jack pressure cells (Wood, 1990b) were installed in the top of column 2 and column 3 to measure the load carried by the stone columns under the treated foundation. Similar 0·2 m diameter cells were installed in the fill below each end of the two foundations. The foundations were cast in-situ after excavation and blinding a prepared surface in accordance with normal building practice. Mesh reinforcement (B283) with a total cross-sectional area of 216 mm² was incorporated into the bottom element of the foundation. The footings were cast in shuttering that was reused to thicken the foundations to study the effect of changes in foundation stiffness on the bending strains induced in the strip foundations (Wood et al., 1996). Load was applied using kentledge comprising 1·2 m concrete cubes, each weighing an average 4 t, and with the bottom row supported at intervals on steel spreader beams placed across the strip so as not to increase foundation stiffness. Each load increment comprising a row of blocks applied a total bearing pressure of 41 kN/m² to the foundation strips. A large-capacity crane was utilized, with adequate reach to avoid any effect on the foundations or supporting soil. Immediately after applying the first load increment levels were taken on both strips and the instrumentation was monitored before a second layer of blocks was added, increasing the foundation pressure to 82 kN/m². Both foundation strips were then monitored at increasing time intervals for a period of one month to observe creep settlement, after which the total average applied foundation pressure was increased to 123 kN/m². The foundations were monitored for a further 6–7 months before the first phase of the trial was concluded by fully unloading both strips.

DISCUSSION: FILL BEHAVIOUR AND FOUNDATION PERFORMANCE
Soil response during treatment

Pressure cells at 0·9 and 1·5 m from the column axis and at depths of approximately 0·4 m, 0·9 m and 1·8 m below foundation level measured changes in lateral (radial) total earth pressure during initial poker penetration and subsequent compaction of the stone column material. In Fig. 8 the increase in total earth pressure measured by cells G1 and G4 during treatment is plotted against poker depth, during both initial penetration and subsequent stone compaction back to the ground surface. During initial penetration no stress increase was measured until the poker reached the level of the cells, but up to 60 kN/m² was registered by cell G4 during further penetration to design depth. The diminishing effect with increasing distance from the column centre is indicated by cell G1. Withdrawing the poker to place charges of stone resulted in an almost immediate return to the pre-penetration stress levels, but similar elevated values were again measured as the stone was compacted. The increases in earth pressures disappeared immediately upon completion of the column, including those measured by the deepest cells G1 and G4, which may have been located in more cohesive fill. Each column took an average 12 min to complete. In contrast, it took 27 min to complete the high-energy column 8. Pore water pressure measurements were not made in the fill, but particle size analyses, which indicate essentially granular soils at the level of the pressure cells, combined with the rapid dissipation of excess pressure observed following pressure cell installations, suggest a high permeability for the fill at the level of these cells.

The measured increase in stress resulting from radial displacement of the granular soil, although not sustained, would result in some drained loading, compression and hence densification of the soil surrounding the columns. Modest stress increases

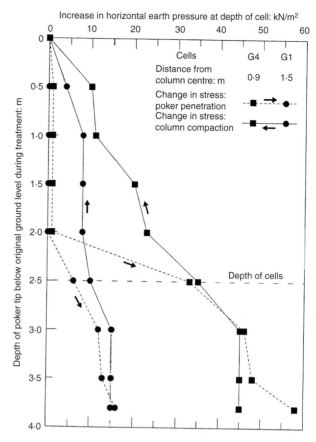

Fig. 8. Lateral stress changes due to poker penetration and stone compaction

were measured at a distance equivalent to 2·5 times the column diameter from the centre of column 2. In granular soils densification would be expected around the periphery of the stone columns owing to the rearrangement of soil particles in response to the vibratory action of the poker and stone compaction. In order to investigate this, dynamic probing was carried out at a range of distances around stone columns. In Fig. 9 dynamic probings at increasing distances from test column 9 indicate a significant increase in blow counts close to the edge of the column, with a more modest increase at 0·6 m from the column centre. Greater increases in blow count are apparent in

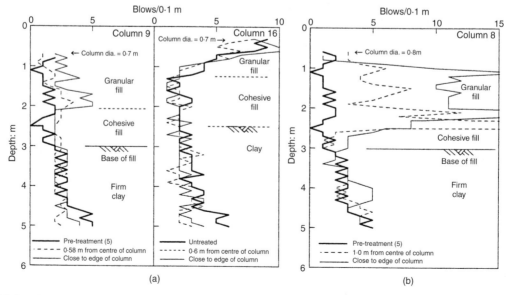

(a)   (b)

Fig. 9. Radial densification of fill around stone columns: dynamic probing

granular fill, but significant increases above pre-treatment values were also measured in the more cohesive fill. Much larger increases, averaging ten times pre-treatment values, were measured in the granular fill immediately surrounding test column 8, which received more compactive energy during treatment and resulted in a larger column diameter. An average five-fold increase was measured in the granular fill at a distance of 1·25 times the column diameter from the column centre. By contrast, probing around column 16, installed using the wet top-feed process, which is designed to minimize disturbance of the surrounding soil, showed no change in resistance in the surrounding soil.

These observations of changes due to column installation in the granular soils surrounding the columns, in particular the direct measurement of increased penetration resistance, confirm densification to at least 0·6 m ($1·7 \times r_0$) from the column centre, and suggest that the use of a diameter ratio $n = 1·5$ in equation (6) is appropriate. No instrumentation was installed around column 8 to measure soil response during treatment, but a much greater effect in terms of post-treatment penetration resistance was measured at a distance equivalent to $n = 2·5$. The increased stress measurements at cell G1 outside column 2 would translate into $n = 5·0$. While it would not be prudent to use this very high figure, the adopted value of $n = 1·5$ is judged to be amply justified in these soil conditions. Both types of measurement confirm the effect of radial modification of the surrounding granular soil as a result of column installation. These observations suggest some validity in the use of K_s rather than K_0 for the calculation of the total pressure in the top of the column for granular soils.

Foundation and soil settlement

The maximum settlement of the 250 mm thick strip foundations on treated and untreated ground is shown in Fig. 10, plotted against time since load was applied. The immediate settlement resulting from the first foundation pressure of 41 kN/m² was similar for both strips. Increasing the foundation pressure to 82 kN/m² produced larger settlement of the untreated strip than the treated strip, and the rate of creep for the untreated strip was also greater after one month. The third foundation pressure of 123 kN/m² induced much larger settlement of the untreated strip. At the end of a further 7 months' monitoring, the maximum settlements of the treated and untreated strips were 16 mm and 26 mm respectively. Both foundations exhibited sagging along their length. Fig. 11 shows the development and distribution of settlement along each foundation measured at the end of each loading period, with a longitudinal section indicating the average change in depth of fill and the position and depth of the stone columns. The plot shows greater differential settlement in the untreated strip after the application of the highest foundation load and greater total settlement where the fill is deepest.

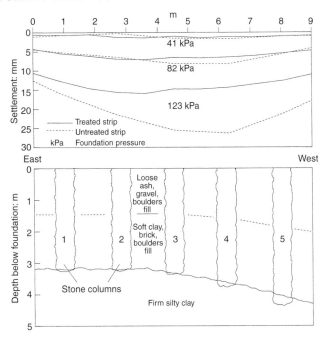

Fig. 11. Differential settlement along foundation strips

The benefits of the stone column treatment in this case are evident, but it is useful to examine the distribution of settlement with depth for the treated and untreated conditions. Results obtained from the borehole extensometers positioned under the centre of each strip are given in Fig. 12 for the two average loading conditions of 82 kN/m² and 123 kN/m². As in Fig. 11, it is apparent that the effect of the stone columns in significantly limiting surface settlement did not occur until the load was increased to 123 kN/m². However, at that load the treated foundation performed slightly better than predicted, and this was due principally to lower compression in the upper granular material. This may be due to an underestimate of the diameter ratio, n, and laboratory tests also indicate significantly higher strength parameters for the ash fill and stone column material

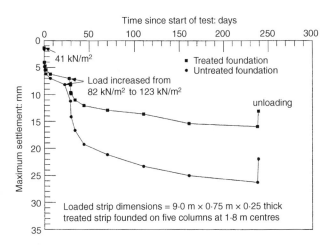

Fig. 10. Maximum settlement along strip foundations

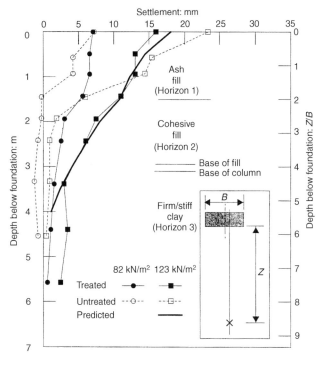

Fig. 12. Settlement profile within foundation soils

(Table 1) than those used to predict foundation performance (Fig. 7). Fig. 12 also shows that the presence of the columns gives rise to much more deep-seated settlements occurring within the cohesive fill and the underlying stiff clay below the base of the columns. The percentage contribution to surface settlement resulting from the maximum applied load of 123 kN/m^2 for the treated and untreated foundations is given in Table 2. Examination of Fig. 12 also shows that there is an increase in the compressibility of the cohesive fill, which may be due to disturbance caused during column installation, load transfer down the stone columns, or a combination of both effects. Load transfer down the stone columns is supported by the fact that there is some 3 mm of settlement observed in the stiff clay beneath the base of the columns, which is absent for the untreated strip. Interpretation of the observed settlements is further complicated by the apparent heave recorded at depth below the untreated strip at an applied load of 82 kN/m^2 but, as shown in Fig. 13, the pattern of settlement with depth is similar even if the observations at an applied pressure of 82 kN/m^2 are taken as the datum.

The principal objective of vibro stone column treatment is to reinforce the existing soil mass to produce a composite structure that, overall, has load-carrying characteristics that are superior to those of similar untreated ground. During the construction of columns stone is densely compacted and fully interlocked with the surrounding soil. The mechanisms that govern stone column behaviour are well documented. Hughes & Withers (1974) describe the similarities with pile behaviour in that, when loaded, they settle and develop end bearing and cohesive resisting stresses up the side. Unlike piles, they comprise compacted cohesionless granular material, which bulges under load and must be supported by lateral stresses exerted by the soil. Greenwood (1991) goes on to explain the differing behaviour of piles and stone columns arising from the hugely different ratio of their stiffness to that of the soil, and also points out that the relative stiffness of stone columns to soil can change signifi-

cantly as load is applied. Hughes & Withers and Greenwood draw the analogy between stone column behaviour and a pressure meter.

Consideration of all of the evidence from this trial would tend to suggest that the deeper-seated settlements associated with the treated strip in this trial are due to the presence of the columns, which encourages load transfer to a greater depth: that is, concentrating the applied load within the column/soil mass and limiting the normal spread that has occurred under the untreated strip. When the observed settlements with depth are compared with the predictions given in Fig. 7 several significant points become apparent. First, for the untreated strip it is clear that the contribution to the surface settlement from the cohesive fill has been overestimated, as illustrated in Fig. 13, and it is apparent the stiffness of this stratum has been underestimated. Second, for the treated strip the agreement between the observed measurements and the prediction for $n = 1.5$ is good. These comparisons are in contradiction to each other, and suggest one of two conclusions. Either the stiffness of the cohesive fill is significantly different under each of the strips, perhaps owing to the effects of the installation of the columns themselves, or the behaviour of the column/soil mass is different from that assumed in the method of settlement prediction, albeit that the prediction of the surface settlement of the treated strip was in good agreement with the observed value.

Although not conclusive it is felt that the balance of the evidence suggests that the presence of the columns leads to an increase in the contribution to the surface settlement from the underlying deeper strata due to load transfer down the stone columns. It is thus important to distinguish between the use of stone columns to increase the bearing capacity of poor soils from their use to limit settlements. Where stone columns fully penetrate soft and more compressible soils, and are toed into an underlying stiffer stratum as in this trial, both the bearing capacity and the settlement performance of the foundation should be improved. Where this is not the situation, as may occur on thick deposits of soft soils when the stone columns are limited to penetrating a partial thickness of the soft soil, the bearing capacity may be improved but a significant proportion of the predicted reduction in surface settlement may not occur. The latter is dependent on the thickness of soft soil remaining below the toe level of the stone columns and the load concentrating effect of the column/soil mass above. To ameliorate this situation the depth of partial penetration treatment should be critically examined. Hughes & Withers (1974) proposed a simple procedure for calculating the minimum length of stone column required to prevent end bearing failure at the toe of the column occurring before bulging failure at a critical depth near the top of the column. This method can be used to estimate the depth at which the vertical stress in the column is zero. Watts & Serridge (2000) describe a field trial of vibro stone columns in soft clay soil in which the efficacy of this procedure is examined and present some evidence to support the theory.

Table 2. Contribution to surface settlement for treated and untreated foundations

Strata	Untreated: %	Treated: %
Granular fill	85	30
Cohesive fill	15	40
Underlying clay	0 (5)*	30

* There is an apparent movement of 1 mm, which is the limit of accuracy of the subsurface settlement measurement, and is judged to be zero.

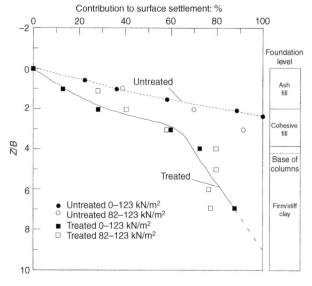

Fig. 13. Percentage contribution to surface settlement

Stress distribution

It has generally been assumed that the ratio of stress in the vibro stone columns to that in the intervening ground is at its maximum when load is first applied: that is, the columns initially carry a high proportion of the total foundation load (Greenwood, 1974). Greenwood (1991) also reported a maximum ratio of column to soil stress of 25 being measured in very soft clay and a ratio of 4 measured in drained silt over applied pressure ranges similar to this trial. In both cases ratios were seen to decrease significantly as the applied load increased. In Fig. 14 stresses measured in the columns and the intervening soil supporting the treated foundation strip are plotted for the three load increments applied. Stresses measured at similar locations under the untreated foundation strip are also plotted for comparison. Changes occurring during periods of constant applied load are indicated by arrows, and reflect a redistribution of stresses that occurred while foundation settlement took place. The ratio of average column stress to stress in

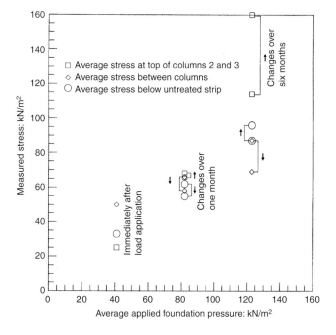

Fig. 14. Vertical stresses beneath the foundation strips

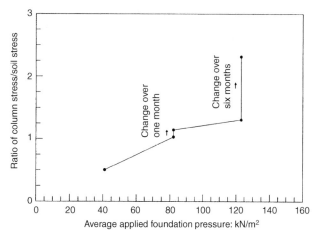

Fig. 15. Column/soil stress ratios with applied loads

the soil between columns is also plotted in Fig. 15, and shows a consistent increase over the range of applied foundation loads, as well as a significant increase as foundation settlement took place at constant load. On first loading the ratio was about 0·5, indicating that a high proportion of the applied load was carried by the intervening soil. At the highest applied load, the ratio increased to a maximum value of 2·5. The increasing concentration of stress in the column resulted in the measured increase in vertical strain in the deeper soils (Fig. 12).

Changes in lateral stress were also measured by the pressure cells alongside the central columns during static loading of the foundations (Fig. 16). The stresses measured at all the cell

locations were significantly in excess of those calculated from normal stress distribution in an elastic medium, with the highest, as a proportion of the applied vertical load, being recorded at 0·9 m from the column centre and 1·8 m below formation level by cell G4. This indicates that the column may be bulging at a depth equivalent to three times its diameter, just below the granular ash fill, and confirms that there was significant stress transfer down the column. This also confirms the necessity, where cohesive soil is located at or close to the formation level, for a check on ultimate column capacity using, for example, the method suggested by Hughes & Withers (1974).

CONCLUSIONS
(a) The trial has demonstrated the ability of vibro stone columns to reduce total and differential settlement of conventional strip foundations constructed on a weak variable fill.

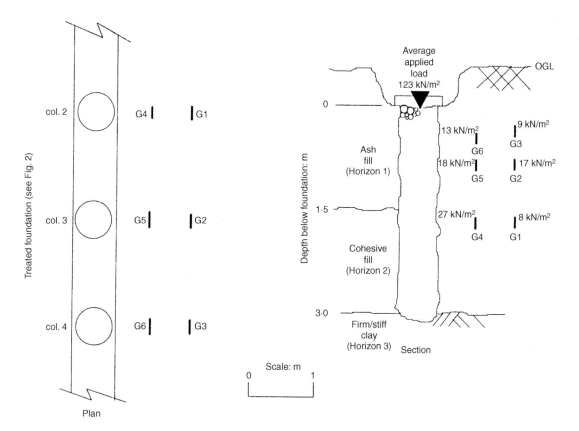

Fig. 16. Lateral stresses measured at applied foundation load of 123 kN/m²

(b) Predicted surface settlements for the treated strip are in close agreement with actual measured performance but settlement measurements with depth suggest that there was significant compression in the lower clay soils, probably as a result of load transfer down the columns.

(c) Partial depth treatment of deep deposits of weak material should be critically examined to ensure settlements will be acceptable.

(d) Stress measurements indicate a much lower proportion of the applied load was carried by the stone columns than predicted by standard analysis.

(e) The ratio of stress in the column/improved zone and intervening soil, P_i/P_s rose with applied load and increased further as settlement of the foundations continued under constant applied load.

(f) Poker penetration and stone compaction affected the ground up to 1·5 m from the centre of the columns that supported the strip foundation. Densification occurred in the granular fill up to a distance equivalent to a diameter ratio, $n = 2·5$ for normally constructed stone columns.

(g) A modified Baumann and Bauer analysis, utilising a value of $n = 1·5$, gave a reasonable prediction of settlement of the treated strip foundation but substantially overestimated the stress ratio P_i/P_s.

(h) The radial effect of column installation was related to the nature of the material, to the level of compaction (workmanship) and to the technique (dry or wet) employed.

ACKNOWLEDGEMENTS

The authors appreciate the assistance provided by Hilary Skinner of BRE in the final drafting of this paper. The BRE research work formed part of a research programme for the Construction Directorate of the Department of the Environment, Transport and the Regions. Bauer Foundations (UK) Ltd carried out the ground treatment and were responsible for construction of the foundations and logistical support. Further financial support was provided by the Faculty of the Built Environment, South Bank University, London. Ahmed Saadi was in receipt of a South Bank University research scholarship. Permission to carry out the trial was given by the site owners, Lancashire County Council.

NOTATION

a $\sqrt{(A_o/\pi)}$, equivalent radius of foundation area per compaction

c_u undrained shear strength of soil surrounding the stone column

h 'critical depth' at which bulging failure of stone column occurs

m_v coefficient of volume compressibility

n diameter ratio of improved zone to stone column

r radius of improved zone

r_o original radius of stone column

u pore water pressure

u_{so} initial pore pressure in soil at depth h

u_s pore pressure in soil at yield at depth h

u_c pore pressure in column at yield at depth h

A_c area of column

A_i area of column/improved annulus

A_o total foundation area per stone column

A_s area of soil between improved zones

D_c stone column diameter

D_i diameter of improved annulus

$E_{Ave.}$ average modulus of deformation for column/improved annulus/unmodified soil

E_c modulus of deformation for stone column

E_i modulus of deformation for stone column/improved annulus

E_s modulus of deformation for soil

H thickness of soil layer

K_s coefficient of earth pressure for unimproved soil (after Baumann & Bauer, 1974)

K_i coefficient of earth pressure for stone column/improved annulus (modified from Baumann & Bauer, 1974)

K_{pc} coefficient of passive earth pressure for stone column

$K_{p(ave)}$ average coefficient of passive earth pressure for stone column/improved annulus

K_0 coefficient of earth pressure at rest for unimproved soil

P average imposed stress on foundation with area A_o

P_c average stress in stone column at foundation level

P_i average stress in stone column/improved annulus at foundation level

P_s average stress in unmodified soil between columns (Baumann & Bauer, 1974)

ϕ'_s angle of internal friction of unimproved soil

ϕ'_c angle of internal friction of column material

ϕ'_a angle of internal friction of improved annulus

γ_s bulk unit weight of unimproved soil

γ_c bulk unit weight of stone column material

σ'_{rc} radial effective stress in stone column

σ_{ro} initial radial total stress in soil prior to column construction

σ_{rs} radial total stress on column boundary

σ'_{vc} maximum vertical effective stress capacity of stone column

REFERENCES

Abbiss, C. P. (1981). Shear wave measurements of the elasticity of the ground. *Géotechnique* **31**, No. 1, 91–104.

Baumann, V. and Bauer, G. E. A. (1974). The performance of foundations on various soils stabilized by the vibro-compaction method. *Can. Geotech. Journal* **3**, No. 2, 509–530.

BRE (2000). *Specifying vibro stone columns*, Report BR391. Construction Research Communication, London.

British Standards Institution (1990). *British standard methods of test for soils for civil engineering purposes: In-situ tests*, BS 1377: Part 9. Milton Keynes: BSI.

Cheney, J. E. (1973). Techniques and equipment using the surveyor's level for accurate measurement of building movement. In *Field instrumentation in geotechnical engineering*, pp. 85–99. London: Butterworth.

DIN (1974). *Subsoil dynamic and static penetrometers: Dimensions of apparatus and method of operation*, DIN 4094: Part 1. Berlin: Deutsches Institut für Normung eV.

Greenwood, D. A. (1974). Vibroflotation: rationale for design and practice. In *Methods of treatment of unstable ground* (ed. F. G. Bell), London: Butterworth. 148–171.

Greenwood, D. A. (1991). Load tests on stone columns. In *Deep foundation improvements: design, construction and testing*, ASTM SPT 1089, 189–209. Philadelphia: American Society for Testing and Materials.

Hughes, J. M. & Withers, N. J. (1974). Reinforcing of soft cohesive soils with stone columns. *Ground Engineering*, **7**, No. 3, 42–49.

National House Building Council (1995). Vibratory ground improvement techniques. *NHBS Standards*, Ch. 4.6. Amersham: NHBC.

NHBC (1988). In text but not in reference list.

Priebe, H. (1995). The design of vibro replacement. *Ground Engineering*, December, 31–37.

St John H. D., Hunt R. J. & Charles J. A. (1989). *The use of 'vibro' ground improvement techniques in the United Kingdom*. BRE Information Paper IP 5/89. Garston: Building Research Establishment.

Watts, K. S. & Serridge, C. J. (2000). A trial of vibro bottom-feed stone column treatment in soft clay soil. *Proc. 4th Int. Conf. Ground Improvement Geosystems*, Helsinki, 549–556.

Watts, K. S., Saadi, A., Wood, L. A. & Johnson, D. (1992). A field trial to assess the design and performance of vibro ground treatment in fill and strip foundations on vibrated stone columns, *Proc. 2nd Int. Conf. Polluted and Marginal Land*, London, 215–221, Engineering Techniques Press.

Wood, L. A. (1990a). The design of foundations on stone columns. *Proceedings of the international conference on construction on polluted and marginal land*, (ed. M. C. Forde), Engineering Techniques Press.

Wood, L. A. (1990b). The performance of a raft foundation in London clay. In *Geotechnical instrumentation in practice*, 323–340. London: Thomas Telford.

Wood, L. A., Johnson, D., Watts, K. S. & Saadi, A. (1996) Performance of strip footings on fill materials reinforced by stone columns. *Struct. Engr* **74**, No. 16, 265–271.

INFORMAL DISCUSSION

Session 3

CHAIRMAN DR KEN BEEN

David Greenwood, *Geotechnical Consulting Group*

Professor David Muir Wood's work seems to indicate, to me anyway, the futility of attempting to simulate by load testing the performance of vibro stone columns when there is a widespread load, such as an oil tank or an embankment foundation, resting on a large array of stone columns. Without expending huge sums of money it is not practical to simulate the constraint that the widespread load imposes on the columns under an embankment or tank. It would be far better to spend the money that would otherwise be spent on testing columns, on supervising the quality of construction and making sure that the columns are built absolutely to specification (the amount of stone that goes in, depth and so on).

Just as an aside, I had to wait twenty years for this model work to vindicate a hypothesis I put forward at our first conference on *Ground Treatment and Deep Compaction* (*Ground treatment by deep compaction*, 1976, Institution of Civil Engineers), to explain an apparent failure of embankment foundations at East Brent. I am grateful to you for completing that work – thank you.

Declan Carney, *Wardell Armstrong*

If the stone columns are constructed in essentially clayey fill, would you have any concerns about water getting into the columns and undermining what you are essentially trying to do, which is to densify the ground?

By constructing stone columns in a clayey soil you will have effectively created sumps, or areas where standing water can remain within the columns. Would you have any concerns about water standing in the columns causing either inundation or softening of the surrounding clayey fills?

Ken Watts

The inclusion of a stone column into the ground, in any ground condition, needs to be critically appraised from this point of view. The problem could in fact just as easily apply to a granular fill, particularly if it is a deep fill where you are, in effect, potentially providing water access into the ground.

I take your point about this problem in soft material. If, of course, the soft ground in which that column is installed is already saturated, I would speculate that the effect would be fairly minimal. If it isn't saturated, of course yes, you have a valid point – providing the water can accumulate in there and stand. It very much depends on the individual ground conditions. Also, bear in mind that the columns will generally be covered over very quickly in a construction process and so it is unlikely that additional water will get down the holes.

It very much depends on whether those columns are above or below the groundwater table. If it is below the water table in a saturated condition, once the columns are installed there would not be a major issue in that respect.

Dr Alan Bell, *Keller Ground Engineering*

I have a comment on Ken Watts' response (above). Practical experience dictates that certainly in the UK there have been problems with deep colliery fill. The stone columns will allow water to get deep into unsaturated non-engineered fills of the kind that Dr Andrew Charles mentioned in his keynote paper (Session 1). These do require a great deal of care, at least in design and in the construction details. I am not aware of any problems with water in purely granular materials, other than extreme flushing can cause stripping of fines, which would probably have occurred over time in any event. When you think about it, the permeability of many granular fills is pretty high anyway, certainly within an order of magnitude or so of many of the stone columns, which if constructed by the wet process, for example, might well have fines contained in them. I think caution is a must for non-engineering fills on the basis of experience within Keller, although this would not always preclude treatment using the vibro process.

Dr Marcio Almeida, *Federal University of Rio de Janeiro*

In 1985 I published a paper in *Géotechnique* (Almeida, M. S. S., Davies, M. C. E., Parry, R. H. G. (1988). Centrifuge tests of embankments on strengthened and unstrengthened clay foundations. *Géotechnique* **35**, No. 4, 425–441.), on centrifuge modelling of stone columns. I had two embankments on stone columns and another one without stone columns. Besides the well-known effects of decreasing settlements and decreasing lateral displacement, it also showed that the great beneficial effect of increasing the strength against stability. For the stone columns the embankment height was much greater. With the much finer measurements of pore pressure and quicker generation of pore pressures due to the drainage effect of stone columns, which have become available over the past fifteen years, have stone columns been used to improve stability of embankments on soft clays? Is there any theoretical analysis behind this?

David Johnson, *Bachy Soletanche*

Back in 1993, when I worked for Bauer Foundations, we installed stone columns beneath a coal stockyard in Liverpool docks, which were the subject of a paper that was presented in the 1994 Delhi conference. The stone columns were primarily used to increase the factor of safety against shear failure, but they were also installed to reduce settlements. Therefore, I would refer you to my paper in the Delhi conference (*Proceedings: Thirteenth international conference on soil mechanics and foundation engineering*, New Delhi, January 1994, Oxford & IBH Publishing (1994)).

Mike Jefferies, *Golder Associates*

My reaction to Kovacevic *et al.*'s paper on compaction grouting was all a bit bleak, to put it mildly, so I would like to give you a brief counter example.

Bennett Dam is one of the larger dams in Canada. It is a 183 m (600 ft) high hydro-power dam on the Peace River in northern British Columbia, close to the Alberta border. It has a 235 km reservoir behind it and it generates about a third of the power for British Columbia.

In 1996, a Japanese tourist was driving across the crest of the dam and dropped his wheel into a small hole in the dam. He went into the dam control office, being a good tourist, and said "excuse me, should there be a little hole in the top of your dam?" It was concluded that this should not be there. The dam engineers sent in a drill rig to find out what was going on. When the drill got down to about 3 m (10 ft), it triggered a sink hole collapse and the top of the dam dropped 15 m (50 ft) in, I think it was, 63 seconds. As you might imagine, this produced general panic in the water controllers' office in British Columbia. It turned out that there were two large sink holes going down about 122 m (400 ft) in the centre core of the dam.

The dam's core was a blended sand-silt material. In terms of permeability and so forth, very much at the silt end of the spectrum as it was supposed to be an impervious core. Contrary to the view that you can only do compaction grouting in roughly free draining materials, we actually carried out compaction grouting to remediate this core in March to June 1997. The analysis we did to set up the compaction grouting was very similar to that presented at this conference. The purpose of the analysis was to understand what would be achieved, and then to control the works. Bearing in mind the size of this dam, the consequence of its failure would be to flood half of Alberta.

We were closely scrutinized by review boards – in terms of what we could do. We were not quite at the stage of having to predict each grout stage, but we had to file a prognosis of what we were going to do, why we were going to do it and what it would achieve. Then we had to analyse each hole as it was compacted to show that our predictions were indeed correct. The analysis that we undertook was the familiar cylindrical and spherical.

We then did a trial in Vancouver, before we went up to the dam, and put inclinometers in the ground to look at the displacements, and based on those displacements we then decided how the ground was moving. It turned out that the standard undrained plane strain cylindrical analysis was actually a very good model. We also used large strain alternatives. We used the Nor sand model, which gives a much better fit to the slightly dilatant materials, and rather than just switching between undrained and drained, we undertook a fully coupled Biot formulation so that we could look at different injection rates, which is of course what we were interested in.

To keep this presentation short I will jump straight to one of our results. One axis is excess water pressure, that is the computed excess water pressure line. By memory we are down at 300 kPa, so very similar stress levels to those you saw presented by Dr Kovacevic. Lightly dilatant fill, 10^{-6} type material, grouting injection rate was 0·25 cu.ft (0·09 m³) per minute. That is quite a slow injection rate for compaction grouting. It is possible to plot the void ratio distribution that we computed for fully drained loading. For this particular rate of 0·25 cu.ft per minute we were getting excess pore pressures in the order of 12 m (40 ft) close to the grout hole. That then gave us reduced effectiveness. The middle line between the two is what then happens when you consolidate.

Now contrary to what you saw earlier, what we are seeing is that, with these slightly dilatant type loose silty sands, you can achieve quite effective compaction even with these excess pore pressures in place. The reason is, if you think of the way these materials behave in a triaxial test, they show S-shaped stress paths. Restricting drainage does not necessarily mean you are not going to be able to move the ground around quite nicely.

So, in summary, while truly undrained loading is not good, we can actually live with surprising excess water pressures and still get very effective compaction. The net result was that the dam was satisfactorily returned to service.

Dr Nebojsa Kovacevic

What struck my mind immediately is that angle of dilation. What was the effect of that angle of dilation?

Mike Jefferies, *Golder Associates*

I think that is the big difference between what you got and what I got.

Dr Nebojsa Kovacevic

It is inevitable, as presented in our paper, that if you have a contractive material, it will be effective during the consolidation stage of the analysis. The point of our paper was to show that compaction grouting works much better in free-draining material than in materials that are less permeable. It works in both situations, but the amount of compaction is quite different.

Mike Jefferies, *Golder Associates*

The objection we have is that your paper paints compaction grouting as a rather ineffective process. Our experience is that it can work quite well. I think some of the difference between what we got and what you got is that you have an artifact in your results because of choosing the Cam-Clay model. The Cam-Clay model has some very restrictive behaviours and I think your computed results depend very much on Cam-Clay. I think that if you ran a slightly better soil model you will actually get different engineering implications.

Dr Nebojsa Kovacevic

The reasons why we used the modified Cam-Clay model were two-fold. First, it clearly distinguishes between compressibility during the first loading and the much stiffer response during unloading and re-loading. This is a very important ingredient for showing that the amount of volumetric strain that was formed during the consolidation stage of undrained loading, at the end of undrained loading, would be much less. Second, it was to induce excess pore water pressure due to undrained shearing. The amount of pore water pressure that can be generated in the modified Cam-Clay model is not solely dependent on the mean effective stress – it depends on the deviatoric stress as well. As such I think the modified Cam-Clay model can be very useful.

Mike Jefferies, *Golder Associates*

Nor Sand does every thing your model does but also does more.

Sam Frydman, *Israel Institute of Technology, Haifa*

Another query on the potential danger of rock or stone columns. One of the applications is in soft clays below railway embankments – this was mentioned by Dr Bell. In this situation there is the possible problem of pumping, in other words, of washing of clay into the voids of the column due to the repeated loading effects of the train or other repeated loads. Does Dr Bell have any comments about this?

Dr Alan Bell, *Keller Ground Engineering*

Again I would have to refer to what experience tells rather than what theoretical problems would suggest. During the upgrade to high-speed status for many of the railways in Germany, vibro was the chosen method. The embankments are relatively shallow, and I am not aware of any problems that have arisen since. Certainly the kind of embankment I was thinking of in my presentation was the larger embankment where there is a large amount of overburden between the railway structure and stone columns. I am not aware of any problems in practice arising from the kind of mechanism that you describe.

Doug Ayres

Concerning railway embankments where there are columns, this is reproduced by the old method of repairing slopes where a trench is dug and a counterfort is made with packed stone. Looking at these over a period of 40 or a 100 year life, clay gradually does work into them. They had two advantages. One is that they took out any water that built up in the ballast pocket and kept the water table down. The other is that they connected points, if there were in fact pockets of more permeable material.

About 50 years ago the fashion was to talk of sand piles applied to railways. Looking with hindsight, these were mainly in areas offered by Professor Peck and others where you were connecting different aquifers in the system and very often in a sub tropical arid climate.

Observation of these counterforts showed that the slips they were put into to cure did in fact stay stable, for about 20 or 30 years, and in fact then built up. Clay washed into this carefully hand-packed stone and clogged it up and the slip then occurred again.

David Greenwood, *Geotechnical Consulting Group*

If the wet technique of vibroflotation is used, the void spaces of stone columns are filled with the coarse, sandier elements of the natural soil. The silts and the finer sands are washed out and only the coarse sand fills the void. If you refer to the discussion in the 1976 conference *Ground Treatment and Deep Compaction* (Institution of Civil Engineers, 1976), you see the answer there. The same question was asked at that conference and answered more comprehensively at that time.

Alan Moxhay, *Roger Bullivant*

Following the presentation by Professor Muir Wood. You will be familiar with the work of Priebe. For many years he has done exhaustive work on the group effects of stone columns and the establishment of settlement reduction factors, which we have been using in design for many years now. Have you any comment on any conclusions you have drawn or are about to draw on your work in terms of group effect and that of Priebe?

Wei Hu

We did some research on Priebe's German work and I suppose there are some empirical results that are presented. From my limited research at the time, I think the group effect that he drew was quite empirical, and there was not much material that we found to support what he proposed. I did try to do some further work on that by looking at flexible footings, which is a quite commonly used method – as Dr Greenwood mentioned earlier. As you know, £35,000 is not much money for larger scale jobs such as that. So I would say that we are familiar with Priebe's work, but I would like to see some more substantial funding to go with it in order to make it more plausible.

Kovacevic, N., Potts, D. M. & Vaughan, P. R. (2002). *Géotechnique* **52**, No. 1, 74–75

WRITTEN DISCUSSION

The effect of the development of undrained pore pressure on the efficiency of compaction grouting

N. KOVACEVIC, D. M. POTTS and P. R. VAUGHAN

M. G. Jefferies and D. A. Shuttle, *Golder Associates*

The authors state that the effect of excess pore water pressures caused by relatively rapid grout injection is that the efficiency of treatment may be reduced substantially. This may be alternatively expressed to say that fully drained conditions appear to be essential for effective densification by grouting. We suggest that this is too strong a conclusion, and possibly a consequence of the chosen soil model and initial conditions. Other simulations suggest that good levels of densification may be achieved under partially drained conditions, and this approach may be more cost-effective and practical than injecting slowly enough to achieve a fully drained grout injection when dealing with silty soils. We offer the following example.

In 1997 two sinkholes developed in the core of Bennett Dam in British Columbia. This dam is some 200 m high, retaining a reservoir (Lake Williston) 235 miles long (380 km), and provides as much as a third of the electrical power to British Columbia. The appearance of the sinkholes caused much concern, instigating immediate intensive investigations and, eventually, the use of compaction grouting to restore the integrity of the dam core. Stewart *et al.* (1997) provide an overview of the occurrence of the sinkholes and their investigation, while Garner *et al.* (2000) discuss the selection of compaction grouting and its implementation as the remediation.

Because of the importance of the dam, and its potential to affect large numbers of people, compaction grouting was required to be controlled and engineered. Further, the results obtained at each stage were to be analysed in near real time so as to obtain assurance that compaction was proceeding as expected. The discussers' company was retained for the prediction and analysis of this compaction grouting.

As in the authors' paper, radial symmetry was adopted for the analysis, with a large-strain formulation. Details of the numerical formulation are given in Shuttle & Jefferies (1998), and an overview of the modelling of compaction grouting is in Shuttle & Jefferies (2000). However, our work differed in two respects: the representation of drainage, and the range of constitutive behaviour modelled. We also examined the issue of radial idealisation, but used the more conventional approach of comparing cylindrical and spherical symmetry rather than the comparison of cylindrical plane strain and plane stress used by the authors. Trials using inclinometers in the influence zone of compaction grouting at Canoe Pass as part of the preparatory work for the dam remediation showed that the cylindrical plane strain idealisation was a good approximation for at least the primary pass of compaction grouting.

We represented drainage as the full spectrum of drained through to undrained using a standard coupled Biot formulation (Smith & Griffiths, 1988). This allowed us to investigate the full range of grout injection rates. We represented the soil being compacted with the NorSand model. This allowed us to investigate loose to compact sands with realistic constitutive behaviour, and in particular including the familiar S-shaped stress paths exhibited by such soils under undrained conditions. Note that Cam Clay is a particular case of NorSand, so that the model used by the authors can be captured by appropriate choice of parameters and initial conditions with the approach presented. It is this choice of initial conditions that leads to rather different results from those of the authors.

In the case of Bennett dam the sinkhole soils (sandy silts) were not as loose as would be represented by normally consolidated modified Cam Clay. Rather, testing showed that dam

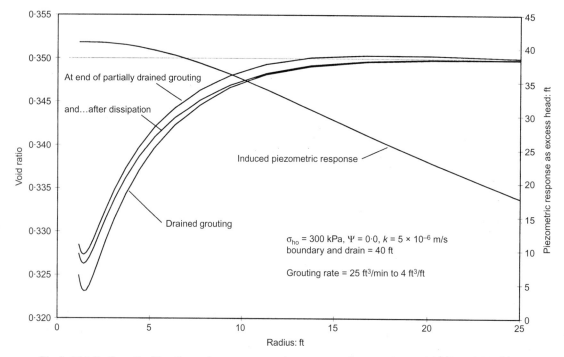

Fig. 9. Distribution of void ratios and excess pore water pressure after grouting to 4 ft³/ft at 0·25 ft³/min

core soils in the sinkhole were essentially at the critical state, consistent with their vertical movement of many metres. This was captured in our simulations. An example of results obtained is shown in Fig. 9, this simulation being for a grout injection rate of $0.25 \text{ ft}^3/\text{min}$ up to a maximum injected volume of 4 ft^3 per ft of grout hole (imperial units are given consistent with North American grouting practice). This injection rate was at the slowest end of the desired operating range.

Figure 9 shows both the void ratio distribution (left-hand axis) and excess pore water pressure distribution (right-hand axis) against distance from the centreline of the grout hole. Three void ratio distributions are plotted: on completion of injection, after consolidation, and what would have been achieved by drained injection. In contrast to the authors' results, compaction is surprisingly efficient, even with a peak excess pore water head of 40 ft. Expressed in terms of the initial vertical effective stress, this is an excess pore pressure ratio of 0.4.

It follows from Fig. 9 that the authors' conclusion that high excess pore water pressures are associated with ineffective compaction is too sweeping. Depending on the initial conditions, material state and material properties, reasonably effective compaction may be achieved despite quite substantial excess pore water pressures. Having said that, we did conclude that grout injection rates as low as $0.25 \text{ ft}^3/\text{min}$ were required at Bennett Dam.

Authors' reply

The authors thank Jefferies and Shuttle for their interesting discussion. They had not been aware of this work. The authors see no significant disagreement between their approach and conclusions and those of Jefferies and Shuttle. The authors considered the limiting cases of drained behaviour, when volumetric compression occurs during injection and compaction grouting is fully efficient, and undrained behaviour, when volumetric compression occurs only after treatment as the soil consolidates. In the latter case treatment is much less efficient. It is perhaps self-evident that intermediate behaviour, in which partial consolidation occurs during injection, will result in compression that lies between the two limits, although this is not stated in the paper. The authors quoted the field data from Boulanger & Hayden (1995) for Sacramento silt (Figs 1 and 8

of the paper), which show quite effective treatment despite the development of excess pore pressure. The authors think that, if the same material assumptions were made, their analyses and that of Jefferies and Shuttle would give similar results. The authors agree that effective treatment can be achieved without full consolidation during treatment. This may be a cost-effective approach. However, if only partial consolidation is occurring during treatment, then a relatively small change in permeability may give a rather larger change in the efficiency of treatment.

Compaction grouting has grown up as a semi-empirical procedure. It is not normally applied except in materials that would largely consolidate during treatment, although this has not been given as a specific limitation to the treatment. This limit has not been put on a qualitative or theoretical basis. Thus there has been a risk that this limit would be unappreciated or forgotten. Indeed, the work done by the authors was stimulated by proposals to use compaction grouting in a loose moraine soil in which significant consolidation during treatment was unlikely. The work of Jefferies and Shuttle and the work of the authors show that such a process can be analysed numerically. The type of material in which it can be effective can then be established quantitatively. The authors believe that this is a much safer way forward than reliance on experience empirically formulated.

REFERENCES
Boulanger, R. W & Hayden, R. R. (1995). Aspects of compaction grouting of liquefiable soil. *J. Geotech. Engng Div., ASCE* **121**, No. 12, 844–855.
Garner, S. J., Warner, J., Jefferies, M. G. and Morrison, N. A. (2000). A controlled approach to deep compaction grouting at WAC Bennett Dam. *Proc. 53rd Can. Geotech. Conf., Montreal.*
Shuttle, D. A. & Jefferies, M. G. (1998). Dimensionless and unbiased CPT interpretation in sand. *Int. J. Numer. Anal. Meth. Geomech.* **22**, 351–391.
Shuttle, D. A. & Jefferies, M. G (2000). Prediction and validation of compaction grout effectiveness. *Proc. Wallace Baker Memorial Symposium, ASCE GeoDenver Conf.*
Smith, I. M. & Griffiths, V. (1988). *Programming the finite element method.* Wiley.
Stewart, R. A., Watts, B. D., Sobkowicz, J. C. and Kupper, A. (1997). *Geotechnical News* **15**, No. 2, 32–42.

Muir-Wood, D. Hu, W. & Nash, D. F. T. (2001). *Géotechnique* **51**, No. 7, 649

WRITTEN DISCUSSION

Group effects in stone column foundations: model tests

D. MUIR-WOOD, W. HU and D. F. T. NASH

D. A. Greenwood, *Geotechnical Consulting Group, formerly Cementation Piling & Foundations*

The model tests and theoretical treatment of stone column groups by Muir-Wood *et al.* confirm the hypothesis outstanding since the 1975 Conference that loading conditions strongly influence the stiffness and strength of stone columns (Greenwood, 1976). Except for colums near the edge of a loaded area that are not uniformly constrained nor wholly vertically loaded, the columns become stiffer and stronger as load is applied. In fact the applied load is the dominant influence on the strength and stiffness of columns. In the case of bulk storage tanks it is the live load that dominates the radial constraint on columns.

As a result, columns in large arrays under wide loaded areas such as embankments, ore stockpiles or oil tanks perform much better that those under narrow strips and pad footings where all columns are close to the edges of the footing and are thus constrained only by unloaded ground.

This fact demonstrates the futility of field load testing a small group of columns in the hope of representing their performance under wide loaded areas. Such tests will always show them to be less stiff and failing under lesser loads than those under large loaded areas. There is generally no practical and economic way of simulating stone column performance under wide loads.

For a comparable test the ambient stresses and excess pore water pressures generated by load must be accurately replicated. In my opinion this is more important for column behaviour at typical area ratios than mutual support from surrounding columns as suggested in the paper.

Money allocated for proving stone columns under wide loaded areas is best spent by supervising construction very closely to ensure that design specifications are consistently met (Greenwood, 1991).

REFERENCES

Greenwood, D. A. (1976). In *Ground treatment by deep compaction*, Institution of Civil Engineers, London, discussion pp. 107–109.

Greenwood, D. A. (1991). Load tests on stone columns. In *Deep foundation improvements: design, construction and testing*, ASTM STP 1089, pp. 148–171. Philadelphia: American Society for Testing and Materials.

Session 4

Al-Khafaji, Z. A. & Craig, W. H. (2000). *Géotechnique* **50**, No. 6, 709–713

Drainage and reinforcement of soft clay tank foundation by sand columns

Z. A. AL-KHAFAJI* and W. H. CRAIG†

A series of centrifuge model tests, at acceleration levels of 105 *g*, is reported, in which loosely placed vertical sand columns and densified sand columns have been used to improve the performance of oil storage tank foundations. A simulated tank of 34 m diameter, with a flexible base, has been subjected to incremental fluid loadings typically to 160 kPa, with foundation drainage permitted. Model preparation included the installation of as many as 572 sand columns in a single foundation. Comparison is made between the experimental results and the analysis of Priebe, which indicates that that analysis will overestimate the improvements if three-dimensional considerations are not taken into account. Comments are also included as to the extent to which model column formation may be deficient.

KEYWORDS: centrifuge modelling; drainage; ground improvement; reinforced soils.

Nous décrivons une série d'essais centrifuges modèles, utilisant des accélérations de 105 *g*, essais dans lesquels des colonnes de sable verticales espacées et des colonnes de sable rapprochées ont été utilisées pour améliorer la performance des fondations de réservoirs de pétrole. Un réservoir simulé de 34 m de diamètre, à base flexible, a été soumis à des charges fluides augmentant par degrés jusqu'à 160 kPa, le drainage des fondations étant permis. La préparation de la maquette comportait l'installation de 572 colonnes de sable dans une seule fondation. Nous comparons les résultats expérimentaux à l'analyse de Priebe (1995); il en ressort que cette analyse surestime les améliorations si certains facteurs tridimensionnels ne sont pas pris en compte. Nous montrons aussi dans quelle mesure cette formation modèle à colonnes peut être déficiente.

INTRODUCTION

The use of compacted sand and vibrated stone columns in the improvement of both fine- and coarse-grained soils has become widespread around the world in the last 30–40 years. Eurocode EC7 (1995) refers only briefly to ground improvement and reinforcement, and notes that in many cases this type of structure should be classified in Geotechnical Category 3, implying that these are not yet considered as conventional types of structure (Simpson & Driscoll, 1998). Stone columns are generally preferred, but sand is used for economic reasons on many sites, where stone, which is more angular and generates higher strength and stiffness, is not readily available in the quantities required. A number of reviews, emanating from different parts of the world, are available that follow the development of both field techniques and associated design methods over the years (Baumann & Bauer, 1974; Barksdale, 1987; Van Impe *et al.*, 1997). One of the most widely used design approaches is that of Priebe (1995).

Field data have been included in many of the reviews, which confirm the likely magnitudes of improvements in performance, though these prove difficult to predict with any accuracy. A number of numerical and experimental studies have concentrated on the performance of a unit cell of soil reinforced by a cohesionless central column and on developing equivalent cohesion/friction and compressibility parameters for the reinforced soil (e.g. Bouassida *et al.*, 1995; Gutub & Khan, 1998; Bergado *et al.*, 1999). This approach is not adopted in the present work, where settlement control rather than bearing capacity is of interest.

Model testing in centrifuges, related to column reinforcement, has been reported on a number of occasions, looking principally at overall benefits and performance under loading up to the collapse of overlying structures such as slopes and embankments, quay walls and sheet pile walls. It should be noted that with few exceptions the centrifuges to date have been used with two-dimensional models. In few of these studies are settlements of prime interest, and indeed they are rarely reported. Modelling of the column structures has generally been

achieved by pouring or tamping sand into pre-formed holes, or by insertion of frozen columns into pre-formed holes. Area replacement ratios have varied enormously, and in one extreme case, with a replacement ratio of 70%, clay columns have actually been inserted into holes bored in a frozen sand block (Kitazume *et al.*, 1998). Almost all the centrifuges used have been of the rotating beam type, although a drum-type centrifuge has been used by Jung *et al.* (1998) with frozen 10 mm diameter sand piles up to 72 mm long installed in containers 165 mm square, primarily for consolidation rate and settlement studies. In a novel development, Ng *et al.* (1998) have reported a device that has been used to form sand columns while a centrifuge is running. This uses an Archimedes' screw to force sand into a clay bed using a support casing that is withdrawn. Modest numbers of piles (18) have been installed in a regular grid before stopping the centrifuge to remove the installation tool and add an overlying structure, which is then subject to test. It is probably not realistic to use this approach for models in which several hundred piles are required.

The present paper deals specifically with three-dimensional, axisymmetric centrifuge models of clay reinforced by sand columns where shear strains are limited and serviceability limit states are more important than collapse limit states. It summarizes data from two series of tests, carried out by Al-Khafaji (1996), in which the settlement performance of a simulated oil storage tank founded on soft clay was improved by the presence of loose sand columns (sand drains) and densified sand columns or sand compaction piles (SCP). Area replacement ratios, a_s, of up to 40% were used. Only a single tank and clay layer geometry has been modelled, and the results obtained must be specific for this case. However, the significance of the results for other scenarios can be seen in broad terms.

Historically, sand drains have been used as consolidation accelerators with small area ratios, and the contribution of reinforcement has been negligible. Barron's (1948) design curves, which still feature in many text books, are commonly presented in terms of a parameter *n*, which is defined as the ratio of the diameter of the cylinder of influence of the drain to that of the drain itself. Typically *n* varies in the range 5–100, implying area replacement ratios of 4% to 0·01%. In practice, sand drains are rarely if ever used today, but the theory of consolidation is applied to band or wick drains of various designs, with even smaller true area replacement ratios and, because of their narrow cross-section, even lower axial load carrying/stiffening capacity. Nevertheless it was felt appropriate

Manuscript received 3 May 2000; revised manuscript accepted 7 September 2000.
* Consulting Engineer, Vancouver, Canada.
† Manchester School of Engineering, University of Manchester, U.K.

to include a test series in which substantially higher area replacement ratios could be utilized with both loose and dense sand columns, to separate the variations caused by differing degrees of densification as a function of column stiffness. The use of sand drains can be considered as a simple soil replacement technique with guaranteed minimum benefits, where the use of densified columns combines the effects of replacement and densification when reinforcement benefits will be higher.

Where the column material is cohesionless and frictional the stiffness is not unique, but varies with relative density, depth and the degree of confinement. In soft fine-grained deposits, for which this construction technique is often appropriate, the strength and stiffness of the host material are also often variable with depth. Many of the design methods for sand and stone columns utilize the relative stiffness between that of the host soil, E_s, and that of the column, E_c. Baumann & Bauer (1974) tabulate possible values of E_s/E_c as 1/8 for sandy silt, 1/16 for clayey silt and 1/25 for silty soft clay, when considering immediate undrained settlement, which will precede settlements associated with drainage. Priebe (1995) presents an analysis for the case of an unlimited load area on an unlimited column grid in terms of the ratio of constrained moduli of soil and column materials, as measured by large-scale oedometer tests, i.e. for drained conditions.

In practice the procedures for installing sand columns in soft clays, which may exhibit some sensitivity, can initially reduce clay strength around the column, though this is counteracted by rapid local drainage once installation is completed and particularly once the clay between the columns experiences additional vertical loading. The concept of a single stress ratio, considered as the vertical effective stress in the column to that in the host soil, involves assumptions that these stresses are uniform at any level and do not vary laterally within and between columns despite ongoing drainage. The stress ratio is also generally assumed to be constant with load level and time and over the depth of the reinforced stratum. Sheng (1986), for example, reports observed ratios between 1·44 and 6 from a range of sites on three continents and values between 1 and 6 used in design. For a field project with a storage tank of dimensions similar to those considered below, Greenwood (1991) reports measured stress ratios varying from 25 to 5, with increasing load levels in a drained situation. All the above assumptions are likely to be simplifications; see also Juran & Guermazi (1988). In the present paper no attempt is made to quantify effective stresses or stress ratios in the ground. Rather there is concentration on point pore pressures and settlements, which indicate, in the first instance, drainage performance at identifiable single points within the foundation volume, and in the second the global performance of the structure—both of which are routinely assessed in field practice.

Priebe's design method utilizes a nominal stiffness ratio. As indicated above, the ratio is determined from laboratory materials testing under fully constrained, one-dimensional conditions. This is not the field situation in most practical cases. In the field the measurement of stiffness and its variation beneath and around a loaded foundation is not undertaken, and no attempt to do so has been made in the model tests. Rather the overall foundation performance is assessed in terms of a back-figured stiffness ratio in order to give guidance on the value that might be applicable in a design situation where a 'Class A' prediction of likely settlement may be required.

THE PRESENT INVESTIGATION

A total of eight tests have been performed, in two series, at a centrifugal acceleration of 105 g to investigate the settlement behaviour of a clay foundation improved using large diameter, closely spaced, loose sand drains and densified sand compaction piles. The initial soft clay foundation was the same for all tests, and was constructed using a bed of reconsolidated clay with a thickness of 200 mm. For each model a specific number of 10 mm diameter loose or dense sand columns were installed in a uniform triangular pattern inside a limited circular foundation

area of 380 mm diameter. The numbers of columns employed were 144, 287, 431 and 572 occupying an area equivalent to 10%, 20%, 30% and 40% of the circular foundation area. A circular model tank 325 mm in diameter, with a flexible base, was placed on top of the foundation and subjected to incremental fluid loading with periods of drainage between increments. Measurements of pore water pressure at different depths and positions in the clay bed were carried out, and settlements were also recorded at several locations in the model tank.

The clay foundation was constructed using Cowden clay consolidated from slurry in a rigid steel box 560 mm square and 460 mm deep. The clay was consolidated under a vertical effective stress of 200 kPa and had a uniform initial undrained shear strength, measured by hand vane, of 26–30 kPa. After consolidation the clay bed was scribed to the desired level and sand columns were installed using techniques involving pouring and vibrating sand into pre-bored holes, described fully by Al-Khafaji (1996) and briefly for the loose columns by Al-Khafaji et al. (1998) and the dense columns by Craig & Al-Khafaji (1997). The model beds were then transferred to the centrifuge, which has a radius of 3·2 m to the face-plate.

Several pore pressure transducers were installed inside the treated area under the model tank centre, at two-thirds of the tank radius, r_t, measured from the centre and under the edge at depths up to 110 mm. Other transducers were installed in columns of sand at the corners of the model box to monitor the ground water table maintained throughout the test runs. A circular plastic ring, 380 mm in diameter and 20 mm high, was placed centrally on the clay surface and used to restrain the tank foundation pad material throughout the test. The material used to form this pad was moist sand, compacted by hand inside the ring.

The circular plastic tank model, which was 325 mm in diameter with a 3·5 mm wall thickness, was placed on the centre of the foundation. Two pipe connections were made near the bottom of the tank shell to allow filling and emptying, i.e. loading and unloading, during tests at 105 g. Fig. 1 shows the general arrangement. The tank base was made using a sheet of 0·2 mm thick aluminium foil to have a flexibility corresponding to typical tank installations in the field. An annular Perspex lid was placed on top of the tank to support displacement transducers at four quadrant points, measuring the settlement and any tilt. Other transducers measured the settlement of the tank base at the centre and at two-thirds radius.

The diameter of the model drains was 10 mm, equivalent to 1·05 m field diameter and the tank equivalent to a field diameter of 34·4 m, with a potential fluid storage depth of 20 m, corresponding to a uniform base loading approaching 200 kPa, for a storage volume of approximately 18 000 m^3.

In each of the model tests summarized here two centrifuge runs were conducted: bedding and main runs. The bedding run was performed with the clay bed having only the sand base for the tank on top, i.e. in the absence of the model tank structure. The aim was to reconsolidate the clay layer under its own weight and that of the tank base. Throughout this run pore water pressures were recorded, and once equilibrium was achieved, at times indicated by theoretical calculations and confirmed by the readings reaching a steady state, the centrifuge was stopped. The main run was carried out the following day with the model tank in place on top of the foundation.

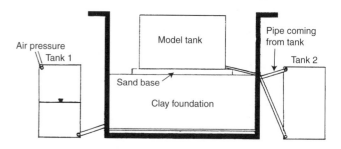

Fig. 1. General arrangement of model with water supply to tank

Additional water pressure transducers measured the tank water level during this run. Two small pneumatic jacks were used to hold the tank in place at rest on the centrifuge arm; these were withdrawn at an acceleration of 21 *g*, allowing the tank to settle freely and possibly tilt, owing to differential movements, from that point onwards. After reaching 105 *g*, with the tank empty, the model was left, as in the bedding run, until equilibrium was reached. Then the tank loading process was started by pumping water from storage tanks, located alongside the model box, to the model tank until a tank base pressure of 40 kPa was reached. The model was left to consolidate under this tank pressure, allowing the full dissipation of excess pore pressures in the foundation. The loading was then increased in stages to 80 kPa, 120 kPa, 160 kPa and sometimes higher, with consolidation at each stage. Ground water level was maintained just above the clay surface throughout the centrifuge runs. Water was fed to the clay surface whenever any reduction of this level was observed from the transducers located in the box corners. While settlements were substantial in all models, tilt rotations were always minimal and are not considered further. This reflects the uniformity that can be achieved under laboratory conditions, but may not be achievable under field conditions.

TEST RESULTS

A typical example of pore pressure records from the tank loading stage of one test, with densified columns, is shown in Fig. 2. The acceleration of the centrifuge was built up to 105 *g* in stages. The equilibration stage at this acceleration was followed by five loading increments that induced initial excess pressures that were allowed to dissipate. The drainage periods varied from test to test according to the area replacement ratio or column spacing. Settlement records from the same test are shown for the period of tank loading in Fig. 3.

From any one test the drained settlement associated with any level of loading was determined at the tank centre, at the edge and at the intermediate point. Comparing data between tests allows an assessment of the benefit of changing the area replacement ratio; extrapolation from the measured data, to the point where $a_s = 0$, allows an assessment of the performance of an unreinforced foundation. It was deemed unrealistic to determine this last information directly, since centrifuge run times close to 24 h were necessary for the tests with the lowest area ratios. Continuous running for periods of upwards of a week would have been needed to approach equilibrium in the absence of any vertical drainage elements in the foundation.

Data collected from the densified column models at all loading levels are shown in Fig. 4. The settlement of the tank centre, S_t, is normalized with respect to the clay layer thickness, T, which is a site-specific parameter in any design study. An alternative would be to normalize with respect to the tank

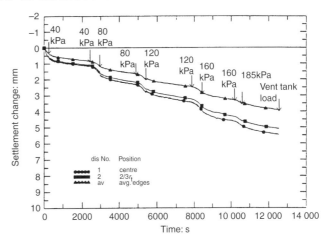

Fig. 3. Settlement changes plotted against time during tank-loading process; $a_s = 30\%$

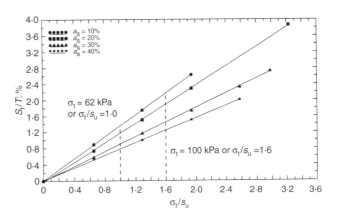

Fig. 4. S_t/T as a function of σ_t/s_u: densified columns

diameter, but this may be a prime design variable. The internal tank pressure, σ_t, can be left as a prime parameter, or can be normalized with respect to some characteristic clay strength. Various possible characteristic strengths are possible, and there is no obvious single value; selection must necessarily be considered arbitrary. In this instance a value of s_u of 62 kPa has been used, being the mean final undrained shear strength at the mid level of the clay layer after consolidation under a tank load of 160 kPa. Fig. 5 shows an example of data obtained for the strength of the clay foundation in the one test shown in earlier figures. The measurements before the test were made in the uniform soil bed before column installation and those after the

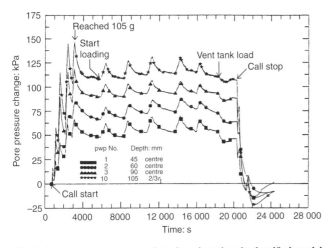

Fig. 2. Pore pressure changes plotted against time in densified model with $a_s = 30\%$

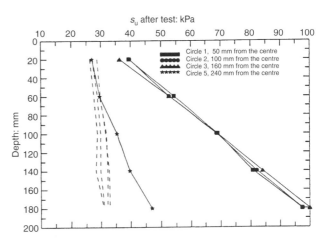

Fig. 5. Undrained strength profiles before and after test: $a_s = 30\%$

test in the zone beneath the foundation, or in the unreinforced and largely unloaded areas outside, when the model was finally stripped down. In Fig. 4 both absolute pressures and dimensionless groupings are shown.

From the information in Fig. 4 and similar data for the sand drain test series it is possible to find a relationship between the tank settlement, the area replacement ratio and the applied fluid stress levels in the tank and the characteristic clay strength in the form

$$\frac{S_t}{T} = f\left(\frac{\sigma_t}{s_u}, a_s\right)$$

or

$$\frac{S_t}{T} = f(\sigma_t, a_s)$$

Fitting a linear relationship to the data appears initially attractive, but this leads to the result that at a certain replacement ratio there will be no settlement (Fig. 6), and this approach is discarded. An exponential relationship also shown in Fig. 6, for the densified columns, has been fitted. For the two test series, at the three points at which settlements were measured, the relationship is of the form

$$\frac{S_t}{T} = A\frac{\sigma_t}{s_u}.e^{(-B.a_s)}$$

or

$$\frac{S_t}{T} = C\sigma_t.e^{(-B.a_s)}$$

where the dimensionless coefficients A and B and the dimensional C are given in Table 1.

For any given radial position the absolute settlement for the unreinforced foundation is found from the value of A or C, while the settlement improvement ratio is computed from the exponential term. In a set of perfect experiments the A, or C, values for the loose and dense columns should be the same at any radial position. However, those computed here for the loose state are consistently 10% higher than for the dense state, with the edge settlement 69% of that at the centre in each case. The ultimate improvement ratio, for 100% replacement, increases

with B and—as is to be expected—is always higher towards the centre. The lower values at higher radii are consistent with increased divergence from the simple concept of using a one-dimensional approach: thus the three-dimensional models are expected to yield lower apparent modular ratios than are consistent with the one-dimensional theory.

Applying the above coefficients, obtained from the practical range of area replacement ratios at the centre of the tank, to the situation where the replacement ratio is 100% will give an estimate of the likely reduction in settlement to be obtained for full replacement. In effect this figure gives an indirect, approximate measure of the mean modular ratio, E_c/E_s, of the drained one-dimensional compression modulus of the replacement sand to that of the parent clay. In the analysis of Priebe (1995) this ratio is a critical design parameter, which is combined with the angle of friction of the column material—a parameter that may not be readily available as it depends on the degree of compaction and the confining stress. In Priebe's analysis the assumption is made that the reinforced soil is of infinite lateral extent—a more restrained condition than that for the tank scenario under consideration here, even considering the central location.

From the data above, the best estimate of the settlement improvement ratio, S_r, for 100% replacement of a foundation of infinite extent for the loose material is 4·2, while for the densified material it is 7·0. Fig. 7 shows a comparison of Priebe's analysis for combinations of a stiffness ratio of 7 with varying friction angles, ϕ, for the column material, showing that the analysis indicates an improvement for the infinitely reinforced layer that is higher than that achieved in the models. The comparison suggests a friction angle of only 30° for the densified columns and an even lower value for the loose columns. It should be noted that no direct measurement of the sand density or friction angle as placed in the models was possible, but the stiffness ratios are broadly in line with what might be expected, allowing for the limited lateral extent of the foundation and the improvement. While the loose columns of sand might be expected to have a friction angle as low as 30°, the densification, to whatever uncertain degree, must raise this considerably, and 35–40° is a reasonable assumption. The apparent lack of higher ratios may be attributable to the consid-

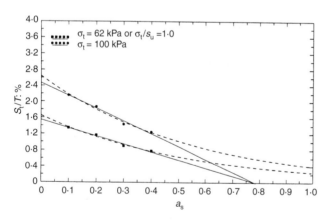

Fig. 6. Best fit for S_t/T as a function of a_s at constant σ_t: densified columns

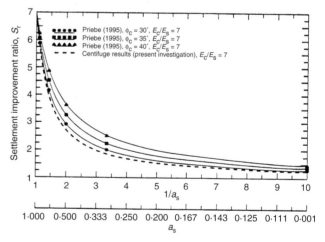

Fig. 7. Comparison between experimental values of S_r and those from Priebe's solution

Table 1.

Settlement position	Loose columns			Densified columns		
	A	B	C: m²/MN	A	B	C: m²/MN
Centre	1·79	1·44	28·9	1·65	1·95	26·6
$2r_t/3$	1·48	1·03	23·9	1·29	1·37	20·8
Edge	1·23	0·96	19·8	1·14	1·20	18·4

erable depth of the clay layer and the columns relative to the tank diameter, and to breakdown of the unlimited load case. As indicated above, the quantitative results are limited to the specific tank and clay layer properties and dimensions used. Any numerical extrapolation to other scenarios must be treated with caution, but the patterns of behaviour observed should still be a useful guide to designers. In Priebe's (1995) example, with a shallower clay layer, only a single settlement, or settlement ratio, was reported, and this is somewhat simplistic. As has been seen here, the settlement for the flexible foundation of the tank has varied considerably between centre and edge, as expected. This is in agreement with data from Greenwood (1991) for a 36 m diameter storage tank at Canvey Island subjected to drained loading on columns 10 m long.

CONCLUDING REMARKS

It is recognized that there remain deficiencies in model preparation where multiple columns are prepared and installed at unit gravity prior to centrifuging. This may have resulted in lower than ideal model stiffness in a manner similar to the discrepancy seen between the behaviour of model piles in sand where the piles are installed either on the laboratory floor, before centrifuging, or while the centrifuge acceleration is applied (Craig, 1984). In this respect the achievement of Ng *et al.* (1998) in installing sand columns in the centrifuge is considerable.

The reported tests involved three-dimensional modelling of a flexible base tank on a uniformly spaced column configuration extending just beyond the tank edge. This is a departure from the essentially one-dimensional basis of most theoretical studies of column-reinforced ground of infinite extent, which have to be modified empirically. It is also a departure from the predominance of previous two-dimensional physical models that have concentrated on failure scenarios. The accent on settlement measurement and analysis has been used to demonstrate the effect of the three-dimensional scenario, and leads to the possibility of using non-uniformly spaced column grids for flexible foundations of this size if limitation of differential settlements is sought.

The internal consistency of these model tests is high, and the benefits of series of tests with repeatable, controlled foundation conditions are clear when seen in the light of widespread scatter in reported field behaviour. If the results err, they will err on the conservative side, but the benefit from showing the effects of varying geometrical and densification conditions is still considerable. A more extensive study, involving variations in clay properties and in tank and layer geometry, may be better undertaken by numerical methods—though this is not a trivial exercise. However, for a 'Class A' prediction the greatest problem will lie in assessing the soil stiffness and its changes. The present data set provides a physical benchmark against which such calculations can be tested.

NOTATION

A	Dimensionless coefficient
a_s	Area replacement ratio
B	Dimensionless coefficient
C	Dimensional coefficient (m^2/MN)
E_c	Stiffness of sand column/pile
E_s	Stiffness of host soil
n	Sand drain radial ratio

r_t	Radius of storage tank
S_r	Settlement improvement ratio
S_t	Settlement of storage tank
s_u	Undrained shear strength of clay
T	Thickness of clay layer
ϕ	Angle of friction of sand
σ_t	Internal tank base pressure

REFERENCES

Al-Khafaji, Z. A. (1996). *Reinforcement of soft clay using granular columns*. PhD thesis, University of Manchester.

Al-Khafaji, Z. A., Craig, W. H. & Cruickshank, M. (1998). Reduction of clay settlement by large closely spaced sand drains. *Proc. Int. Conf. Centrifuge 98*, Rotterdam: Balkema, **1**, 861–866.

Barksdale, R. A. (1987). *State of the art for the design and construction of sand compaction piles*, US Army Corps of Engineers, Report REMR-GT-4. Washington DC.

Barron, R. A. (1948). Construction of fine-grained soil by sand wells. *Trans. ASCE* **113**, 718–742.

Baumann, V. & Bauer, G. E. A. (1974). The performance of foundations on various soils stabilized by the vibrocompaction method. *Can. Geotech. J.* **11**, No. 4, 509–530.

Bergado, D. T., Lin, D. G. & Nakamura, M. (1999). Evaluation of silty sand as a material for soil compaction piles and applications. *Ground Improvement* **3**, No. 1, 7–19.

Bouassida, M., de Buhan, P. & Dormieux, L. (1995). Bearing capacity of a foundation resting on a soil reinforced by a group of columns. *Géotechnique* **45**, No. 1, 25–34.

British Standards Institution (1995). *Eurocode 7: Geotechnical design: Part 1, General rules, together with UK national application document*, DD ENV 1997-1:1995. Milton Keynes: BSI.

Craig, W. H. (1984). Installation studies for model piles. *Proceedings of the symposium on application of centrifuge modelling to geotechnical design*, pp. 441–456. Rotterdam: Balkema.

Craig, W. H. & Al-Khafaji, Z. A. (1997). Reduction of soft clay settlement by compacted sand columns. *Proc. 3rd Int. Conf. Ground Improvement Geosystems: Densification and Reinforcement*, 218–224.

Greenwood, D. A. (1991). Loads tests on stone columns. In *Deep foundation improvements: design, construction and testing*, ASTM STP 1089, pp. 148–171. Philadelphia: American Society for Testing and Materials.

Gutub, M. Z. A. & Khan, A. M. (1998). Shear strength characteristics of Madinah clay with sand compaction piles. *Geotech. Testing J.* **21**, No. 4, 356–364.

Jung, J. B., Moriwaki, T., Sumioka, N. & Kusakabe, O. (1998). Consolidation behavior of composite ground improved by soil compaction. *Proc. Int. Conf. Centrifuge 98*, Rotterdam: Balkema, **1**, 825–830.

Juran, I. & Guermazi, A. (1988). Settlement response of soft soils reinforced by compacted sand columns. *J. Geotech. Engng. Div., ASCE* **114**, No. 8, 930–943.

Kitazume, M., Shimoda, Y. & Miyajima, S. (1998). Behavior of sand compaction piles constructed from copper slag sand. *Proc. Int. Conf. Centrifuge 98*, Rotterdam: Balkema, **1**, 831–836.

Ng, Y. W., Lee, F. H. & Yong, K. Y. (1998). Development of an in-flight sand compaction piles (SCPs) installer. *Proc. Int. Conf. Centrifuge 98*, Rotterdam: Balkema, **1**, 837–843.

Priebe, H. J. (1995). The design of vibro replacement. *Ground Engng* **28**, No. 10, 31–37.

Sheng, C. W. (1986). Some experiences in construction and design of stone column reinforced ground. *Proceedings of the international conference on deep foundations*, Beijing, pp. 1.1–1.6.

Simpson, B. & Driscoll, R. (1998). *Eurocode 7: a commentary*. London: Construction Research Communications Ltd.

Van Impe, W. F., De Cock, F., Van der Cruyssen, J. P. & Maertens, J. (1997). Soil improvement experiences in Belgium: part ll. Vibrocompaction and stone columns. *Ground Improvement* **1**, No. 3, 157–168.

Slocombe, B. C., Bell, A. L. & Baez, J. I. (2000). *Géotechnique* **50**, No. 6, 715–725

The densification of granular soils using vibro methods

B. C. SLOCOMBE,* A. L. BELL* and J. I. BAEZ†

The method of densifying in-situ sandy soils using deep vibration (vibro) was first developed in Germany over 60 years ago. Since then the technique has been extended to improve a wide range of soil types for structures of up to 30 storeys in height. A major development in the vibro market worldwide over the last 25 years has been the treatment of soils to resist the effects of liquefaction during seismic events. It is well known that the presence of silt and clay fines inhibits the physical in-situ densification of sandy soils. Experience has shown that a maximum fines content of about 15% would normally permit the performance of vibro compaction methods without the addition of stone aggregate. However, the development of new vibrators and modified construction techniques has enabled sands with significantly higher fines content to be treated. The paper describes how densification of granular soils has progressed with the development of new equipment, the greater appreciation in the manner in which soils respond to both vibration and displacement, and the importance of workmanship during treatment. A number of case histories including in-situ test results, long-term monitoring of ageing effects and references to treated sites already subjected to earthquakes are presented.

KEYWORDS: case history; earthquakes; ground improvement; in-situ testing; sands; vibration.

La méthode qui consiste à densifier sur place des sols sableux en les soumettant à des vibrations profondes (au vibro) a été mise au point en Allemagne il y a plus de 60 ans. Depuis, la technique a été élargie et elle est maintenant employée pour améliorer de nombreux types de sol appelés à supporter des bâtiments pouvant avoir jusqu'à 30 étages. Le traitement de sols pour leur permettre de résister aux effets de la liquéfaction pendant les secousses sismiques est un développement majeur dans le marché mondial du vibro de ces 25 dernières années. On sait que la présence de limons et de fines d'argile empêche la densification physique in-situ des sols sableux. L'expérience a montré qu'une teneur maximum en fines d'environ 15% permet normalement d'employer un compactage au vibro sans avoir à ajouter d'agrégats pierreux. Cependant, l'apparition de nouveaux vibrateurs et de techniques de construction modifiées permet de traiter des sables ayant une teneur en fines beaucoup plus élevée. Cet exposé montre comment la densification de sols granuleux a progressé avec le développement de nouveaux équipements et une meilleure appréciation de la manière dont les sols réagissent aux vibrations et aux déplacements ; il montre aussi l'importance des compétences humaines pendant le traitement. Nous présentons plusieurs d'histoires de cas avec le résultat d'essais effectués sur place, le suivi à long terme des effets de vieillissement et des références aux sites traités qui ont déjà subi des tremblements de terre.

INTRODUCTION

The vibro compaction technique was developed by Johann Keller GmbH, with loose sands beneath the first structure, a building in Berlin, being densified in situ in 1937. Development then proceeded in Germany and the USA in the 1940s. The system was introduced to Britain and France in the 1950s. The technique has now been performed in both developed and developing economies worldwide.

The principle of the process is based on particles of non-cohesive soil being rearranged into a denser state by means of dominantly horizontal vibration, produced by specially built electrically or hydraulically powered depth vibrators. The resultant reduced void ratio and compressibility, and increased angle of shearing resistance, then permit the adoption of higher imposed design loadings at lesser settlement and increased seismic resistance. In-situ densification from horizontal vibration has been consistently reported to be more efficient, to a higher relative density over a wider range of soils, than densification using vertical vibration. Similar equipment can be used for the construction of vibro stone columns. These can be used to enhance densification in finer granular soils, or to reinforce clayey soils.

VIBRO EQUIPMENT

Vibrations are generated close to the tip of the depth vibrator, and are produced by a series of rotating internal eccentric weights mounted on a shaft driven by a motor. The original Berlin vibrator was of relatively low power, and was suspended

from a well-drilling platform. Vibrator development has been intensive over the last 20 years, with the aim of greater efficiency in densifying the soils at wider compaction centres, as well as longer working life and reduced breakdown/maintenance costs. Centrifugal forces of over 300 kN are currently in use at frequencies of 20–30 Hz, in some cases with variable frequencies and with yet more powerful vibrators under development. These are highly sophisticated machines with state-of-the-art materials.

The vibrator is usually connected via a flexible coupling to extension tubes for the required depth of treatment and suspended from a crawler crane. Purpose-built base machines incorporating a pull-down facility to ease penetration of the vibrator into the ground have also been developed. These include bottom-feed delivery systems (see Figs 1(a) and 1(b)) and, since the 1980s, computer-based instrumentation packages that provide continuous records of the construction of the compaction locations. More recent developments include digital data acquisition and modem transfer of information from site to office, as well as servo-controlled automatic construction operations.

Depths of treatment of up to 56 m have been achieved to date. Vibro compaction can also be performed from barges/platforms over water, in which case vibrators are sometimes connected in tandem for greater efficiency.

BASIC TECHNIQUE

The basic vibro compaction technique uses water jetting to assist penetration of the vibrator to the required depth. In unsaturated soils, densification is achieved by temporarily reducing the friction between soil particles and then allowing them to redeposit under gravity and vibration into a more compact condition. In saturated soils, densification is achieved by increasing pore water pressure and reduction of shear strength, together with cavity expansion and vibratory action. The me-

Manuscript received 1 June 2000; revised manuscript accepted 7 September 2000.
* Keller Ground Engineering, Coventry, UK.
† Hayward Baker Inc., Santa Paula, USA.

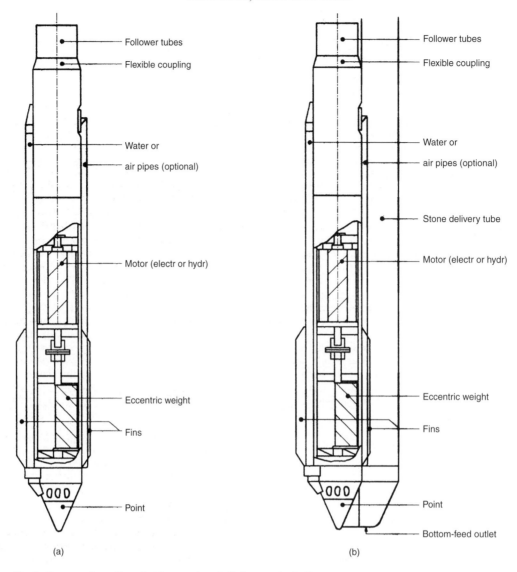

Fig. 1. Cross-sections through (a) normal and (b) bottom-feed vibrators

chanisms of such densification have been studied in detail by Baez & Martin (1992). The vibrator is then lifted a short distance and densification is repeated up to the desired level, normally ground level. This process of densification results in a lowering of the ground surface. The construction operation can be accelerated by adding granular material, sometimes stone, from the surface. The use of stone or other infill also assists in maintaining site levels. More detailed description is given by Moseley & Priebe (1993).

The densification effect reduces with increasing radius from the vibrator (Greenwood & Kirsch, 1984), such that area treatment is attained by suitable grids to achieve overlapping compaction from adjacent vibrator locations. The degree of densification is dictated primarily by the total energy input, i.e. compaction centres, vibrator power, and time spent at each horizon, as well as the particular sand characteristics and experience of the site operatives. The resultant profile of reasonably uniform relatively density, i.e. increasing SPT/CPT values with depth, is well suited to seismic protection.

It was found in the late 1930s that the presence of significant fines severely inhibited the radial densification effect. Experience has shown that sands of total fines content of up to 15% (grains finer than 0·06 mm) and/or clay and fine silt content (particles smaller than 5 μm) of less than 2% will respond to treatment without the addition of stone. However, modified construction technique with higher-powered vibrators has proved to be capable of densifying soils of up to 45% total fines and 5% clay content, by the addition of stone to assist transfer of

energy and provide enhanced drainage, as illustrated in the case histories.

DENSIFICATION MECHANISM

Baez & Martin (1992) have described densification mechanisms during the installation of stone colums (in sands at Monterey, California) using the vibro replacement technique. The loose to medium dense saturated silty sands were susceptible to liquefaction during a 0·2 g seismic event. Treatment was performed to provide bearing pressure and settlement performance as well as to prevent liquefaction. Fig. 2 illustrates the typical pore water pressure and vibration time histories together with the respective depth location of the vibrator throughout construction. This clearly shows the increase in both vibration and excess pore water pressures for each re-penetration of the vibrator during construction of the stone columns. The instrumentation was located at lateral distances of 0·9, 1·5 and 2·1 m and depth of 4·3 m while treating to 5·5 m depth. The vibrator used was a Keller bottom-feed S-type of 200 kN centrifugal force and 125 kW motor operating at 30 Hz.

Vertical and horizontal geophones recorded steady-state particle accelerations of 1·7 g and 0·6 g at the same 4·3 m depth at 0·9 and 2·1 m lateral distances. The vertical vibrations also became higher than the horizontal during the re-penetration of the stone column for its construction in a series of short lifts. The condition of liquefaction (taken as the ratio of excess pore pressure to effective overburden pressure of 1·0) was observed

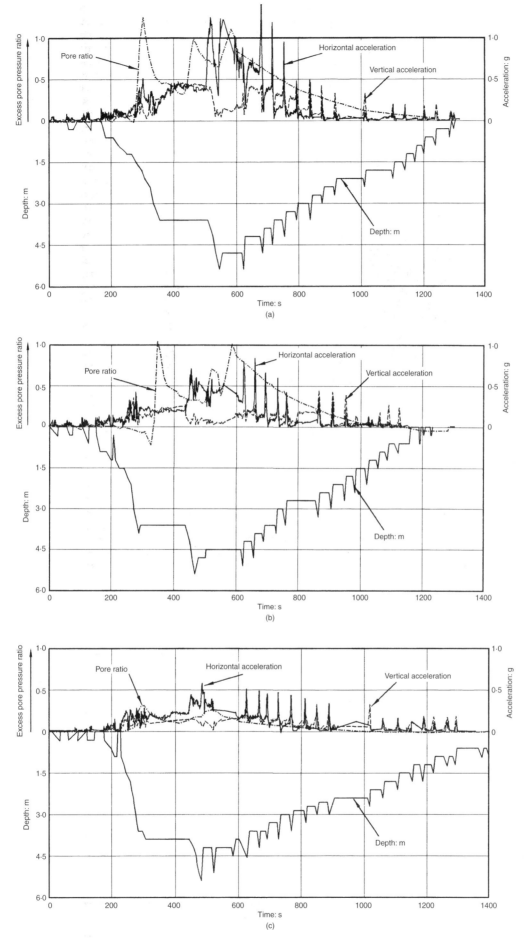

Fig. 2. Time histories of vibration, excess pore water pressure ratio and vibrator depth at (a) 0·9 m, (b) 1·5 m and (c) 2·1 m from stone column

at the distances of 0·9 and 1·5 m for periods of about 65 and 30 s respectively, while the excess pore pressure ratio at 2·1 m was 0·45. On completion of each stone column installation the excess ratio had dissipated to less than 0·1 at each distance. This suggests that the stone column had already begun to perform as a drain.

Post-treatment CPT testing was performed one day, as well as one year, after densification at the same lateral distances of 0·9, 1·5 and 2·1 m to correlate to the instrumentation. These have been plotted in Fig. 3(a) as being located at the centroid of equivalent triangular grids of 1·61, 2·65 and 3·67 m respectively. The average one-day values recorded respective increases of 60%, 40% and 4% with further improvements with time closer to a single column (see Fig. 3(a)). These results have been both averaged and normalized to compare with a pre-treatment CPT value of 10 MPa for ease of assessment.

The contract proceeded with CPT tests at the centroids of 1·8, 2·4 and 3·0 m triangular grid spacings. These recorded one-day increases of 190%, 150% and 100% at the respective spacings. It was not possible to test the groups one year later as the intended building facility was already constructed on top of the columns. The trials clearly demonstrated that mechanisms of densification included controlled vibrator-induced liquefaction in surrounding soils, as well as the benefits of enhanced drainage and group effects. The typical grading of the silty sands and added stone are illustrated in Fig. 3(b).

Even when sands are considered to be well suited to densification, it can sometimes be more economic to add stone to accelerate productivity and enhance results. For example, vibro treatment at Miami Airport beneath 10 structures did not utilize stone. For a further structure, a six-storey parking garage, the alternative of 50–60% of the usual number of compactions but

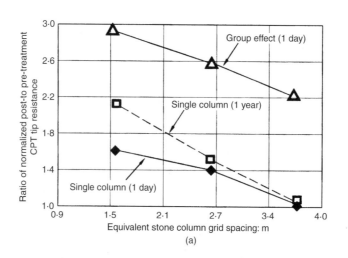

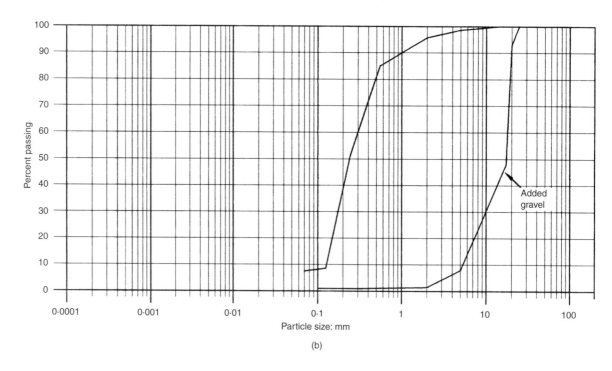

Fig. 3. (a) Effect of time on the penetration resistance (single stone column) and group effect; (b) particle size distribution

with added stone was adopted. The post-treatment SPT results in every case exceeded the specified minimum value by a factor of at least 4. Even with the extra cost of importing stone, the cost of this composite scheme was about 75% of the conventional approach (unpublished internal report).

TESTING THE TREATED GROUND

There are various methods for specifying the requirements of the post-treatment characteristics of the soils. These can range from a safe bearing capacity with associated settlement criteria, through relative density profiles or seismic protection with appropriate factor of safety (usually within the range 1·05–1·5), to limitations on dynamic settlement arising from the design earthquake. It therefore follows that the post-treatment test method should be appropriate to the specified criteria.

The most common methods are in-situ testing by standard penetration test, electric cone penetration test, relating both to conventional geotechnical theory, and/or by load tests. The SPT test is anything but standard, with differences in equipment and performance around the world. It is also heavily dependent upon the care exercised by the driller in controlling the drop height, maintenance of water level in the borehole, and location of the test with respect to the drill casing. Nevertheless, its simplicity in execution and large amount of experience already achieved ensures its continued use.

The CPT test has gained wider use since it is less influenced by the operator and can provide greater detail for thin layers of trapped clay or silt when using a piezocone. However, it can be misled by the presence of thin clay layers, particularly in loose soils, where the CPT tends to 'sense' the clay layer in front of the cone while the friction sleeve is still within the granular soil. Corrections to both SPT and CPT should be applied for fines content, particularly for detailed seismic analysis.

Schmertmann (1978) reported that using the conventional equivalent Young's modulus of $E = 2\cdot5\ q_c$ for square foundations would overpredict the settlement potential of vibro-densified sands. This is considered by the authors to be due to the generation of high lateral earth pressures during vibro compaction, which reduce lateral movement and thus settlements.

Schmertmann recommended for vibro-compacted sands that moduli of about 200% of those suggested by the conventional $E = 2\cdot5\ q_c$ correlations with CPT results be adopted. Some confirmation of this view was obtained at a large project reported by Bell et al. (1986), in which subsequent large-scale loading was carefully monitored for settlements. However, this multiplier cannot be relied upon in all circumstances because of the uncertainty in estimating the over-consolidation ratio for a sand, as illustrated below. Research using calibration chamber data has suggested a wide range for this multiplier: see section 5·5·3 of Lunne et al. (1997).

When time and finance permit it is always useful to correlate the in-situ test results with load tests, particularly when stone columns are constructed. Full-scale foundation loading tests to 250% of working load are sometimes performed but are relatively slow and expensive. As a result there is increasing use of shear wave velocity, geophysical and dynamic loading tests. These methods are valid for low-strain elastic moduli but do not provide the important consolidation characteristics of clayey soils.

SEISMIC CONSIDERATIONS

Seed (1996) states that there is 'recognition of an increasingly broad variety of soil types potentially vulnerable to liquefaction. Liquefaction is no longer an issue only for sandy soils, though sandy soils continue to represent a majority of observed field cases of seismically induced liquefaction'. Mitchell et al. (1995) state that 'there is clear evidence that ground improvement can provide significant protection against earthquake-induced liquefaction and ground failure and that settlements and lateral displacements of treated ground can be reduced to tolerable levels.' Seismic aspects and design methodologies for the mitigation of soil liquefaction, including performance after strong earthquakes, have been discussed in detail by Baez (1995).

Vibro compaction was performed to densify sands beneath the Aswan Dam in Egypt in the 1950s. This project was later positively assessed for resistance against liquefaction. As such it was the first project to receive effective seismic protection. The first in the USA was a waste water treatment works in Santa Barbara in California in 1972, which was subjected to a 0·3 g seismic event in 1978. This structure was being monitored for settlement, and the earthquake had no effect. The first in Britain was a gas terminal at Barrow-in-Furness in 1981. Six vibro-treated sites in California were subjected to the 1989 Loma Prieta earthquake without distress, while adjacent untreated areas liquefied (Mitchell & Wentz, 1991). Three structures on stone columns were subjected to the 1994 Northridge event without ground distress or sign of liquefaction (Hayden & Baez, 1994). Vibro projects in Australia and Greece have also been subjected to significant earthquakes, again without any distress.

The prime factor in resisting liquefaction during the design event is densification. However, a significant bonus when adding stone to assist the densification operation is provided by the resultant higher-permeability column acting as a relative drain to dissipate excess pore pressures. This mechanism of reducing the rate and magnitude of pore water pressure increase during an earthquake is described by Seed & Booker (1976). However, their theory assumed that perfectly drained boundaries are available within the matrix of the stone column, and this may not always be the case. Drainage designs are evaluated in terms of resulting excess pore pressure ratios, which should be limited to less than 0·1–0·4 depending on the final density of the soil in question (Yoshimi & Tokimatsu, 1991). Furthermore, while drainage may mitigate the occurrence of liquefaction, in the absence of densification it can still yield significant vertical deformations.

Ultimately, the engineer is concerned not only with liquefaction but also with its consequences for the structure in question. This is usually evaluated in terms of deformation (settlement or lateral spread), or loss of shear strength (bearing capacity, or slope failure). Although the aim of the vibro stone columns is to achieve a non-liquefiable condition of all cohesionless soils, the presence of the stone column and its ability to carry loads and reinforce the soil may permit pockets of unimproved material to remain after treatment, while not compromising the integrity or ultimate performance of the structure.

As described earlier, the introduction of the vibrator into the ground generates excess pore water pressures. These dissipate rapidly in free-draining low-fines sands. Higher fines contents do not allow such rapid dissipation, and the formation of stone columns therefore has the further benefit of accelerating pore water dissipation by providing a short drainage path during construction. There have also been projects where overall permeabilities have been increased and accelerated dissipation has been introduced by first installing prefabricated wick drains prior to vibro stone columns. The wick drains may permit a drained loading behaviour, thus achieving volume change or densification of interbedded silts and sands (Luehring et al., 1998).

CASE HISTORIES

Vibro compaction has always been an 'observational method', with continual monitoring of power input, induced settlement, consumed quantities of sand or stone and visual response during densification combined with post-treatment testing. The following examples, which all utilize the 120 kW, 200 kN force Keller S-type vibrator with water-flushing unless stated, have been chosen to demonstrate the performance of some difficult sites, or to illustrate a particular aspect of the technique. Typical grading analyses of the sands for each site are given in Fig. 4. Papers by Raison et al. (2000) and Slocombe et al. (1995) provide other useful case histories.

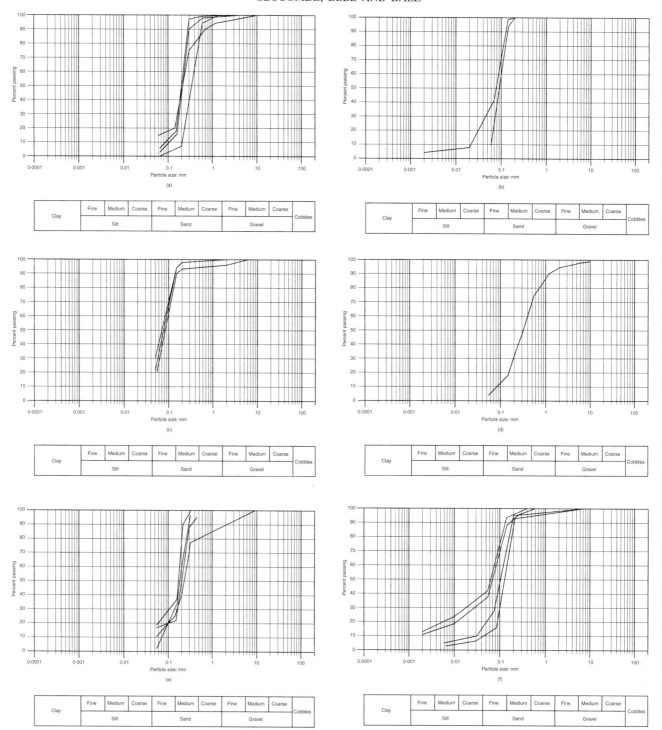

Fig. 4. Particle size distributions for case histories: (a) Great Yarmouth; (b) East Anglia; (c) Swanscombe; (d) Sizewell B; (e) Hartlepool; (f) Trinidad

Great Yarmouth

A new gas-fired power station was to be built on typically about 5·0 m depth of loose to medium dense wind-blown sands underlain by apparently competent sands. The main turbine house included relatively heavy and dynamic loadings. These sands, while being relatively low in fines content, had a local reputation for being difficult to compact by conventional roller methods. Analysis of the soils suggested that a target relative density of 75% using the Jamiolkowski *et al.* (1985) approach would achieve the desired total and differential settlement performance. However, part of the contract included the performance of more intensive pre-treatment CPT investigation, which revealed that the sands below the proposed treatment depth were not as competent as expected. Initial vibro compaction trials without the addition of stone had confirmed that a 2·4 m

triangular grid could compensate for the weaker deeper soils, as illustrated in Fig. 5.

East Anglia

Successful advance trials were performed to confirm the post-treatment settlement potential of a site underlain by about 6·0 m depth of loose to medium dense silty to very silty sands for a power station development. Groundwater occurred at about 2·5 m depth, but water flushing was prohibited owing to slight contamination of the granular fill and the proximity of the site to a river. The trials therefore used an 80 kN force 50 Hz bottom-feed vibrator with gravel backfill in view of the presence of thin clay layers. Trials were performed at 1·5 and 2·1 m grids.

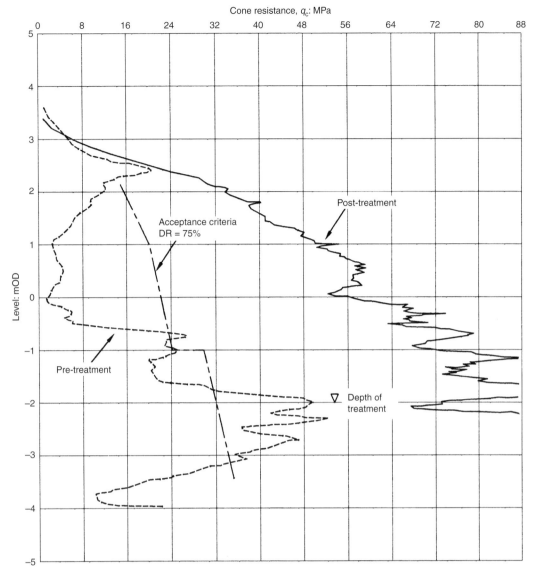

Fig. 5. Comparison of pre- and post-treatment CPT results at Great Yarmouth

The sands were relatively single-sized with total fines content of up to 30% and clay content of 5–10%. As such, only limited densification, even at close grid centres, was anticipated. Even so, the pre-treatment relative density of about 40–45% was increased to about 55–60% at the centroid of the closer grid of almost 900 mm diameter stone columns, increasing to 60–65% at about 0·5 m from the edge of the stone columns. The expected composite structure of stone column surrounded by a ring of relatively dense sand surrounded by less densely compacted sands was therefore confirmed by the CPT testing.

Zone loading tests on a flexible base 3·6 m × 3·63 m were performed to test loading of up to 390 t, or 300 kN/m² imposed pressure. Maximum settlements of 40·5 and 24·5 mm were recorded for areas treated at wider and closer grids respectively. These results implied, by analysis of the multiplier of the respective averaged q_c values to obtain equivalent modulus and the numbers of stone columns beneath the test bases, that the closer grid had imparted a degree of over-consolidation/increase in lateral earth pressure, while the wider grid was too wide for this effect to be significant.

Two sets of continuous surface wave system tests (Matthews et al., 1996) were also performed on untreated and the closer grid of treated ground. The aim was to consider the effect of the inclusion of the stone columns. A comparison between CSWS shear modulus of the untreated/treated sands suggested a 200–300% increase. However, this would have underestimated the zone test settlements as the dynamic modulus was low-

strain. There is, however, considered to be significant potential in these and other dynamic methods for assessing the settlement potential of sands subjected to vibro compaction. The CSWS test results together with a Fugro seismic cone test on the untreated sands are given in Fig. 6.

Swanscombe, Kent

A backfilled quarry was to be reclaimed for a large shopping development. The 13 m depth of fill comprised loose becoming very loose with depth Thanet Sand material that was the original overburden to chalk. This sand was single-sized, very fine, with total fines contents of 25–35% and CPT friction ratio 2%, locally 3·5% where more clayey. A future rising water table would have triggered an estimated 7·5% collapse settlement.

Advance trials were therefore performed by vibro compaction, with and without stone, and dynamic compaction with post-treatment testing by CPT, dilatometer and zone loading tests. It was not expected that vibro compaction without stone would be capable of achieving significant densification. However, in attempting to use stone to enhance the densification, the quantities of added material became far in excess of what was considered to be economic. The dynamic compaction using 15 t weight results were comparable to 5·0 m depth to vibro compaction with stone, the CPT values recording an enhancement of about 500% on the pre-treatment results (see Fig. 7).

It had been expected that a close grid of vibro compaction

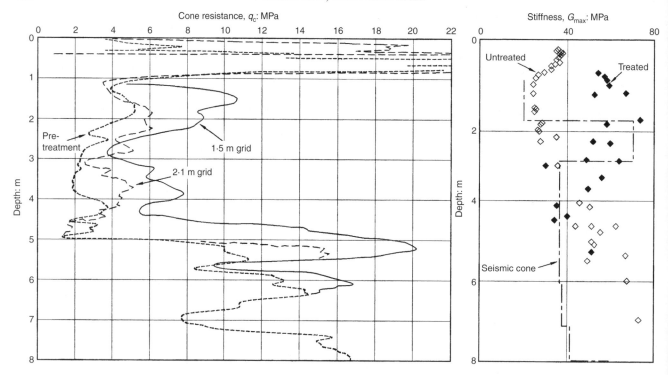

Fig. 6. Comparison of pre- and post-treatment CPT results and CSWS and seismic cone results at East Anglia trial

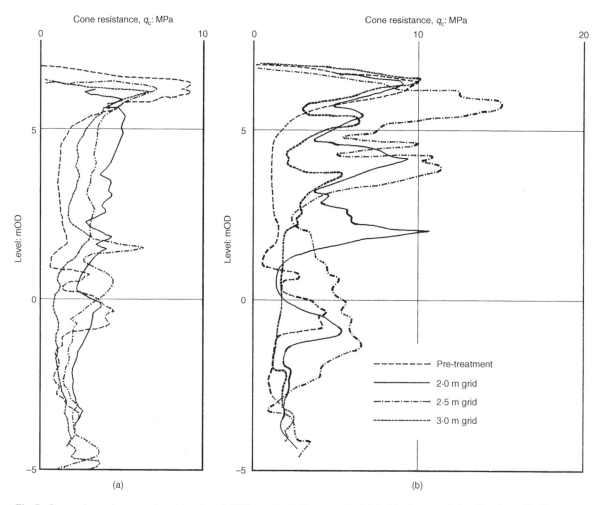

(a)

(b)

Fig. 7. Comparison of pre- and post-treatment CPT results at Swanscombe trial: (a) vibro sand densification; (b) vibro stone columns

stone columns should have been able to provide reasonable settlement performance for the anticipated building loads. However, the zone loading tests delivered a settlement performance equivalent to only $E = 2.5\ q_c$, even though the dilatometer tests suggested lateral earth pressure values of up to 3 to suggest the sands were being densified adequately. It is considered that the prime reason for this poorer than expected performance was the very low pre-treatment density of the Thanet Sand. Closer than normal compaction grid centres would therefore be required. It was concluded that the Thanet Sand fill was not untreatable, but that the cost of vibro treatment would be too high. The development proceeded on the basis of limited dynamic compaction plus driven piles to all major structures.

Sizewell B

Britain is usually considered to be a country of relatively low seismicity. However, the Sizewell B power station contract required advance treatment of two distinct areas, one being clean sands of low density and the other of interbedded sands and clays, to provide adequate factor of safety against potential liquefaction. Planning restrictions, however, did not permit the importation of stone, so the sand/clay area was treated at a closer grid of 2·4 m, the clean sand area at 2·8 m, and on-site sand backfill.

Figure 8 illustrates the comparison between post-treatment test results within the respective areas. The tests within the area of clean sands recorded consistent densification to well in excess of the specified requirements. In contrast, the tests in the interbedded clay and sand area occasionally recorded less than the specified values in sands immediately beneath the clay layers, but with even higher q_c values in zones of greater than about 0·5 m thickness of clean sands.

Hartlepool

Two contracts have also been performed within existing nuclear power plants at Hartlepool and Heysham. The new

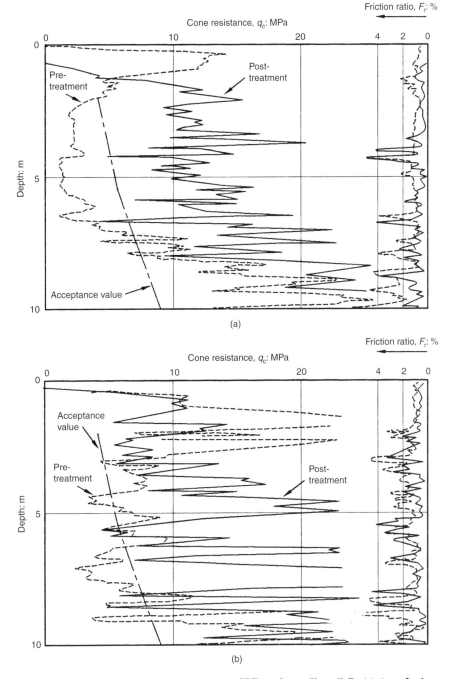

Fig. 8. Comparison of pre- and post-treatment CPT results at Sizewell B: (a) Area 2, clean sands; (b) Area 1, interbedded clays and sands

structures were to be located relatively near to the main reactor buildings. The vibration levels from the S-type vibrator could have triggered the reactor safety mechanisms, so the 80 kW dry bottom-feed vibrator plus stone was used on both projects.

Figure 9 illustrates (note the change in scale above $q_c = 10$ MPa) the comparison between pre- and post-treatment results for sands of relatively poor suitability for densification, and the attempt to penetrate a CPT down a stone column. These test results clearly revealed that the use of lower-powered vibrators at closer centres could still achieve the specified densification requirements, with the added bonus of very dense stone columns to further reduce the settlement potential.

Trinidad

A new LNG plant was to be developed on up to 12 m depth of hydraulically reclaimed ground. The process area, of 70 kN/m^2 average loading, was treated by vibro compaction with the addition of stone to provide adequate bearing capacity, settlement performance to not exceed 50 mm long-term and mitigation of liquefaction of soils during the design seismic event.

Detailed analysis of the pre-treatment borehole and CPT results revealed about 50% of the fill volume to be of an often very silty sand and 50% to be predominantly cohesive. Surcharge by 4·0 m stockpiles of stone was performed to limit post-treatment settlements arising from the cohesive deposits to those specified. The ability of the vibro stone columns to act as drains and accelerate consolidation was employed. Advance trials confirmed the design rate of settlement such that the average surcharge duration was reduced to 4–6 weeks from commencement of building each 25 m square stockpile area.

The sands often contained total fines contents of up to 45%,

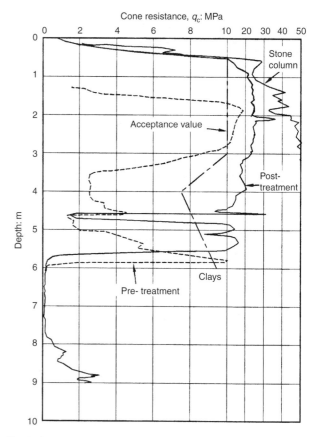

Fig. 9. Comparison of pre- and post-treatment CPT results at Hartlepool

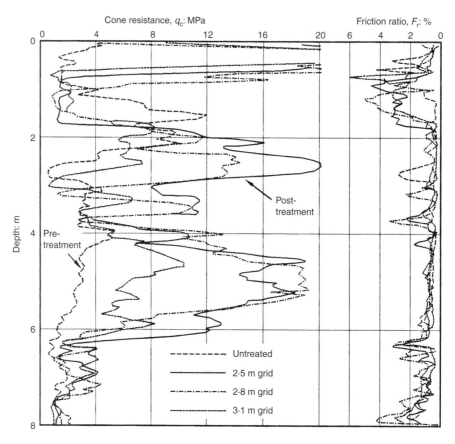

Fig. 10. Comparison of pre- and post-treatment CPT results at Trinidad

Table 1. Trinidad analysis

Classification	1–5	6	7	8	9	10
Total fines: %	>45	35–45	25–35	15–25	5–15	<5
Clay content: %	>12	9–12	6–9	3–6	0–3	0
% of soils	36	26	19	12	6	1
% pass by CPT	100	60	65	75	97	100
% pass by SPT/Lab	100	94	82	83	100	100

with clay contents of up to 15%, and were normally relatively loose and susceptible to liquefaction. A sliding scale of acceptance SPT N1(60) values of 23 at 15% fines reducing to 10 at 35% fines and corresponding CPT values was developed and agreed as a suitable means to ensure that potential liquefaction could be designed out of the scheme for the wide range of soils. Fig. 10 illustrates the comparison between pre-treatment and post-treatment trials for 2·5, 2·8 and 3·1 m grids in an area of relatively clean sands, with friction ratio less than 1% and total fines of less than about 15%.

Post-treatment testing was performed by an initial 25 m grid of piezocone tests and using the Robertson *et al.* (1986) classification for soil behaviour types. An initial 75 m grid of boreholes with SPT tests at 1·5 m depth intervals was also performed. The majority of these tests met the specified requirements. Those that did not were then subjected to boreholes with continuous SPTs. These were then reassessed on the basis of SPT values and laboratory tests in comparison with the sliding scale.

A number of borehole and piezocone tests were also performed pre- and post-surcharge. These did not reveal any measurable improvement in the success ratios. This confirms that, in seismic terms, it is necessary to subject sands to vibration rather than surcharge in order to achieve densification.

Detailed analysis of the post-treatment results was based on reducing the CPT and SPT information to 100 mm thick layers, classifying each and then appraisal of each layer. A statistical analysis of many thousands of tests considered the soils within six classifications, as illustrated in Table 1. Overall about 95% of all tests met the required sliding scale design criteria at the first attempt in what is believed to be the most demanding soils ever attempted for vibro compaction. Not only were the sands of relatively high silt and clay content, but they often occurred in relatively thin layers trapped between clays, which would further dampen any densification effect.

CONCLUSIONS

The degree of densification resulting from deep vibratory treatment is considered to be a function of many factors, including soil type, silt and clay content, uniformity of soil grading, soil plasticity, pre-treatment density, vibrator type and power, grid spacing and type of added material e.g. stone. Sands of single-size grading or higher silt and clay content are more difficult to treat. Sands that occur in relatively thin layers trapped between clays are also more difficult to treat and require particular attention. Carbonate sands tend to crush and require closer grid centres than silica sands in order to achieve the desired degree of densification.

The frontiers of vibro compaction are continually being extended by the development of equipment and experience. The case histories have described a range of factors that can influence the use of the technique. There is considerable evidence to confirm that the addition of stone can enhance the degree of densification. It is considered that the bottom-feed equipment that is sometimes used to add stone may produce results that are superior to the water-flushing technique, but there is insufficient evidence at present to confirm such superiority. However, the bottom-feed technique is vastly more positive in ensuring that the required stone column diameter is being constructed, particularly at depth.

Liquefaction mitigation by vibro stone columns has shown itself to perform well during major earthquake events. The design and methodology incorporates densification as the primary means of increasing liquefaction resistance, but increased drainage and reinforcement also assist in controlling excess pore pressure development and deformations.

REFERENCES

Baez, J. I. (1995). *A design model for the reduction of soil liquefaction by vibro-stone columns.* PhD dissertation, University of Southern California, Los Angeles, CA.

Baez, J. I. & Martin, G. R. (1992). Liquefaction observations during installation of stone columns using the vibro-replacement technique. *Geotech. News Magazine*, September, 41–44.

Bell, A. L., Slocombe, B. C., Nesbitt, A. M. & Finey, J. T. (1986). Vibro compaction densification of a deep hydraulic fill. *Proceedings of the Conference on Marginal and Derelict Land*, Glasgow, 791–797. London: Thomas Telford.

Burland, J. B. & Burbridge, M. C. (1985). Settlement of foundations on sand and gravel, *Proc. Inst. Civ. Engrs, Part 1* **78**, Dec., 1325–1381.

Greenwood, D. A. & Kirsch, K. (1984). Specialist ground treatment by vibratory and dynamic methods. *Proceedings of the Conference on Piling and Ground Treatment*, London, 17–45. London: Thomas Telford.

Hayden, R. F. & Baez, J. L. (1994). State of practice for liquefaction mitigation in North America. *Proc. 4th US–Japan Workshop on Soil Liquefaction: Remedial Treatment of Potentially Liquefiable Soils*, PWRI, Tsukuba City.

Jamiolkowski, M., Ladd, C. C., Germaine, J. T. & Lancelotta, R. (1985). New developments in field and laboratory testing of soils. *Proc. 11th Int. Conf. Soil Mechanics and Foundation Engineering, San Francisco* **1**, 57–153.

Luehring, R., Dewey, R., Mejia, L., Stevens, M. & Baez, J. (1998). Liquefaction mitigation of a silty dam foundation using vibro stone columns and drainage wicks: a test section case history at Salmon Lake Dam. *Proc. 1998 Ann. Conf. Association of State Dam Safety Officials, Nevada*, 719–729.

Lunne, T., Robertson, P. K. & Powell, J. J. M. (1997). *Cone penetration testing in geotechnical practice.* Blackie Academic & Professional, London.

Matthews, M. C., Hope, V. S. & Clayton, C. R. (1996). The use of surface waves in the determination of ground stiffness profiles. *Proc. Inst. Civ. Engrs. Geotech. Engng* **119**, Apr, 84–95.

Mitchell, J. K. & Wentz, F. L. (1991). *Performance of improved ground during the Loma Prieta earthquake*, Report No. UCB/EERC-91/12. Berkeley: Earthquake Engineering Research Center, University of California.

Mitchell, J. K., Baxter, C. D. P. & Munson, T. C. (1995). *Performance of improved ground during earthquakes*, ASCE Geotechnical Special Publication No. 49, pp. 1–36. American Society of Civil Engineers.

Moseley, M. P. & Priebe, H. J. (1993). Vibro techniques. In Ground improvement (ed. M. P. Moseley), 1–19. Blackie Academic and Professional.

Raison, C. A., Slocombe, B. C., Bell, A. L. & Baez, J. I. (1995). North Morecambe Terminal, Barrow, ground stabilisation and pile foundations. *Proc. Third Int. Conf. on Recent Advances in Geotechnical Earthquake Engineering and Soil Dynamics, St Louis, Missouri* **1**, paper 3.02, 187–192.

Robertson, P. K., Campanella, R. G., Gillespie, D. & Greig, J. (1986). Use of piezocone data. *Proc. ASCE Speciality Conf. In Situ '86: Use of In Situ Tests in Geotechnical Engineering, Blacksburg*, 1263–1280.

Schmertmann, J. H. (1978). *Guidelines for cone penetration test, performance and design*, Report FHWA-TS-78-209. Washington, DC: US Federal Highway Administration.

Seed, R. B. (1996). Recent advances in evaluation and mitigation of liquefaction hazard, *Proceedings of the conference on ground stabilisation and seismic mitigation: theory and practice*, Portland, Oregon.

Seed, H. B. & Booker, J. R. (1976). *Stabilization of potentially liquefiable sand deposits using gravel drain systems*, Report No. EERC 76-10. Berkeley: University of California.

Slocombe, B. C., Bell, A. L. & May, R. E. (2000). The in-situ densification of granular infill within two cofferdams for seismic resistance. *Proceedings of the Conference on Compaction of Soils, Granulates and Powders*, Innsbruck, pp. 33–42.

Yoshimi, Y. & Tokimatsu, K. (1991). Ductility criterion for evaluating remedial measures to increase liquefaction resistance of sands. *Soils Found.* **31**, No. 1, 162–168.

Renton-Rose, D. G., Bunce, G. C. & Finlay, D. W. (2000). *Géotechnique* **50**, No. 6, 727–737

Vibro-replacement for industrial plant on reclaimed land, Bahrain

D. G. RENTON-ROSE,* G. C. BUNCE† and D. W. FINLAY*

This paper describes the improvement by means of vibro-replacement of hydraulically placed fill, recently constructed to form a reclaimed platform for a coke calcining plant in Bahrain. The improved fill will directly support some of the principal structures constructed with shallow foundations. The alternative would be to support these structures by means of bored piles extending into the bedrock, some 15–20 m below. The initial density/stiffness properties of the as-placed calcareous sand fill are presented, together with observed improvement with time after placement, prior to treatment. The pre-tender design method is described, from which confidence to proceed with the approach was derived. Relevant construction details are presented and, in particular, the validation testing regime, which included pre- and post-construction cone penetration tests and five large zone load tests (up to 11·5 m² loaded area). The improvement achieved by the treatment works is compared with the initial improvement predictions as well as with recalculated predictions using the tender design approach, but employing observed stone column and treated fill properties. This review indicates that the design approach adopted was acceptable, slightly underestimating the improvement achieved by a factor of around 1·3.

KEYWORDS: bearing capacity; case history; footings/foundations; ground improvement; settlement; stiffness

Cette étude décrit l'amélioration d'un sol par replacement vibratoire d'un remblayage hydraulique, récemment construit pour regagner sur la mer une plate-forme destinée à une usine d'embouteillage de Coke à Bahrain. Le remblai amélioré supportera directement certaines des structures principales à fondations peu profondes. L'autre solution consisterait à soutenir ces structures au moyen de piles forées, se prolongeant dans la roche de fond, entre 15 et 20 mètres en dessous. Nous présentons les propriétés initiales de densité/rigidité du remblai de sable calcaire tel qu'il est placé ainsi que l'amélioration observée dans le temps après la mise en place et avant le traitement. Nous décrivons la méthode conceptuelle de pré-soumission, d'où nous déduisons que cette approche est fiable. Nous présentons des détails de construction pertinents et, en particulier, le régime d'essais de validation qui comportait des essais de pénétration au cône avant et après construction et cinq essais de charge sur grandes zones (la surface chargée ayant jusqu'à 11·5 m²). Nous comparons l'amélioration obtenue par les travaux de traitement aux prévisions initiales et aux prévisions recalculées utilisant la méthode conceptuelle de présoumission, mais en employant les propriétés observées des colonnes de pierre et du remblayage traité. Cette étude indique que la méthode conceptuelle adoptée est acceptable, sous-estimant légèrement, par un facteur d'environ 1·3, l'amélioration obtenue.

INTRODUCTION

A major coke calcining plant is under construction on reclaimed land in Bahrain, for Aluminium Bahrain BSC (ALBA), with a capital cost of around US$500 million. Hyder Consulting Limited (Hyder) was appointed by ALBA in 1998 to review piled foundation proposals, with a view to achieving more economical designs. From this review, an alternative approach was proposed for some structures, comprising shallow foundations bearing onto fill materials improved by vibro-replacement techniques.

In order to develop outline designs for tender purposes, it was necessary to determine design parameters prior to the construction of the reclaimed platform. As such, designs relied heavily on the reclamation works specification and on published data, for similar works. This necessitated extensive large-scale pre-works trials to be written into the foundation contract, for verification of performance of the foundation contractor's final designs.

This paper describes the ground treatment design, and discusses relevant construction issues. The suitability of the construction approach, as demonstrated by validation testing, is presented. The performance of the improved ground, as indicated by cone penetration testing (CPT) coupled with zone load test data, is compared with that predicted by a recognized design approach. Finally, the appropriateness of the design basis is discussed, by considering the level of conservatism of the performance predictions for treatment in the conditions encountered.

FOUNDING CONDITIONS

The site for the coke calcining plant comprises a newly reclaimed platform, just offshore from Sitrah (Fig. 1).

The reclamation fill was locally won sea-dredged calcareous sand and gravel, comprising typically fine to coarse sand and fine to medium gravel, which included a significant shell content (5–40%, of shells of medium gravel size and larger). However, intact shells tended to be encrusted or filled with finer particles weakly cemented to them, rendering them more like a 'gravel'. The specified fines content (< 0·074 mm) of < 10% was generally achieved, with typically 1–4% recorded. Locally, however, silt pockets were encountered at the base of the fill. These are thought to represent trapped silt that had been transported in suspension away from a point placement and deposited, prior to rapid coverage during further hydraulic filling.

A grading envelope is given in Fig. 2. Any modification to this grading envelope by the breakdown of the shell 'gravel' during the improvement process was not investigated. However, it is likely that some of the shells would have been broken during ground treatment, locally to the injection point, with a loss of some of the finer materials cemented to them. The impact of this would have been to reduce the medium gravel content, with a corresponding rise in the proportion of sand and fine gravel.

Typically, 6 m of fill was placed by hydraulic techniques within a bunded area, to produce a platform at +3·8 m chart datum (CD), around 1·5 m above high water level of +2·3 m CD. Dredged fill was placed by means of a large-diameter pipe discharging into the reclamation area from a height of around 1–2 m above sea level. Filling was generally carried out working from the seaward end, to control the flow of discharge water/suspended fines. Water discharging from the reclamation area was required to pass through a silt trap and enter open water with no more than 2% suspended fines.

The as-placed fill was dense to very dense within and above the intertidal zone; below this level it became loose. The under-

Manuscript received 4 May 2000; revised manuscript accepted 25 July 2000.
* WSP Environmental Limited formerly Hyder Consulting Limited.
† Hyder Consulting Limited.

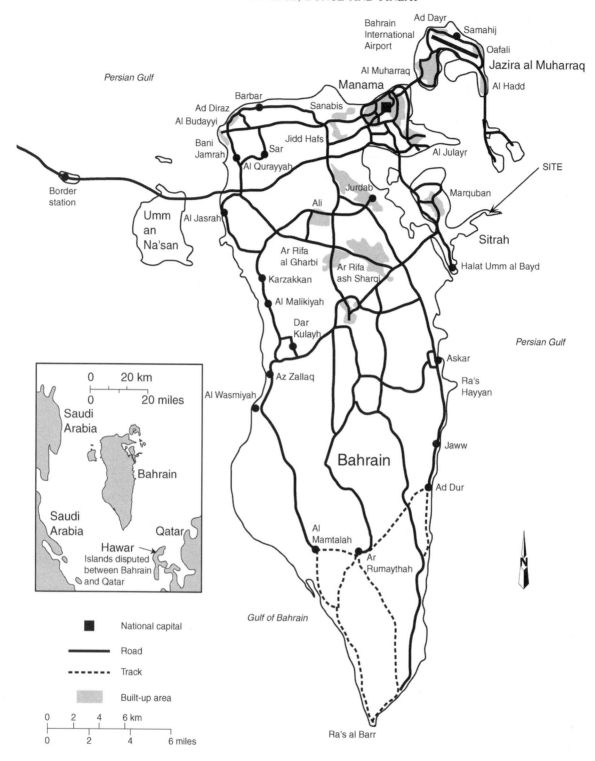

Fig. 1. Map of Bahrain

lying marine deposits comprised around 2 m of loose to medium dense calcareous silty sand, over a caprock layer (generally
< 1 m thick), above medium dense silty calcareous sands. The
solid formations below consisted of very weak to weak calcisil-
tite over calcilutite. The typical material profile is presented in
Fig. 3.

FOUNDATION ASSESSMENT AND DESIGN
Many of the structures will apply considerable founding loads
and/or are settlement-sensitive. These structures will be supported by bored cast in place piles, socketed into bedrock.

However, a limited number of structures, although supporting
significant loads, were deemed to be relatively insensitive to
post-construction settlement. For these structures, shallow foundations on improved ground were assessed to be viable. The
design and performance of treatment works for the green coke
storage building (GCSB) are the subject of this paper. The
GCSB comprises a partitioned structure measuring 240 m by
70 m (approximately) within which 80 000 t of coke will be
stored.

Owing to the restricted contract programme, outline design
was required prior to the reclamation works. The likely engineering properties of the reclamation fill were predicted from

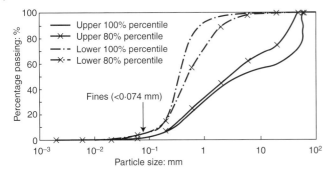

Fig. 2. Reclamation fill grading envelope

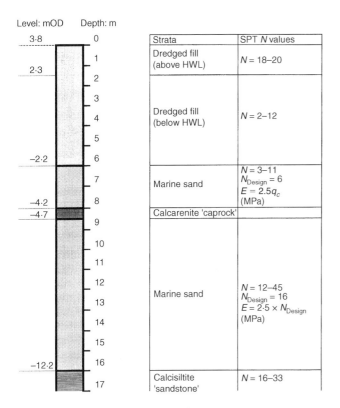

Fig. 3. ALBA site: typical strata profile

the specification requirements relating to grain size and grading, and using published data for similar materials elsewhere. It is notable that much of the published information reviewed related to fills where the fines content was at or above the limit of 10% specified for the ALBA site. Based on this information, therefore, higher q_c values might be expected for the ALBA site. Information from the recently constructed Hidd Power Plant was also considered (provided by Keller, Middle East), where the fill materials used were obtained from the same borrow area as for the ALBA site, with reported fines contents of <5%. From this review, a pre-treatment CPT tip resistance (q_c) of 2·5 MPa for materials below the intertidal zone was selected for the design of improvement measures. Above this level, a sharp increase in penetration resistance was generally reported, frequently with q_c values of 5–8 MPa.

During the tendering period for the piling/ground treatment contract, quality control CPT data for the reclamation fill material became available. The results of this testing, typically carried out within one month of placement and sometimes within a week, are summarized in Fig. 4. Also shown are the CPT results just before ground treatment some 7–8 months after fill placement. These data indicate that the initial q_c assessment was somewhat conservative, with long-term values in fill below the intertidal zone of around 4 MPa. This is

probably as a result of the very low fines content recorded for the reclamation fill.

From reference to Fig. 4, it is apparent that a time-dependent improvement of penetration resistance has been demonstrated. CPT q_c values recorded 7–8 months after placement have typically increased by between 1·5 and 2 times compared with values recorded a few weeks after placement. It is likely that the majority of this improvement takes place shortly after placement, as the permeable fill drains and consolidates.

The following relationships for estimation of soil stiffness were adopted for the outline design of improvement works:

(a) E (MPa) $= 2\cdot5q_c$ in fill/marine sand (Beckwith & Hansen, 1982; Meigh, 1987).
(b) E (MPa) $=$ SPT N in marine sand (Stroud, 1989; Clayton, 1993) for normally consolidated sands.
(c) E (MPa) $= 2\cdot5$ SPT N sand below caprock (Stroud, 1989; Clayton, 1993), for lightly overconsolidated sands at small shear strain (say 0·25%).

The predicted stiffness profile from which required improvement factors for the hydraulic fill and underlying silty marine sand were determined is indicated in Fig. 3. Improvement factors were derived from the ratio of predicted foundation settlements arising from (a) unimproved and (b) improved materials. Settlement criteria agreed with ALBA's process contractor for subsequent construction works were used to determine the required level of improvement for the different foundation areas.

Outline design for ground treatment was performed using the method given by Priebe (1995), which allows for consideration of the densification of fill between columns, after treatment, coupled with the improvement due to the stone columns themselves. Priebe's approach is empirical, developed from the back-analysis of many case histories, within a range of soil types. Nominally 1 m diameter stone columns were selected for design, based on discussions with Middle Eastern contractors for the anticipated types of plant available in the region. These were assumed to have a constrained modulus of 120 MPa, with an internal angle of friction of 40° from reference to published information and test data provided by Keller (Bahrain) for similar projects.

The initial step in deriving the enhancement of the ground's deformation modulus due to the stone columns themselves requires the derivation of a basic improvement factor (n_0), for an infinitely loaded area. From reference to the ratio A/A_c and the friction angle of the stone columns, a value of n_0 is obtained, assuming the stone columns are incompressible ($A =$ effective area improved by each column; $A_c =$ cross-sectional area of the column). To allow for the actual compressibility of the stone columns, a reduced improvement factor (n_1) is obtained from consideration of the ratio D_c/D_s, where $D_c =$ constrained modulus of stone column and $D_s =$ constrained modulus of soil. A third step is then applied to allow for the increasing radial constraint to the stone column with depth, which yields a final improvement factor n_2.

If the design process considers the estimated ground modulus prior to treatment, the assessed improvement will be due to the contribution of the stone columns themselves. In granular soils, the installation process normally results in densification of the ground between columns. To take account of this densification in Priebe's approach, a factor of 2 was applied to the assessed soil modulus for untreated hydraulic fill. This was considered to be realistic based on published case history data for similar conditions. Using this enhanced fill modulus for the entire area of soil between the columns an overall improvement value (n_2) was obtained for the treated ground.

Predicted settlement of the improved fill, based on Priebe's infinite analysis approach, was coupled with the calculated settlement in the in-situ material below, derived from a linear elastic analysis using the VDISP computer program (Oasys, 1997). VDISP assesses the settlement response of a loaded flexible foundation bearing on linear elastic materials. Corrections were made to these predictions to allow for the size of the

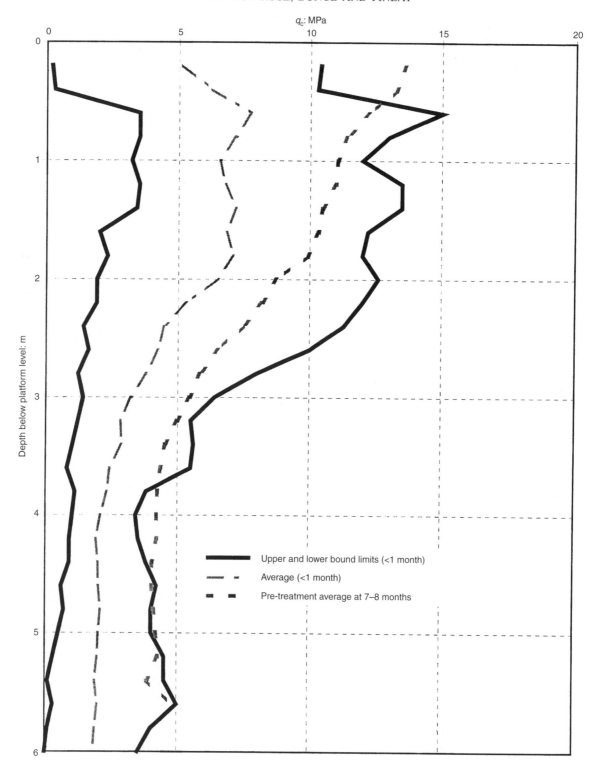

Fig. 4. Reclamation CPT tip resistance (q_c profile) (<1 month after filling)

foundations (i.e not infinitely wide) and for the fact that, owing to their thickness, they will act as virtually rigid footings.

The settlement of the improved ground predicted from Priebe's approach was compared with predictions using the method of proportioning column and soil stiffness after Greenwood & Kirsch (1983). In addition, immediate settlement predictions using Priebe's approach, considering the effects of densification only, were compared with estimates derived from the methods of Burland & Burbridge (1985) and Schmertmann et al. (1978), based on the q_c–SPT (N) correlations above. These comparisons showed that Priebe's approach predicted larger settlements, possibly by a factor of around 1·5. Priebe's approach was therefore considered a suitably cautious approach to follow for

tender design purposes, with a comprehensive testing regime required to verify the designs proposed under the foundation contract. This testing regime comprised:

(a) pre- and post-construction CPTs
(b) zone load tests
(c) plate bearing tests.

Table 1 gives the limiting settlements for the various foundation pressures under the GCSB, based on the outline design approach described above, together with improvement values for treated ground predicted to be required.

Table 1. Pre-tender (outline) design of column arrangement and performance

Location	Footing width: m	Gross foundation pressure: kPa	Assumed triangular grid spacing: m	Assumed improvement due to densification	Improvement due to stone columns	Overall improvement	Settlement above caprock: mm
Side wall	8	110	2·15	2 (fill) 1·25 (marine sand)	1·9	3·5	15
Outer divider wall	19	140	2	2 (fill) 1·25 (marine sand)	2·4	4·4	19
Central divider wall	19	200	2	2 (fill) 1·25 (marine sand)	2·4	4·4	27·5

CONSTRUCTION

Stone columns were constructed at two locations on the site, with around 3750 columns at the GCSB and around 550 columns at the potable water tank. All these columns were constructed to caprock, which was generally at a depth of between 7·5 and 8·0 m below platform level. This level was easy to identify, by means of an audible 'rattling' whenever the vibroflot made contact with the caprock layer. The stone columns were constructed in triangular grids with centre-to-centre spacings of between 1·5 and 2·1 m. These spacings corresponded to approximate treated grid areas (A) of 1·94 m² and 3·82 m² respectively. Based on these grid areas, the total area treated on site was approximately 15 000 m².

The vibroflots were fed with stone from 360° excavators using supplies that were generally delivered and tipped adjacent to the area to be treated. The 'wet' placement system was used for this site, with the vibroflot penetrating through the dredged fill and marine sand to caprock level by means of vibration and water jetting. Upon reaching the caprock, the water flow was reduced and stone was introduced around the annulus. From preliminary trials it was agreed that the vibroflot must penetrate the full profile to caprock and then be raised to the surface before repenetrating to caprock and commencing the placement of stone. It was also agreed that the maximum lift of the

vibroflot should be no more than 1–1·5 m between compactions of the stone.

A daily target of 40 columns per flot was required for completion within the contract period. This target was further increased owing to breakdowns. However, productivity rates were regularly in excess of 50 columns for an 11 h day and nearly always more than 40 columns per day per rig, which allowed the programme to be recovered. This rate was significantly in excess of the 25 columns per day per rig identified at tender stage. Two vibroflots were used for the majority of the construction period, completing the 4300 columns in slightly under 2 months.

Automated monitoring data were not available from the rigs, and hence visual inspection was carried out during construction operations. The main criteria recorded, along with reference and location details, were boring and packing durations and pressures, and the quantity of stone per column. The quantities were assessed by counting the number of excavator buckets used, having already determined the volume and weight of stone carried in an excavator bucket from site trials.

Equivalent as-constructed column diameters were cross-checked by comparing the quantity of stone delivered to site with the number and depth of constructed columns. An approximate stone column diameter of 940 mm was estimated from

Fig. 5.

Fig. 6.

the total weight of stone delivered and a bulk density of 1.5 t/m^3, assuming a wastage of 15%. This corresponded well with site observations of excavated columns.

TESTING REGIME

An extensive testing and validation programme was undertaken for the ground treatment works. This comprised pre-treatment and post-treatment CPTs, zone tests, plate bearing tests, and material acceptability testing. Prior to the installation of stone columns, CPTs were performed on a 15 m square grid

in order to record pre-treatment q_c profiles with depth. Over 100 pre-treatment CPTs were carried out at the treated areas.

Five zone tests were carried out on trial stone column installation areas, outside the works, at design column spacings of 1·5, 1·9 and 2·1 m. The maximum test pressures for the zone tests, which were approximately twice the design pressures, ranged from a pressure of 200 kN/m^2 for a spacing of 2·1 m to a pressure of 450 kN/m^2 for a spacing of 1·5 m. The zone tests were carried out using an 11 m^2 steel reinforced bearing plate that transferred the applied load to a triangular grid of three columns. A stiff spreader system was used to distribute the applied loads fairly uniformly to the steel bearing plate. The actual bearing areas for the different spacings were varied by constructing a 100 mm thick layer of blinding concrete beneath the steel plate, to the exact dimensions required. The area of the blinding, which was always less than the steel bearing plate, was required to be nA, where n is the number of columns under the test slab and A is the treated grid area as defined by Priebe (1995). The slab was required to extend at least 0·5 m beyond the edge of the stone columns being tested. Fig. 5 illustrates the calculation of the area and shape of bearing plates required to satisfy the nA requirement.

Load tests were carried out at a depth of 1·5 m below platform level, to simulate actual founding levels. During excavation for the zone tests, as-constructed spacings and diameters of the stone columns were checked. These checks proved the column diameter to be generally acceptable, at over 900 mm, although a limited number of the columns were slightly out of position.

The zone tests were carried out by means of maintained loading, applied in two cycles, in general accordance with the *ICE Specification for Ground Treatment* (1987). The first cycle went up to gross design pressure and the second cycle to the maximum test pressure. Load increments and decrements were maintained until the rate of settlement had reduced to less than 0·5 mm/h. The test pressures along with allowable and actual settlements are shown for the GCSB zone tests (zone tests Nos 2–5) in Table 2. The maximum permitted settlements, with respect to the performance of the structures, were derived using Priebe's approach considering three columns being loaded.

The zone tests indicated very little recovery following the loading cycles. The graph for zone test 3 in Fig. 6 clearly shows this. This phenomenon was expected at design stage and consid-

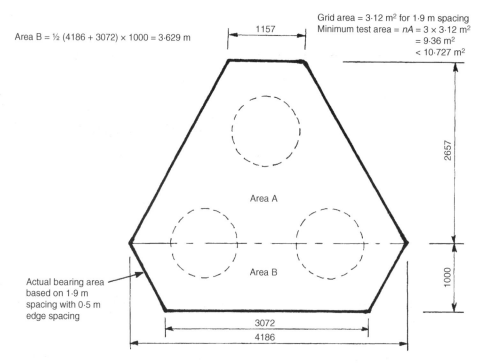

Fig. 7. Calculation of test area for Zone test 5

Table 2. Zone load test performance criteria

Zone test no.	Gross design pressure: kN/m²	Max. test pressure: kN/m²	Design column spacing*: m	Maximum permitted settlement of test at pressures stated†: mm		Observed settlement at pressures stated: mm	
				GDP	MTP	GDP	MTP
2	110	200	2·1	7	15	3·7	9·5
3	250	450	1·5	17	33	5·8	16·4
4	110	200	2·1	7	15	2·2	8·3
5	195	340	1·9	13	25	5·7	13·7

GDP = gross design pressures; MTP = maximum test pressure.
* Column arrangements proposed by contractor.
† Settlements based on pre-tender analysis.

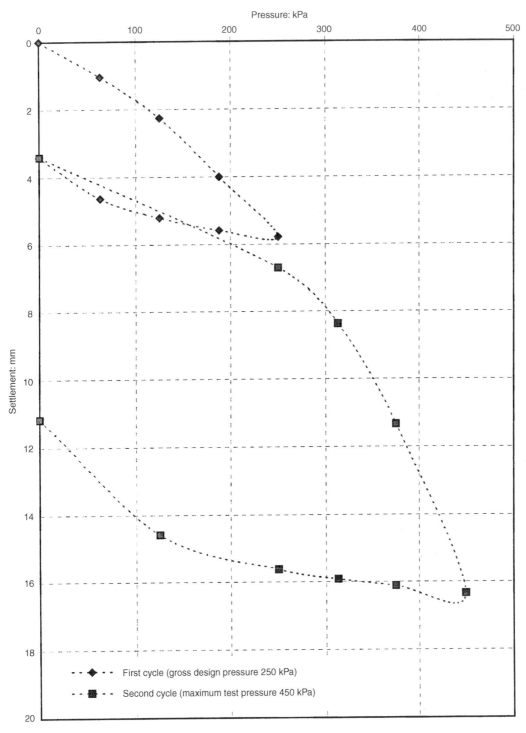

Fig. 8. Zone test 3 pressure–settlement plot

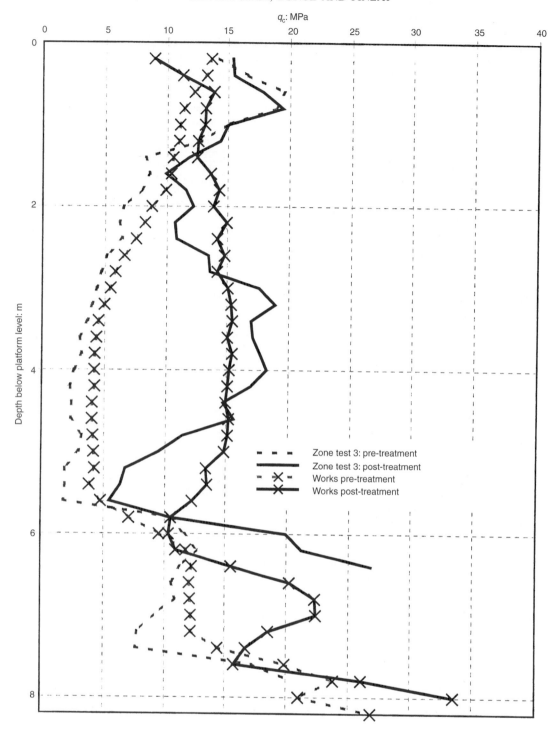

Fig. 9. Pre- and post-treatment reclamation CPT tip resistance for zone test 3 (average q_c profiles)

Table 3. Plate bearing test performance

Maximum test pressure: kPa	Maximum test load: kN	No. of tests	Range of settlements: mm	Average settlement: mm
334	94	8	0·74–2·95	1·2
425	120	1	1·64	1·64
500	141	11	1·53–3·18	2·1

Plate diameter – 600 mm

ered for the GCSB. This building is to be used to store green coke in chambers, separated by divider walls, which will be supported on stone columns. To prevent the potential for differential settlement, it was recommended that adjacent chambers be loaded simultaneously. The average pre- and post-treatment cone

resistance (q_c) values for zone test 3 are illustrated in Fig. 7, along with the average pre- and post-treatment cone resistance (q_c) values for the works columns as a comparison.

The monitoring of the settlement of the zone tests was carried out using an independent reference frame. The settle-

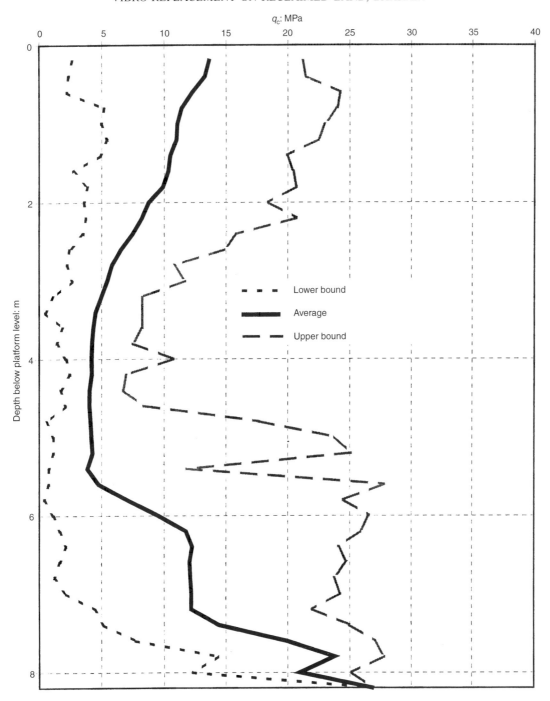

Fig. 10. Pre-treatment reclamation CPT tip resistance (q_c profiles) (85 results)

ment monitoring points were positioned midway between the centre and the edges of the steel bearing plate, as agreed with the contractor.

Following completion of the construction of the stone columns, post-treatment CPTs were carried out at a frequency of one test per 50 m^2 of ground treated. The CPTs were undertaken equidistant from three columns in the triangular grid layout. A total of 246 post-treatment CPTs were carried out.

Further validation testing was also undertaken in the form of plate bearing tests directly on a number of stone columns. The columns tested were chosen on the basis of a number of criteria such as depth, quantity of stone placed, and difficulties with the construction sequence. This included some columns that were 'boxed in' owing to the sequence of construction of surrounding columns, as these typically used less stone in their construction. It was recognized, however, that this was a function of the greater densification of the ground prior to installation, which

would increase constraint to radial displacement of the columns when load tested.

Maximum plate test pressures of 330–500 kN/m^2 were applied to the columns, which corresponded to around 30–50% of the predicted ultimate capacity considered at design stage. An allowable settlement limit of 6 mm was assessed under these pressures, with actual settlements recorded as 0·74–3·18 mm for the range of columns tested. Table 3 shows the range and average settlements for the plate bearing tests for each maximum test pressure.

During construction material acceptability testing of the delivered stone was undertaken. The stone compliance testing included percentage fines, aggregate crushing value, and effective angle of internal friction. The testing frequency was one set of tests per 250 columns. Monitoring of the stone at the quarry source was also found to be required by the contractor, to ensure compliance with the specification.

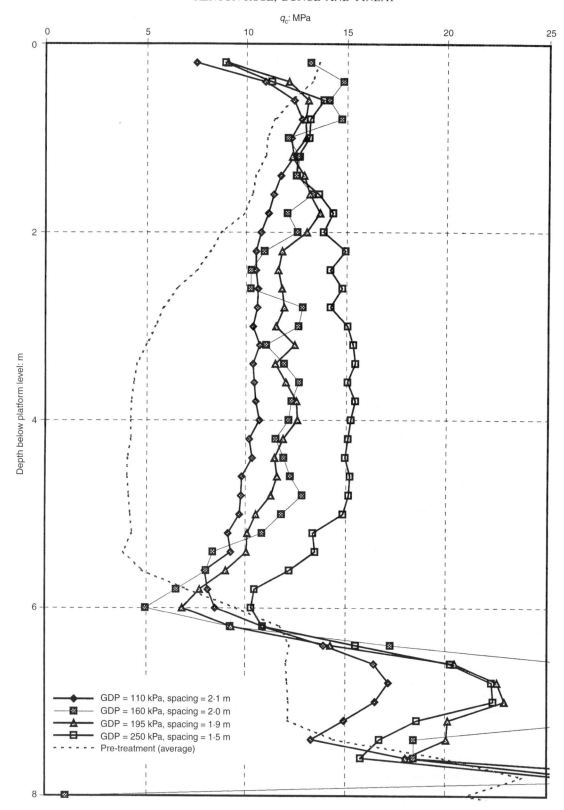

Fig. 11. Post-treatment average reclamation CPT tip resistance (q_c profiles) for all stone column spacings

COMPARISON OF TREATMENT WORKS WITH DESIGN

The assessment of the treated ground comprised a review of the performance of the five zone tests, as well as all the cone penetration tests and the plate bearing tests, from which the stone column constrained modulus was determined.

All zone tests were satisfactory, with those at the GCSB showing actual settlements at maximum test pressure of between 50% and 63% of the permitted settlement, based on preliminary design parameters.

An assessment of the adequacy of the densification (improvement) between columns was made, by comparing the average post-treatment cone resistance (q_c) values with the target value calculated by the contractor to suit his design column arrangement. Post-treatment CPTs were performed at the geometric centre of the triangular column layout. Required average target q_c values were estimated from settlement calculations. With values ranging from 6·5 MPa for a column spacing of 2·1 m, to 14 MPa for a column spacing of 1·5 m, the majority of the

Table 4. Comparison of design (target) and observed CPT resistances (above caprock)

Gross design pressure: kN/m²	Design column: spacing: m	Pre-treatment q_c: MPa	Target average q_c: MPa	Observed average q_c: MPa[†]	Improvement due to densification between columns	No. of results
250	1·5	5–10 (1·5–3 m and 5·5–6 m)* 4 (3–5·5 m)* Fill profile average = 5·6	14	14	2·5	24
195	1·9		8	12	2·1	49
160	2·0		7	12	2·1	11
110	2·1		6·5	12	2·1	118

* Depth below platform level.
[†] Between approximately 1·5 m and 6 m below platform level.

Table 5. Comparison of predicted and observed zone test performance

Zone test no.	Gross design pressure: kPa	Outline design settlement (1): mm	Recalculated settlement (2): mm	Observed settlement (3): mm	Ratio (2)/(3)
2	110	7	4·8	3·7	1·3
3	250	17	6·1	5·8	1·1
4	110	7	4·1	2·2	1·9
5	195	13	5·9	5·7	1·0
				Average	1·3

averaged cone resistance values exceeded the target values. However, a small number were marginally below target values.

A review of the plate bearing tests on columns, where actual settlements were all less than 53% of allowable values, indicated that the average 'as-built' stone column constrained modulus was around 140 MPa. This compared with the assumed design value of 120 MPa. This higher 'as-built' modulus, coupled with the better than predicted performance of the zone tests, was used in recalculations to give reduced target q_c values. All post-treatment q_c values exceeded the reduced target, justifying previously marginal failures as being acceptable.

The pre-treatment CPTs were used to assess the degree of improvement. Fig. 8 illustrates the range and average of pre-treatment cone resistance values. It is notable that the upper bound values and average values have risen significantly compared with the values obtained after less than one month from placement, as presented in Fig. 4. However, the lower bound values have not increased by the same degree. Fig. 9 illustrates the averaged pre- and post-treatment cone resistance values for all the areas with column spacings of between 1·5 and 2·1 m. Generally, the construction of the stone columns improved the cone resistance within the fill by a greater margin at depths where q_c values were lowest prior to the treatment. Looking in more detail, however, it appears that a slight to significant reduction in q_c value occurs just at the base of the fill, which may be associated with the development of elevated pore pressures within silty marine sands/localized silt pockets, generated during stone column installation.

A very marked increase in q_c values is also observed within the marine sand close to the seabed, where average pre-treatment values were relatively high, at around 12 MPa. Furthermore, the degree of improvement within the marine sands does not correlate directly with stone column spacing, with columns on a 2·0 m grid indicating the greatest degree of improvement, followed by those on a 1·9 m grid. Only the 2·1 m grid spacing shows less improvement in the marine sand than for the closest grid spacing of 1·5 m. However, the reliability of data at this depth is reduced, since a number of CPTs terminated at the top or within this horizon, thereby distorting the 'true' average resistance.

Testing of the stone during the site works indicated the initial design angle of friction of 40° to be acceptable.

OBSERVATIONS AND CONCLUSIONS

A significant time-dependent improvement in untreated fill materials is indicated in Fig. 4, with improvement by a factor of 1·5–2 comparing q_c results obtained within a month of placement with those at around 6 months later. From reference to Fig. 9, it appears that in the silty materials at the base of the fill this improvement may be reversed, at least in part, shortly after the installation of stone columns. This is possibly due to the short-term elevation of pore pressures within these fine materials. It is important to recognize the potential for time-dependent strength gain in similar hydraulic fills since pre-treatment q_c values are used as the basis for determining the level of improvement required and, consequently, the required spacing of treatment points. The time required to dissipate placement pore pressures will depend on the fines content of the fill. For this site, owing to the relatively coarse nature of the fill, dissipation was assessed to have been essentially complete, based on data from surface settlement plates. Further time-dependent improvement may be expected within calcareous soils, however, with the development of weak cementation.

Pre- and post-treatment cone penetration test q_c profiles are plotted against depth in Figs 8 and 9 to give an indication of the improvement factor achieved by the different stone column arrangements. Table 4 identifies the number of tests in each area of the GCSB and the stone column spacing, together with target and observed q_c values.

The post-treatment CPTs indicated an improvement factor of around 2·1–2·5 on the pre-treatment CPTs, within the fill below the intertidal zone. In the surface deposits above the intertidal zone densification was negligible, mainly owing to the fact that the pre-treatment q_c values were already high, as a result of consolidation and terrestrial placement/compaction in this zone. Compared with the outline design assumption of pre-treatment q_c of 2·5 MPa in fill, an improvement by a factor of around five times would be attributable. However, this assumed low initial q_c value is indicated to be conservative based on Fig. 4, being typical of unconsolidated fill prior to primary consolidation.

The higher, post-consolidation q_c values recorded prior to treatment are considered to reflect the grading and low fines content of the fill, providing a greater pre-treatment density and susceptibility to densification than it was possible to confidently assume for fill materials that would comply with the reclamation specification.

At around existing seabed level, approximately 6–6·5 m below platform level, the post-treatment CPTs indicated a reduction of q_c values in the soft silts on the existing seabed and localized silt at the base of the fill. It was expected at the design stage that little or no densification would be induced by treatment works in these materials owing to their fines content, and this was allowed for in the calculated average target q_c values required. The observed reduction in q_c, however, is considered to reflect short-term increases in pore water pressures, with values expected to return to at least pre-treatment levels once pore pressure dissipation is completed.

A recalculation of the predicted performance of the zone test has been made using the observed stiffnesses of the stone columns and soil between columns, the latter derived from the CPTs carried out actually at the zone test location. CPT q_c profiles at zone test positions were similar to the average profiles in Fig. 9, but showed local variation. Consequently, recalculated settlements for the same founding pressure and stone column spacing show small differences. The findings of the reassessment are given in Table 5, for tests carried out at the GCSB.

Comparison of the recalculated estimations of zone test settlements using Priebe's method with observed settlements indicates that the design approach may slightly overestimate the predicted foundation settlement. Recalculated settlements are 1·0–1·9 times observed movements, being an average of 1·3 times greater for the four tests considered. This observation is in general agreement with the outline design review, where Priebe's method was indicated to be slightly conservative compared with other estimation approaches. However, this level of accuracy, which errs on the safe side, is considered appropriate for such a geotechnical design where many variables exist, including the ground conditions and the uniformity of stone column installation.

While the design methodology for determining improvement due to installation of stone columns combined with densification between them is indicated to be satisfactory, the sensitivity of the column arrangement to the pre- and post-treatment soil properties is clearly demonstrated. For this case study, where designs were required prior to in-situ test data being available, it is apparent that higher q_c values and wider column spacings could have been adopted, with the design load–settlement criteria still being satisfied. However, the conservatisms demonstrated in this project are deemed to reflect the necessary level of caution required for a design and construct contract with such a restricted programme.

REFERENCES
Beckwith, G. H. & Hansen, L. A. (1982). Calcareous soils of the south western United States. In *Geotechnical properties, behaviour and performance of calcareous soils* (eds K. R. Demars and R. C. Chaney), ASTM STP 777, pp. 16–35. Philadelphia: American Society for Testing and Materials.
Burland, J. B. & Burbridge, M. C. (1985). Settlement of foundations on sand and gravel. *Proc. ICE, Part 1* **78**, Dec., 1325–1381.
Clayton, C. R. I. (1993). *The standard penetration test (SPT): methods and use*, Funders Report CP/7. London: Construction Industry Research and Information Association.
Greenwood, D. A. & Kirsch, K. (1983). Specialist ground treatment by vibratory and dynamic methods. In *Advances in piling and ground treatment for foundations*. London: Thomas Telford. pp. 17–45.
ICE (1987). *Specification for ground treatment*. London: Thomas Telford.
Meigh, A. C. (1987). *Cone penetration testing: methods and interpretation*, CIRIA Ground Engineering Report: In-situ testing. London: Butterworths.
Oasys Limited (1997). *VDISP program* (vertical displacement analysis).
Priebe, H. J. (1995). The design of vibro replacement. *Ground Engng*, Dec., 31–37.
Schmertmann, J. H., Hartman, J. P. & Brown, P. R. (1978). Improved strain influence factor diagrams. *Proc. Am. Soc. Civ. Engrs* **104**, No. GT8, 1131–1135.
Stroud, M. A. (1989). The standard penetration test: its application and interpretation. *Proceedings of the geotechnology conference on penetration testing in the UK*. London: Thomas Telford. pp. 29–49.

Croce, P. & Flora, A. (2000). *Géotechnique* **50**, No. 6, 739–748

Analysis of single-fluid jet grouting

P. CROCE* and A. FLORA†

An overall analysis of single-fluid jet grouting is presented. Current concepts and design rules are first reviewed and compared. Alternative approaches to the analysis of single-fluid jet grouting are proposed and implemented with experimental data and observations taken from recent and well-documented case histories. In-situ observations and measurements as well as laboratory tests are presented, indicating the influence of both treatment parameters and original soil properties on treatment effects. On the basis of the experimental evidence, the mechanical phenomena induced by jet grouting are discussed.

KEYWORDS: ground improvement; grouting

Nos présentons une analyse générale de la cimentation par injection à un seul fluide. Nous passons d'abord en revue et nous comparons les concepts actuels et les règles de formulation. Nous proposons et essayons d'autres manières d'analyser ce type de cimentation en utilisant des données expérimentales et des observations provenant d'histoires de cas récentes et bien documentées. Nous présentons les observations et les valeurs relevées sur place, ainsi que les essais en laboratoire, indiquant l'influence des paramètres de traitement et des propriétés originales des sols sur les effets du traitement. En nous basant sur les preuves fournies par les expérimentations, nous analysons les phénomènes mécaniques provoqués par la cimentation par injection.

INTRODUCTION

Currently adopted jet grouting techniques can be classified according to the number of fluids injected into the subsoil:

(*a*) grout (single-fluid system)
(*b*) air + grout (double-fluid system)
(*c*) water + air + grout (triple-fluid system).

The fluids are injected into the soil at very high speed through small-diameter nozzles placed on a grout pipe, which is rotated at a constant rate and slowly raised towards the ground surface. The jet propagates radially with respect to the treatment axis and the injected mortar solidifies underground.

The final result of jet grouting is then a cemented soil body of quasi-cylindrical shape, which is usually called a jet column. By appropriately arranging several columns in the subsoil, it is possible to solve different geotechnical problems (e.g. increasing bearing capacity and reducing settlements of new or existing foundations, supporting open and underground excavations, or creating water cut-offs for dam foundations).

At the design stage, however, there is still a relative degree of uncertainty arising from the lack of reliable methods for predicting the diameter of the columns and the mechanical properties of the cemented soil (soilcrete). It follows that most jet grouting projects are first planned on the basis of subjective rules of thumb, and are then specified on site by means of field trials. More satisfactory design methods should thus be developed.

Each jet grouting system has its own advantages and limitations, and thus all systems deserve equal attention. The present paper is devoted to the analysis of the single-fluid system, which generally produces columns that are not as large as those obtained by the double- and triple-fluid systems. However, the single-fluid system is the simplest, the most flexible and the most economical. Therefore, it often provides the most convenient solution to many projects. For instance, the single-fluid system is the jet grouting system routinely used for provisional tunnel support.

In the following, current concepts and design rules from published papers are first reviewed and compared. Alternative design approaches are then proposed and implemented with experimental data and observations taken from recent and well-

documented case histories. In-situ observations and measurements as well as laboratory tests are presented, in order to describe the influence both of treatment parameters and of original soil properties on treatment effects. On the basis of the experimental evidence presented, possible hypotheses on the mechanical phenomena induced by jet grouting are discussed, and design rules are presented.

CURRENT CONCEPTS

The first works published on jet grouting date back to the 1970s, and in the last 20 years numerous papers have been produced. It seems, however, that most findings are still confined to the experience of individual authors, although in a few cases collective views have been reported (e.g. Covil & Skinner, 1994). Current concepts are reviewed in the following, pointing out three key issues that appear rather controversial:

(*a*) the phenomena induced by jet grouting (i.e. the interactions between grout and natural soils)
(*b*) the main variables affecting treatment results (jet column diameter and soilcrete properties)
(*c*) the soil types suited for single-fluid jet grouting.

With regard to the first issue, it is generally assumed that jet grouting destroys the native soil structure, separating the soil particles, which are then mixed and cemented with the jetted grout (Miki, 1985; Tornaghi, 1989; Shibazaki, 1991; Kauschinger *et al.*, 1992; Bell, 1993; Covil & Skinner, 1994). However, Kauschinger *et al.* (1992) and Covil & Skinner (1994) point out that, for single-fluid jet grouting, the degree of replacement is rather low. In the case of gravel, Miki (1985) and Bell (1993) observe that the grout flows through the soil pores without destroying the soil structure, thus implying that for coarse-grained materials permeation becomes the dominant mechanism.

Concerning the second key issue (the main variables affecting treatment results and their possible relationship) various suggestions have been made. In particular, the influence of soil properties on jet column diameter (D) has been investigated, considering three variables: grading, relative density, and undrained shear strength s_u. Relative density has been estimated by SPT blow count (N). Some published relations (Botto, 1985; Bell, 1993) suggest that D decreases with decreasing soil grain size (Fig. 1), while other relations (Miki & Nakanishi, 1984; Tornaghi, 1989) indicate that D decreases with increasing N and increasing s_u (Figs 2, 3).

In order to account for the influence of treatment parameters, several empirical formulae have been proposed (Covil & Skinner, 1994). Such formulae, however, do not seem to have a

Manuscript received 22 May 2000; revised manuscript accepted 7 September 2000.
* University of Cassino, Italy.
† University of Napoli, Federico II, Italy.

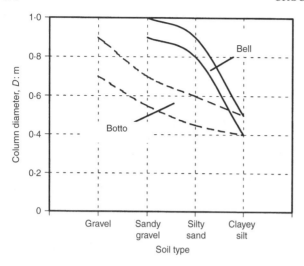

Fig. 1. Column diameter ranges against soil type according to Botto (1985) and Bell (1993)

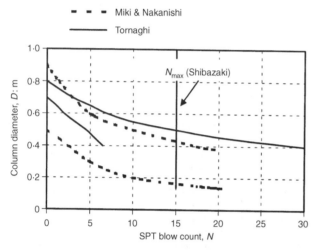

Fig. 2. Column diameter ranges in granular soils against SPT blow count according to Miki & Nakanishi (1984), Tornaghi (1989) and Shibazaki (1991)

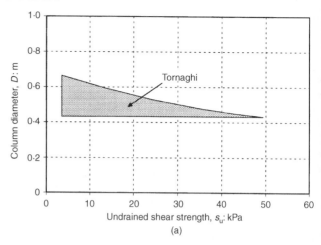

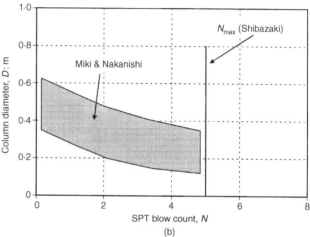

Fig. 3. Column diameter ranges in fine-grained soils: (a) against undrained shear strength (Tornaghi, 1989); (b) against SPT blow count (Miki & Nakanishi, 1984; Shibazaki, 1991)

clear physical meaning. One notable exception is provided by Tornaghi (1989), who suggests correlating column diameter with treatment energy per unit of column length. The energy is calculated at the pump (E_p) by the following formula:

$$E_p = \frac{pQ}{v} \tag{1}$$

where p = grout pressure, Q = grout flow rate, and v = monitor lifting rate.

With reference to the mechanical properties of soilcrete, it is pointed out that they are generally evaluated by means of uniaxial compression tests, and only seldom with direct shear (Mongiovì *et al.*, 1991; Croce *et al.*, 1994) or triaxial tests (Croce & Flora, 1998). Many uniaxial strength data have been published (e.g. Shibazaki, 1991), and some expected ranges of values are given in Table 1. The coefficient of permeability of soilcrete has not been investigated, and this may be explained by the fact that the overall permeability of cut-offs is affected mainly by their discontinuities.

Finally, with regard to the third issue (the soil types suited for single-fluid jet grouting), literature indications are quite divergent. In fact, some authors believe that jet grouting is applicable to a wide range of soils (e.g. Covil & Skinner, 1994), while others express some doubts with respect to fine-grained soils (e.g. Bell, 1993). Other authors (Kanematsu, 1980, as reported by Covil and Skinner, 1994; Shibazaki, 1991) claim that single-fluid jet grouting is applicable only for very low SPT values (cohesive soils $N < 5$–10, sandy soils $N < 15$), but do not provide any explanation for such sharp limits.

DESIGN APPROACHES

Previously observed uncertainties are not surprising if the complexity of the phenomena associated with the production of a jet column is considered. In particular, the following three stages can be identified (Fig. 4):

(1) water–cement mixing and grout pumping
(2) mechanical interaction between soils and grout
(3) delayed soilcrete hardening.

In the first stage, cement powder and water are mixed in various prescribed proportions to form a suspension (grout), which is

Table 1. Expected uniaxial compressive strength of soilcrete, according to various authors

Natural soil type	Bell (1993) q_u: MPa	Miki (1985) q_u: MPa)	Shibazaki (1991) q_u MPa
Clay	0·5 to 8	<5	10
Silt	4 to 18		
Sand	5 to >25	5–10	30
Gravel	5 to >30		

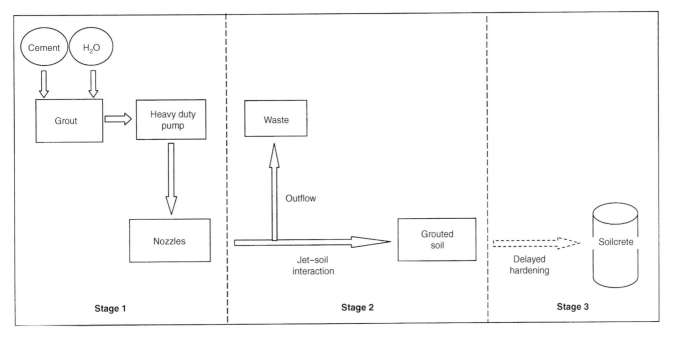

Fig. 4. Sketch of the production stages of a jet column

then subjected to high pressure as it is pumped to the monitor. The grout is conveyed through flexible tubes and then through the hollow metallic injection stem, eventually reaching the monitor nozzles. Therefore the grout is ejected at very high speed, but some hydraulic head is lost between the pump and the nozzles.

The second stage is characterized by the fact that only a part of the injected grout is absorbed by the soil, while the remaining flows towards the ground surface, passing through the annular space bounded by the perforation hole and the injection stem. Measured percentage of outflow, composed of grout and some eroded soil, can vary between 0 and 80% (Kauschinger *et al.*, 1992). Such a large range of outflow shows that erosion and replacement may be produced to varying degrees.

The third stage consists of the cementing process due to the solidification of the grout retained by the ground. This is a time-dependent phenomenon, which derives from hydration of the cement particles by the water contained in the grout and/or in the natural soil. In this final stage, soil grain size distribution and mineralogy will influence the strength of the soilcrete.

Two possible alternative approaches are shown in order to improve present design methods. The first and simplest approach consists in relating column diameter with treatment energy and with some simple soil parameters (e.g. grain size, N, S_u). This is an empirical procedure, which has the advantage of giving practical information for design purposes, but provides no insight into the mechanism. It is suggested, however, that the energy should be taken at the nozzles (E_n) rather than at the pump, in order to account for the losses along the circuit and through the nozzles. In particular, with reference to the unit length of column L:

$$E_n = \frac{m v_n^2}{2L} \qquad (2)$$

where m is the grout mass delivered in the time interval Δt corresponding to treatment length L, and v_n is the grout velocity at the nozzles. The mass is calculated as $m = \rho_g Q \Delta t$, where ρ_g is the grout density and Q is the grout flow rate. Expressing v_n as the ratio between Q and the overall nozzle cross-section, it follows that

$$E_n = \frac{8 \rho_g Q^3}{\pi^2 M^2 d^4 v} \qquad (3)$$

where M = number of nozzles, d = diameter of nozzles, and v = monitor lifting rate.

With respect to the current definition of treatment energy (equation (1)), the above formulation (equation (3)) has the advantage of accounting for grout density and nozzles dimension, which can change from case to case, rather than for grout pressure, which is almost constant (typically 45 MPa).

The second approach for pursuing more satisfactory design methods relies on theoretical modelling of the mechanical phenomena induced by jet grouting. The latter approach has been attempted by Croce & Mongiovi (1995), who have modelled the complex interaction between jet flow and soil by subdividing the jet column in two coaxial zones: an inner zone, where the soil grains are displaced by the grout and an outer zone, where the grout flows through the soil pores without moving the grains. However, such hypotheses should be experimentally verified.

Unfortunately, observation of the interaction between the grout and the natural soil during the process of column formation is not feasible, and possible mechanical phenomena (e.g. permeation, erosion, mixing, replacement) should thus be assessed by careful examination of the physical features of the columns after treatment has been performed. This procedure can be integrated by a method, proposed herein, that allows the experimental determination of the percentages of absorbed grout and removed soil. The method is based on the continuity condition applied to the volume of jet column per unit length ($V = \pi D^2/4$). In particular, considering first the mechanical effects of grouting (end of stage 2) the following relation holds:

$$V = \frac{\alpha}{\delta(n + \beta - n\beta)} V_j \qquad (4)$$

where V_j is the volume of injected grout per unit length of treatment, α is the volumetric percentage of grout retained by the subsoil, β is the volumetric percentage of soil removed by the jet action, δ is the percentage of soil pores filled with grout, and n is the original soil porosity.

Considering then the hardening process (stage 3), the volume of column per unit length can also be expressed as:

$$V = \frac{k\alpha}{n + \beta - n\beta - n_{sc}} V_j \qquad (5)$$

where k is the volumetric ratio between hydrated cement and retained grout, and n_{sc} is the soilcrete porosity.

The parameter k can be readily determined via the following relation, which is obtained by applying mass continuity from grouted soil to soilcrete:

$$k = \frac{(1 + r_s)\rho_g}{(1 + r_g)\rho_{hc}} \quad\quad\quad (6)$$

where r_s is the stoichiometric water/cement ratio (≈ 0.30), r_g is the water/cement ratio of grout, ρ_g is the grout density, and ρ_{hc} is the hydrated cement density (≈ 25.5 kN/m³). The derivation of equations (4), (5) and (6) is reported in Appendix 1.

The injected grout volume, V_j, is derived from known treatment parameters ($V_j = Q/v$), while n, V and n_{sc} should be measured by experimental investigations performed respectively on natural soils, column shape and soilcrete characteristics. In particular, V can be obtained by direct measurement of D along the jet column, and n_{sc} can be determined on undisturbed soilcrete specimens retrieved by the column.

It follows that equations (4) and (5) contain three unknowns (α, β, δ). In general, however, it can be assumed that the grout completely fills the soil voids ($\delta = 1$) at the end of stage 2, with the exception of treatments performed on clean gravel (Chiari & Croce, 1991). Therefore the percentages of absorbed grout (α) and removed soil (β) can be determined by experimental investigations, providing basic information on the mechanical phenomena induced by jet grouting.

CASE HISTORIES

The experimental investigations reported herein are taken from three well-documented case histories, located in Italy:

(a) case A – the Fadalto viaduct in the Eastern Alps (Croce et al., 1990; Mongiovi et al., 1991)
(b) case B – the Polcevera river levees in the city of Genoa (Croce et al., 1994)
(c) case C – the Vesuvius tunnel near the city of Naples (Croce & Flora, 1998).

For case A, two field trials were executed on a steeply inclined slope made of calcareous debris (angular sandy gravel). Each of the two trials comprised 166 vertical treatments, 17–23 m deep, arranged on a rectangular grid plan at 1·10 m centres. These trials were performed in order to investigate the feasibility of deep massive foundations made of soilcrete, but results were not conclusive, and it was then decided to build shaft foundations made of concrete, employing jet grouting only to sustain the shaft walls during construction.

Case B concerned a field trial of 19 columns, 13 m deep, arranged in a circular pattern, in order to allow excavation of an exploration shaft after treatment. The subsoil comprised recent alluvial deposits, subdivided into a superficial layer (thickness 7·0 m), made of heterogeneous sandy gravel, and a deeper layer of clayey silt.

Case C involved the construction of a field trial of seven columns, 10 m deep, positioned in a single row, at 1·20 m centres, in a deposit of pyroclastic silty sand.

Accurate in-situ measurements and laboratory tests were carried out in all cases on both natural and cemented soils. Available data include soil properties, treatment parameters, column dimensions (Table 2) and mechanical characteristics of the cemented material (Table 3). In Table 2, data on a published case history (Bianco & Santoro, 1995) are reported as well (case D), and used in the discussion of treatment effectiveness. This case history (the Rio Matzeu viaduct in Sardinia) reported on a trial field consisting of several isolated vertical treatments performed in two soil types (sandy silts and sandy gravels).

Detailed information on each case is provided in the quoted references.

Direct observation of the jet columns, as well as testing, proved that single-fluid jet grouting is quite effective for coarse-grained soils (Fig. 5(a)), but that treatments carried out on fine-grained soils (Fig. 5(b)) had no measurable effects.

The diameters of the jet columns measured after excavation are plotted in Fig. 6 (cases B and C). In case B (Fig. 6(a)), all treatments were performed with the same parameters, but a random variation of diameter is observed as a consequence of the strong heterogeneity of the alluvial soil. In case C (Fig.

6(b)) the shape of the columns is much more regular, because the pyroclastic sand deposit is rather homogeneous, but the diameter of each column changes depending on the adopted combination of treatment parameters. The mechanical properties of soilcrete are excellent (Table 3), with values of the uniaxial compressive strength q_u in general accordance with the ranges reported in the literature (Table 1).

DISCUSSION

Current design rules can be checked by comparing the relations previously reported in Figs 1 and 2 with the experimental data taken from the four case histories presented. In particular, measured column diameters are plotted against soil type in Fig. 7, confirming the trend of diameter reduction with decreasing soil grain size. However, the data fall within ranges wider than those proposed by Botto (1985) and Bell (1993). On the contrary, it can be seen from Fig. 8 that the column diameter for granular soils cannot be simply correlated with SPT blow count, and that large diameters can be obtained even for very dense materials. This finding contradicts the rules suggested by Miki & Nakanishi (1984), Tornaghi (1989) and Shibazaki (1991).

With regard to the proposed energy approach, a preliminary comparison is made between energy at the pump, E_p (equation (1)), and energy at the nozzles, E_n (equation (3)). The values obtained are reported in Fig. 9, showing that about 20% of energy is lost in most cases before reaching the ground. In case A the energy loss was larger because the pump was further from the injection stem.

The average values of the column diameters are then plotted against energy at the nozzles (Fig. 10) for the relevant soil types and N values. Note that energy is almost constant for all cases except case C, where the treatment parameters were intentionally changed over as wide a range as possible. At the same energy, it is confirmed that the column diameter increases with increasing soil grain size while, for case C, data interpolation shows a linear increase in diameter with respect to energy.

The interaction between soil and grout has been investigated by observing the features of the jet columns. A typical shaft wall (case A) made of overlapped jet columns is shown in Fig. 11. From the picture it can be seen that the soils subjected to treatment have retained their original structure, characterized by inclined strata parallel to the ground slope. In particular, the sandy gravel mass (debris) has been fully cemented while the clean gravel strata have not been completely filled by the grout. Thin strata of fine-grained soil have not been cemented at all.

These features confirm that, for gravelly soils, jet columns are produced by soil permeation rather than by mixing. However, in clean gravel the grout may flow into the subsoil without filling the soil pores because soil permeability is too high. Therefore, in clean gravel, permeability is not substantially reduced by jet grouting although high soilcrete strength may be achieved.

In case C direct observation of the jet columns was not helpful in understanding the grouting mechanism because the subsoil was quite homogeneous, and the natural soil structure could not be observed. However, the percentages of removed soil and retained grout for two columns were obtained by applying the new experimental method proposed (equations (4)–(6)). In particular, V was calculated from the column diameters, measured at different depths, and n_{sc} was determined on undisturbed soilcrete specimens retrieved at corresponding depths. V_j was calculated for each column from the recorded grout flow rate and monitor lifting rate. Constant values were assumed for the other parameters: $\delta = 1$, $n = 0.4$, $r_m = 1.0$, $r_s = 0.3$, $\rho_g = 1.47$ g/cm³, $\rho_{hc} = 2.55$ g/cm³.

The results are plotted in Fig. 12, showing that a large amount of soil has been removed by the jet action ($0.30 < \beta < 0.60$), and that most of the injected grout has been absorbed by the soil ($0.65 < \alpha < 0.90$). This is in agreement with the assumption that, for sandy soils, jet columns are produced mainly by soil mixing and replacement. Furthermore,

Table 2. Data on natural soil, treatment parameters and jet columns from the three selected field trials (cases A, B and C) and from a case history (D) taken from literature (Bianco & Santoro, 1995)

Case	Natural soil						Treatment parameters							Jet column	
	Type	N	γ: kN/m³	c': kPa	ϕ': °	s_u: kPa	r_g	p: MPa	Q: 10^{-3} m³/s	v: 10^{-3} m/s	ω: rpm	M	d: 10^{-3} m	Number	D: m
A	SG	50	—	—	—		1·2	45	1·8	5·00	10–15	2	2·8	—	—
B	SG	25	—	—	—		1	45	1·8	5·50	20	2	2·4	1, 2, 18, 19	1·15
C	SS	15	15	55	35		1	45	1·4	5·70	15	2	2·0	0	0·66
							1	45	2·5	5·00	7·5	1	3·8	1	0·96
							1	45	2·4	6·70	15	1	2·6	2	0·75
							1	45	2·5	5·00	11	1	3·8	3	0·97
							1	45	2·4	6·70	10	2	2·6	4	0·71
							1	45	2·5	4·00	9	1	3·8	6	0·95
D₁	SG	Refusal	18·5	0	37		0·9	48	1·7	8·00	10–20	2	2·2	1	0·84
							0·9	50	1·7	8·00	10–20	2	2·2	2	0·95
							0·9	50	1·7	8·00	10–20	2	2·2	3	1·00
							0·9	45	1·6	7·10	10–20	2	2·2	4	1·09
							0·9	45	1·6	5·70	10–20	2	2·2	5	1·14
D₂	SS	60	19·5	20	28	160	0·9	48	1·7	8·00	10–20	2	2·2	1	0·45
							0·9	50	1·7	8·00	10–20	2	2·2	2	0·5
							0·9	50	1·7	8·00	10–20	2	2·2	3	0·55
							0·9	45	1·6	7·10	10–20	2	2·2	4	0·45
							0·9	45	1·6	5·70	10–20	2	2·2	5	0·65

A = Fadalto; B = Polcevera; C = Vesuvius; D₁ and D₂ = Rio Matzeu; SG = sandy gravel; SS = silty sand; N = average SPT value; γ = soil unit weight; c' = soil cohesion; ϕ' = soil friction angle; s_u = soil undrained shear strength; r_g = water/cement ratio by weight; p = grout pressure; Q = grout flow rate; v = monitor lifting rate; ω = rotational speed; M = number of nozzles; d = diameter of nozzles; D = average column diameter.

Table 3. Main characteristics of soilcrete in the three field trials from in-situ and laboratory tests

Field trial	v_{p1}: m/s	v_{p2}: m/s	E_{50}: MPa	q_u: MPa	ϕ: °	c: MPa	k: 10^{-9} m/s
A	2000–3500	2500–5000	5000–20000	18–26	65	1·8	—
B	2000–3000	2600–3600	2000–6700	11–30	52	2·1	1–6
C	—	2600–3200	2000–8000	4–15	32	2·2	—

v_{p1} = average value of velocity of compression waves from cross-hole tests; v_{p2} = average value of velocity of compression waves measured on specimens in laboratory; q_u = uniaxial compressive strength; E_{50} = average value of Young's modulus in uniaxial compressive tests at $q = 0.5q_u$; ϕ = friction angle at peak; c = cohesion at peak; k = Darcy's coefficient of permeability measured on specimens in laboratory.

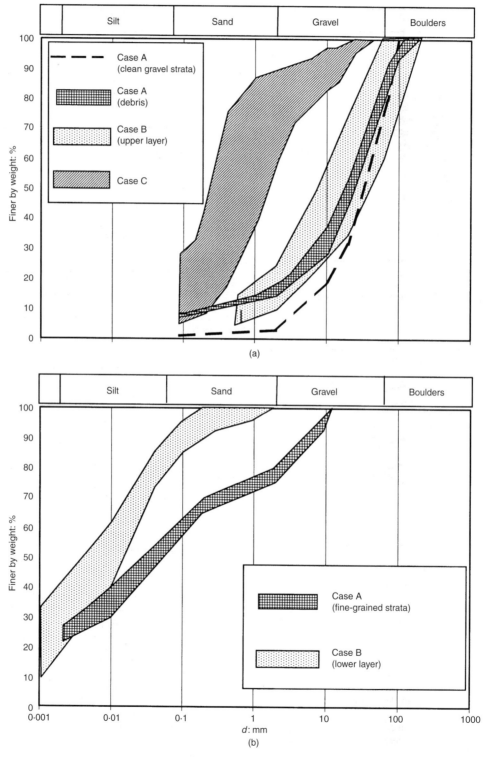

Fig. 5. Grain-size distributions of natural soils for (a) effective and (b) ineffective single-fluid jet grouting treatment

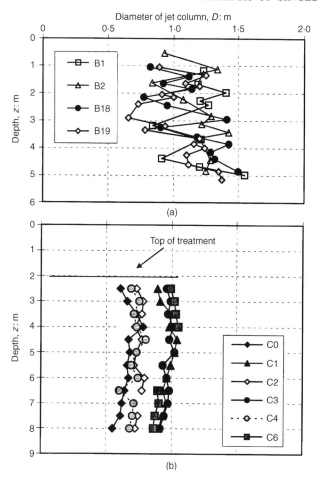

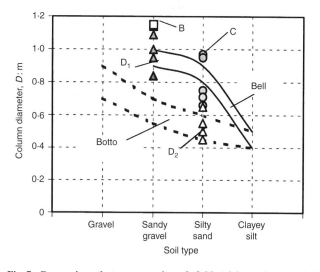

Fig. 6. Measured diameter of jet columns against depth for (a) case B and (b) case C

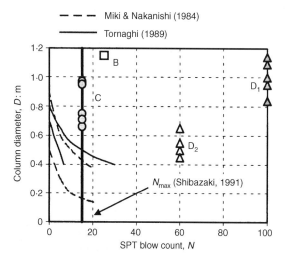

Fig. 8. Comparison between results of field trials and suggested column diameter ranges as a function of SPT blow count

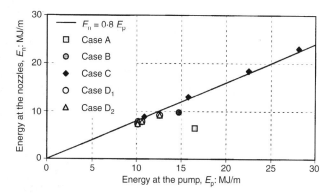

Fig. 9. Comparison between the energy at the pump (E_p) and at the nozzles (E_n)

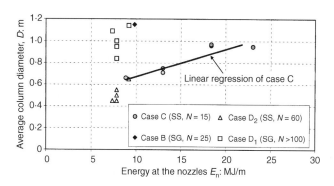

Fig. 10. Average column diameter against energy at the nozzles

Fig. 7. Comparison between results of field trials and suggested column diameter ranges as a function of soil type from Botto (1985) and Bell (1993)

both grout absorption and soil removal decrease with depth. This trend can be explained by considering that sandy soil resistance to cutting increases with depth, owing to the increase in shear strength.

With respect to soilcrete mechanical properties, only uniaxial compressive strength q_u is considered, as very few data are available from other laboratory tests.

In case A, the occurrence of large gravel particles caused erratic results for tests carried out on cylindrical specimens

recovered from borings, and so compression tests were performed on cubic specimens with 20 cm side. In particular, three groups of specimens were prepared: the first group was made only with water–cement mortar, the second group was taken from a large block of soilcrete retrieved on site (undisturbed soilcrete), and the third one was prepared by hand-mixing mortar and natural soil (artificial soilcrete), being careful to obtain the same unit weight of the undisturbed specimens. The results of these uniaxial compression tests are plotted in Fig. 13, showing good agreement between undisturbed and artificial soilcrete, with average strength corresponding to a low-quality concrete. Mortar specimens show a lower unit weight and a larger value of q_u.

Results of uniaxial compression tests from case C are plotted in Fig. 14 together with available data from another field trial

Fig. 11. Shaft wall made of overlapped jet columns (case A), showing the persistent natural soil structure (inclined strata parallel to slope)

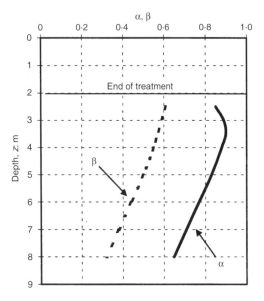

Fig. 12. Case C: retained grout fraction (α) and removed soil fraction (β) against depth

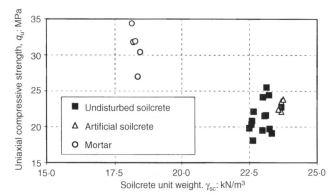

Fig. 13. Case A: influence of composition and preparation on the uniaxial compression strength of cubic (20 cm side) specimens

Table 1, with the exception of some very low values of q_u for the Porto Tolle soilcrete. For case C, soilcrete unit weight is quite low, owing to the very low specific weight of the pyroclastic silty sand. Yet q_u values are relatively high, reaching the higher values reported for the Porto Tolle site.

(Porto Tolle, from Perelli Cippo & Tornaghi, 1985). In both cases, compressive strength increases with soilcrete unit weight (Fig. 11(b)), and soilcrete unit weight increases with depth. The strength data are in general agreement with the ranges given in

CONCLUSIONS

Direct observation of several jet columns and extensive in-situ and laboratory testing has proved that single-fluid jet grouting is very effective for coarse-grained soils, while treat-

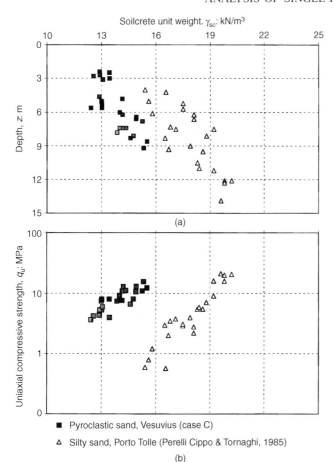

Pyroclastic sand, Vesuvius (case C)

Silty sand, Porto Tolle (Perelli Cippo & Tornaghi, 1985)

(b)

Fig. 14. Relationships between soilcrete unit weight and (a) depth, (b) uniaxial compressive strength

ments performed on fine-grained soils do not produce measurable effects. Previously reported trends of diameter reduction with decreasing soil grain size have been confirmed, even though the data fall in their ranges wider than those proposed in the literature. It has also been proved that column diameter for granular soils cannot be simply correlated with SPT blow count, and that large diameters can be obtained even for very dense materials.

The diameter of jet columns has been related with treatment energy calculated by a new formulation (equation (3)) to account for the losses along the circuit and through the nozzles. By applying equation (3) to the reported field trials it was shown that about 20% of energy is lost before reaching the ground, but energy loss may increase if the pump is far from the injection stem. For a given energy at the nozzles, it is confirmed that the diameter of jet columns increases with increasing soil grain size while, for a given soil, an increase of energy produces a corresponding increase in diameter. However, the effectiveness of energy increase should depend on the grouting mechanism, which in turn depends on soil type.

In-situ observations have confirmed that, for gravelly soils, jet columns are produced by soil permeation rather than by mixing. This explains why, for coarse-grained soils, N is not a relevant parameter, and thus the approach based on SPT blow count may be misleading.

Treatment is very effective in sandy gravel, but in clean gravel the grout may flow into the subsoil without filling the soil pores because soil permeability is too high. Therefore, in clean gravel, permeability is not substantially reduced by jet grouting, although high soilcrete strength may be achieved.

On the contrary, for sandy soils, jet columns are produced mainly by mixing and replacement. The latter mechanism has been detected by interpreting experimental measurements of column diameter and soilcrete porosity with a new method based on the continuity condition applied to the volume of jet column per unit length.

APPENDIX 1. DERIVATION OF EQUATIONS (4), (5) AND (6)

With reference to Fig. 15, the relevant volumetric ratios are

$$n = \frac{V_{v1}}{V} \tag{7a}$$

$$n_{sc} = \frac{V_{v3}}{V} \tag{7b}$$

$$\alpha = \frac{V_g}{V_j} \tag{7c}$$

$$\beta = \frac{V_{s1} - V_{s2}}{V_{s1}} \tag{7d}$$

$$\delta = \frac{V_g}{V_{v2}} \tag{7e}$$

$$K = \frac{V_{hc}}{V_g} \tag{7f}$$

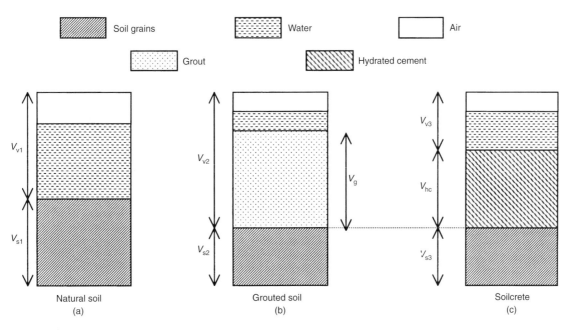

| Soil grains | Water | Air |
| Grout | Hydrated cement |

Natural soil
(a)

Grouted soil
(b)

Soilcrete
(c)

Fig. 15. Schematic composition of (a) natural soil, (b) grouted soil and (c) soilcrete

where V_j is the injected grout volume.

Equation (4) is derived by applying volume continuity from natural soil to grouted soil.

In particular, from equations (7a) and (7d):

$$V_{s2} = (1 - \beta)(1 - n)V \qquad (8)$$

From equations (7c) and (7e):

$$V_{v2} = \frac{\alpha V_j}{\delta} \qquad (9)$$

Combining equations (8) and (9):

$$V = \frac{\alpha}{\delta(n + \beta - n\beta)} V_j \qquad (4)$$

Equation (5) is derived by applying volume continuity from grouted soil to soilcrete.

In particular, from equations (7c) and (7f):

$$V_{hc} = K\alpha V_j \qquad (10)$$

Considering that $V_{s2} = V_{s3}$, and applying equations (7b), (8) and (10), it follows that

$$V = \frac{K\alpha}{n + \beta - n\beta - n_{sc}} V_j \qquad (5)$$

Equation (6) is obtained by applying mass continuity from grouted soil to soilcrete. The relevant mass (m) ratios are

$$r_g = \frac{m_w}{m_c} \qquad (11a)$$

$$r_s = \frac{m_{ws}}{m_c} \qquad (11b)$$

where the subscripts have the following meaning: w = grout water, c = grout cement (dry), ws = stoichiometric water.

The mass ratio between hydrated cement (hc) and grout (g) is

$$\frac{m_{hc}}{m_g} = \frac{1 + r_s}{1 + r_g} \qquad (12)$$

Therefore, considering equation (7f), it follows that

$$K = \frac{\rho_g}{\rho_{hc}} \frac{1 + r_s}{1 + r_g} \qquad (6)$$

REFERENCES

Bell, A. L. (1993). Jet grouting. In *Ground Improvement*, (ed. M. P. Moseley), pp. 149–174. Blackie.

Bianco, B. & Santoro, V. M. (1995). L'importanza dei campi prova e delle sperimentazioni nella progettazione dei trattamenti colonnari. L'esempio delle fondazioni del viadotto Rio Matzeu della nuova SS 131 variante nei pressi di Cagliari. *Associazione Geotecnica Italiana, XIX Congresso Nazionale di Geotecnica*, Pavia, Vol. 1, pp. 81–88 (in Italian).

Botto, G. (1985). Developments in the techniques of jet-grouting. *XII Ciclo di Conferenze di Geotecnica*, Torino, reprint by Trevi.

Chiari, A. & Croce, P. (1991). Some limits to the use of jet-grouting. *Proc. 4th Int. Conf. Piling and Deep Foundations, Stresa*, 201–204.

Covil, C. S. & Skinner, A. E. (1994). Jet grouting: a review of some of the operating parameters that form the basis of the jet grouting process. In *Grouting in the ground*, pp. 605–627. London: Thomas Telford.

Croce, P. & Flora, A. (1998). Jet-grouting effects on pyroclastic soils. *Rivista Italiana di Geotecnica*, XXXII, No. 2, 5–14.

Croce, P. & Mongiovì, L. (1995). Modellazione teorica della gettiniezione. *Associazione Geotecnica Italiana, XIX Congresso Nazionale di Geotecnica, Pavia*, **II** (in Italian).

Croce, P., Chisari A. & Merletti T. (1990). Indagini sui trattamenti dei terreni mediante jet grouting per le fondazioni di alcuni viadotti autostradali. *Rassegna dei Lavori Pubblici*, XXXVII, No. 12, 249–260 (in Italian).

Croce, P., Gaio, A., Mongiovì, L. & Zaninetti A. (1994). Una verifica sperimentale degli effetti della gettiniezione. *Rivista Italiana di Geotecnica*, XXVIII, No. 2., 91–101 (in Italian).

Kanematsu, H. (1980). High pressure jet-grouting method, Doboku Sekoh, Civil Construction, **21**, No. 13.

Kauschinger, J. L., Perry, E. B. & Hankour, R. (1992). Jet grouting: state-of-the-practice. In *Grouting/soil improvement and geosynthetics*, ASCE Geotech. Spec. Publ. No. 30, Vol. 1, pp. 169–181.

Miki, G. (1985). Soil improvement by jet grouting. *Proc. 3rd Int. Geotechnical Seminar on Soil Improvement Methods, Singapore*, 45–52.

Miki, G. & Nakanishi, W. (1984). Technical progress of the jet grouting method and its newest type. *Proceedings of the international conference on in situ soil and rock reinforcement*, Paris, pp. 195–200.

Mongiovì, L., Croce, P. & Zaninetti, A. (1991). Analisi sperimentale di un intervento di consolidamento mediante gettiniezione. *II Convegno Nazionale dei Ricercatori del Gruppo di Coordinamento degli Studi di Ingegneria Geotecnica del CNR, Ravello* (in Italian).

Perelli Cippo, A. & Tornaghi, R. (1985). Soil improvement by jet grouting. In *Underpinning* (Eds S. Thorburn and J. F. Hutchison), pp. 276–292. Surrey University Press.

Shibazaki, M. (1991). The state of art in jet-grouting. *Proceedings of the symposium on soil and rock improvement in underground works*, pp. 19–46. Società Italiana Gallerie.

Tornaghi, R. (1989). Trattamento colonnare dei terreni mediante gettiniezione (jet-grouting). *XVII Convegno Nazionale di Geotecnica, Taormina* (in Italian) 193–203.

INFORMAL DISCUSSION

Session 4

CHAIRMAN MR KEN WATTS

David Greenwood, *Geotechnical Consulting Group*

I would like to begin by congratulating Professor Croce on tackling a terribly difficult subject, and one which really needs attention. I think our other academic friends here could well spend some time on this territory, which is a difficult one.

I would like just to draw attention to figure 10 in Croce and Flora's paper, which relates energy input to diameter. I would recall that, admittedly in soft soils, the Japanese have been producing columns of 5 m to 7 m diameter with single grout jets unaided by air or separate jetting. Although you produce a linear relationship over a relatively short range, I think their energies are somewhere around eight times what you depict in the graph. So we are in quite a different range. Surely it is common sense that the energy curve, it will be a curve, it will not be linear as you eventually get out to the bigger diameters. I certainly agree that you cannot correlate the penetration testing with diameter because, I think, in the sandy gravels we get such a scatter. Perhaps it is the penetration test that is at fault rather than anything else.

You did not mention plasticity index for the finer grained soils, which I think must have some relevance to the diameter that you can get with a given energy level.

On the question of controlling the soilcrete mix, I think you are constrained very strongly by the fact that you first of all have to erode the column. That gives you a certain necessary flow rate and momentum from the jet. If you are using a single jet, the grout mix is constrained because it has to be sufficiently fluid. The rheological properties and proportions are all fixed to some extent at one end of the range by the jetting requirement. If you have an additional water cutting jet and air, you then have a further complication of the ratios between the cement quantities and the water, and so on. Also, using the volume balance that you showed, I think that yes, you can get close to the sort of mix that comes out in the column and the effluent.

I would suggest that where you have a gap-graded material, or perhaps one with a fair bit of gravel in it, you will not have sufficient effluent velocity or space at the annulus around the jet stem to shift the heavier particles. You are inevitably going to get a separation between the fines and the coarser particles, which are going to sink a bit in thick, not viscous, fluid in the hole. They will not sink right to the bottom, but the heavier particles will sink. I think you can exercise some control by putting a casing in the borehole and determining the velocity, to some extent, of the effluent coming out, which determines, to some extent, what size of particle you can lift, knowing the particle size distribution in the ground.

That of course alters the volume balance, which is very significant. Where you tend to wash out nearly all the material and virtually replace it, you have a much more economical consumption of cement and a more economical job. For example, in concrete you might have 200 to 250 kg/m^3, typically in jet grouting you might have 500 kg/m^3. Where you have sandy gravels you might well go up to 700 to 750 kg/m^3. So it is quite a significant point to be made, trying to control the composition of the mix by adjusting the size of the effluent exit annulus.

Professor Paulo Croce

I agree with most of the observations that you have made. Clearly in Italy we usually use lighter equipment than that usually used in Japan. But this is very useful because in this way we can provide new applications or make existing ones easier on a routine basis. For example, we can use it in tunnelling.

I agree with most things you have said. Clearly I was not suggesting that there is a linear relation between jet energy and diameter. In that case in my paper, I only have a limited data because finding enough reliable data is a very big problem. Actually, we worked hard on these three particular cases. When you get data from the literature or from other colleagues, these data are usually not complete, they can be questionable. You do not get out anything that is meaningful. Every single case means a lot of direct work. I do not have enough data for the moment.

My suggestion is that in the future, by collecting good reliable data, we should be able to specify the shape of those curves. They will clearly only be empirical, but could be very helpful in design. I think that the shapes of these curves will be very different according to the different grain-size distribution because with different distributions there will be different kinds of mechanism. If we have different kinds of mechanism, the relations will be different. We may have linear relations for one particular material, curved relations for another - different kind of shapes.

Dr Alan Bell, *Keller Ground Engineering*

Professor Croce, I thought your contribution to a problem that has taxed many minds in jet grouting over the past few years was very interesting. It is very interesting that you indicate the triple system, for which, I think, you have shown some of my data from seven years ago. At that time, the column diameters we were getting with triple systems were quite small, and in your paper you too point out that it is very important to take account of the losses in the system to get the true nozzle energy. You may be interested to know that during the past few years we have spent a lot of time looking at how to eliminate and reduce those losses, so that our efficiency at the nozzle has been substantially improved. Currently we are achieving column diameters of approximately 0·6 m and above in silty clays of medium plasticity.

I notice that for single fluid systems you feel that there is a restriction. For triple fluid systems we seem to be getting somewhere. Do you believe that the relationships that you have developed will apply to triple fluid systems?

Professor Paulo Croce

I had to restrict myself to single fluid jet grouting because I had a chance to work on several projects in Italy where single fluid jet grouting was used. In fact in Italy single fluid is the routine method. I do not have personal experience on the three fluid method. I have, of course, seen in the literature many observations and many details that tell us that with triple fluid systems it is possible to get very large columns, even for fine grained materials. I remain doubtful until I can see it for myself because I have not seen any literature giving me all the data and all the information that I need. I rely mainly on my personal experience, and I hope to be able to undertake similar research for double fluid and triple fluid systems.

Alec Courts

Dr Renton-Rose, you mentioned a time dependent improvement in the cone penetration results prior to treatment. Did you look for or find any similar time dependent variation in the post treatment testing?

David Renton-Rose

Regarding the time dependency, from the outset we obviously looked at the information bearing in mind the limited information we had. There was quite a clear time dependency for the as-placed material, which seemed to correlate in terms of the strength gain or densification with settlement monitoring plates on the site that were installed as part of the reclamation process itself.

In terms of post-treatment time-dependent improvement, I must admit I do not think we have the information because of the nature of the construction site, as is always the problem. A lot of the settlement monitoring information has now been damaged or removed. The correlation with the consolidation response of the ground is not there. I have to say also that I am not personally involved with the scheme anymore and it was not something the client was particularly interested in. I suspect that we would struggle to get funding to do some longer term CPT testing to validate that.

Having looked around some of the literature, we would expect to see, particularly for calcareous sands as we were dealing with here, see some strength gain with time due to ongoing slower rates of consolidation, and also a cementation build-up, particularly in the upper inter-tidal zone. I do not have any information for that particular site to support this conjecture.

Len Threadgold, *Geotechnics*

I would like to draw on a theme that was touched on by Bill Craig and David Renton-Rose in relation to settlements and tolerable differential settlements. We have discussed making settlements tolerable, which obviously implies a specification drawn up by others with whom we interface, namely structural designers and the like, who are going to specify that either differential settlements should be limited to a given amount or that total settlements should be limited to an amount. It was refreshing to learn that on the site that David Renton-Rose talked about there was flexibility in this regard and that non-sensitive structures did not require the same intensity of treatment. I think there is a need for us to interface with designers better than we currently do.

In relation to Bill Craig's contribution, he was talking about identifying the amount of differential settlement in order to define the spacing required to achieve a given amount of settlement differential. Was the design modified to tie in with the amount that could be achieved, or were the designers inflexible, as in fact I find most tend to be, as far as settlement is concerned?

Bill Craig

In relation to the oil storage tanks, there was no specific design being studied. What we wanted to find out is whether we can contribute to the debate on what the relative differential settlement was across, in our case, a tank of 35 m diameter. There clearly is a differential settlement.

One thing we deliberately did not consider was the problem of variability on the site. We were looking at an idealized foundation. One of the issues that perhaps has not been very widely tackled at this symposium is that of variations within a site and the fabric of the soils on the site. We deliberately, as an academic study, played with a uniform soil, albeit one in which the stress levels were field stress levels and where the stress gradients were appropriate by using the best modelling technology we had. But there was no particular site-specific predictor.

David Renton-Rose

You touched on a very sensitive issue there in terms of the interaction with other parties, particularly liability for performance of the treated ground. We spent a lot of time, I personally spent a lot of time going backwards and forwards to satisfy large process contractors' desires to prove the concept, to apply treated ground as a direct substitution for piling. I think a lot of people here would recognize that it is not a straight swap between the two approaches.

To give a little background, a lot of negotiation in fact was carried out with the process contractor's own academic support, based in Germany. We finally persuaded them that some of the criteria being put forward were not meaningful. There were some settlements; the major treatment process was actually based around a large storage facility, which was serviced by rail facilities. The rail facilities are clearly settlement-sensitive, but not to the degree that the process contractor believed. So, by a process of taking their own design support along with us, we actually managed to get some relaxations to the original, very stringent, settlement tolerances, which we would never have got. I have to say, the treatment performed better than expected but not that well. We actually managed to get some relaxation and were able therefore to achieve a satisfactory approach.

David Baker, *Balfour Beatty*

A comment on David Renton-Rose's experience. Some years back we worked on a major power station in Dubai on calcareous soils. This was a design-and-construct project. We did not do zone load testing to monitor vibro compaction. We used only a minimum CPT design line as the criteria for accepting the compaction. Part way through the job there was concern because there were discrete thin layers where the improvement was doing nothing at all. We were just getting nowhere near the line. Furthermore, there were very high friction ratios in the CPT, which would normally indicate organic cohesive layers, of which the boreholes had shown no evidence. If anyone ends up in that situation in the future, do not panic. On this particular site we had the benefit of having an open cut for the main pumping station for which all the strata were exposed. Careful examination down the entire slope showed that there were discrete shell layers, which is not surprising in that environment. If you think about a CPT in those conditions, you get much more crushing under the cone, and the effect on reducing the cone resistance is far more than the effect on reducing the friction. So if you find yourself in those circumstances, that is the probable explanation.

A quick observation on site investigation practice. We had shell and auger boreholes and we paid much care and attention to how they were done. Undersize shells; very careful lifting; keeping the water topped up. But in some of the strata we could not get the SPTs to correlate with the CPTs with standard shell and auger holes. The only way we could achieve that was by small diameter holes with mud flush. So, dare I say it, there may be cases where you might look at a shell and auger investigation and decide you need ground improvement when really you do not.

Dr Angus Skinner, *Skinner & Associates*

Regarding Slocombe, Bell and Baez's paper *The densification of granular soils using vibro methods.* I was very surprised at the limited extent of the liquefaction phenomena that was observed in the three different periods of liquefaction. If I am reading the diagram correctly, the diameter of the liquefied zone that you are measuring is only about 1·5 m to 3 m around the vibrator itself, at that particular depth. There are three phases of liquefaction, but it is a very limited extent of liquefaction. Is that right? It is not clear what the extent of your instrumentation was.

Dr Juan Baez

The instrumentation was located at 0·9 m, 1·2 m and 1·5 m distance. In the 0·9 m and the 1·2 m locations we found significant liquefaction and changes in the penetration resistances, and we have improvement. In the 1·5 m location we did not have as much, but the caveat in those graphs is that they are for single columns. If you look at Figure 3 in the paper, the difference between the single column and the group effect would indicate that in fact the diameter of effective improvement would go beyond 1·5 m. It is perhaps a little misleading to say that at 1·5 m we did not get improvement. By using the group effect we do get improvement.

Dr Angus Skinner, *Skinner & Associates*

The sting in the tail now. The fact is you said there "may" well have been some isolation of your input energy by virtue of the liquefaction. If it is zoned, does in fact the zoning liquefaction prevent liquefaction away from those zones?

Dr Juan Baez

The profile was fairly homogeneous in terms of being a fairly clean sand and as you can see the excess pore pressures dissipate very, very quickly. In fact, by the end of construction of the stone column we have excess pore pressures only of the order of 10%. So I do not suspect that there were any isolation effects from the partial liquefaction. Having said that, we in fact have, in some situations, approached the jobs by forming enclosing patterns rather than forming advancing patterns so we can allow generation and dissipation of pore pressures to take place before hitting it once again.

Renton-Rose, D. G., Bunce, G. C. & Finlay, D. W. (2002). *Géotechnique* **52**, No. 10, 764

WRITTEN DISCUSSION

Vibro-replacement for industrial plant on reclaimed land, Bahrain

D. G. RENTON-ROSE, G. C. BUNCE and D. W. FINLAY

D. A. Baker, *Balfour Beatty Major Projects*

In 1991/1992 Dutco Balfour Beatty carried out both the design and the construction of the civil works for the Jebel Ali G gas turbine and multistage flash desalination station in Dubai. Vibro-compaction was used to densify loose calcareous sand in the power island area. Rather than use zone load testing, acceptance criteria for the ground improvement were based on static cone penetrometer tests (CPT) using a defined minimum cone resistance as a function of depth, corresponding to a minimum relative density. For certain discrete layers it proved impossible to meet the minimum cone resistance criteria. The CPT results showed that these layers had high friction ratios of 3 to 4, which would normally indicate clay. Clay layers were not expected in this environment, however, and had not been identified in boreholes. Fortunately, a nearby deep open excavation for the cooling water pumping station revealed that these zones were actually deposits consisting almost entirely of shells. High friction ratios result in these materials because the reduction in cone resistance due to crushing is far more marked than the reduction in the friction on the penetrometer sleeve. The ground improvement was therefore accepted and has performed well.

Another important observation relates to the pre-densification ground investigation. Standard penetration tests (SPTs) carried out in shell and auger boreholes did not provide results compatible with CPT results. This was despite extremely careful execution and supervision of the shell and auger boring, following all the rules such as use of an undersize shell, slow withdrawal of the shell, and keeping the water level topped up. Compatible results could only be achieved using smaller-diameter boring with mud flush support. The lesson is to be cautious of using standard penetration tests in shell and auger boreholes to determine that ground improvement is required. Without other more reliable data such as static cone penetrometer testing, ground improvement could be implemented unnecessarily.

Authors' response

It is interesting to note Mr Baker's observations of zones of high CPT friction ratio in loose calcareous sand, improved by vibro-compaction, which were subsequently proven to be shell layers rather than cohesive soils, as might have been expected. Since we were aware of certain restrictions on the application of accepted soil classifications based on friction ratio in calcareous soils, only tip resistance was used as a quality control measure on the ALBA site.

Where low q_c values were recorded, further investigation by wash boring techniques was carried out to determine whether they were related to predominantly shell layers that had not consolidated after placement, or to soft cohesive soils. Sampling within an SPT split spoon, to enable recovery, proved the presence of calcareous silt. Low q_c values were recorded at various locations across the site. In areas improved by vibro-replacement, low q_c values were recorded both prior to treatment (7 to 8 months after fill placement) and shortly after improvement, although only at the interface of the former sea bed and overlying fill. It was therefore concluded that localised pockets of silt had become trapped during the hydraulic filling process. Fortunately, these pockets were of limited extent in plan and typically <200 mm thick, as well as being generally outside settlement sensitive areas. Consequently, it was considered unnecessary to carry out additional improvement in these areas.

Croce, P. & Flora, A. (2001). *Géotechnique* **51**, No. 10, 905–906

WRITTEN DISCUSSION

Analysis of single-fluid jet grouting

P. CROCE and A. FLORA

D. A. Greenwood,

I congratulate Professor Croce and his colleague on tackling the analysis of the jet grouting process. This topic is ripe for study.

Concerning the diameter of columns resulting from jet erosion, the key question is the correlation of soil properties derived from site investigation with diameter. The need for initial field testing to confirm design assumptions based on empirical knowledge is often prohibitively expensive.

The correlation with penetration tests and undrained shear strengths is poor. In gravelly soils the scatter of penetration resistance is large, and averages are unrepresentative of the resistance to the physical effects of jetting out the matrix of finer soils. In cohesive soils surely plasticity index is a significant guide in conjunction with undrained strength?

Regarding Fig. 10 of the paper relating diameter to energy input, the linear relation shown applies only over a small range of energies. The Japanese have demonstrated columns of 5–7 m diameter using energy values 6–10 times those of the range indicated. The loss of energy at large diameters must result in a declining relationship.

These large diameters have been achieved using single-jet techniques. The air used in double or triple systems loses its apparent effects at such a range except for airlifting effluent from the bore.

Moreover, the large diameters are achieved using jet pressures the same as those in first-generation machines with increased volumes of jet fluids. This confirms that momentum rather than pressure is the key to erosion.

On the first ever jet grouting contract in 1962 for an underground sewage pumping station at East Pentire, Newquay, UK, Cementation used a single grout jet at 10 bar pressure with an 8 mm diameter orifice. Column sizes of 800 mm were achieved in a uniform medium gravel with about SPT $N = 20$. The effluent was collected, hydrocycloned to remove coarse particles and recirculated with added cement until the correct quantity was achieved. The benefits in terms of fewer environmental and disposal problems and less waste of cement are obvious. This process used about half the cement used with current techniques. It is astonishing that these practices are not used today.

Concerning the control of the soilcrete composition, this is constrained by the necessity of having to compromise with the jetting requirements. It is essential that the jet be sufficiently fluid, and that it has sufficient momentum to be effective. This limits grout material proportions. When separate air and water jets are employed, the ratios of material volumes are altered further.

Using a material input/output balance as illustrated by the authors in the appendix to the paper it is also important to know the cavity volume (and hence its diameter) accurately, as otherwise significant errors arise in the material proportions calculated for the soilcrete.

Furthermore, when jetting in soils containing coarse sands and gravels there may be insufficient velocity in effluent rising up the annular space around the drill pipes to raise the coarser particles. They necessarily remain in the bore, sinking by hindered settlement in the soilcrete slurry. Hence they do not take part in the fluid circulation and should be excluded from the balance.

Control can sometimes be exercised on the particle sizes remaining in circulation, depending on the particle size distribution of the soil. By drilling (and possibly temporarily casing) a borehole of a chosen diameter the up-flow velocity can be predetermined. By reference to the particle size distribution the proportions retained and settling can be controlled to some extent by choosing an appropriate velocity and hence hole diameter. Regard must be taken of the lifting capacity rendered by the rheological properties of the fairly thick effluent (which is essentially the same as the in-situ soilcrete without the heavier particles).

A. L. Bell, *Keller Ground Engineering*

In his presentation, Professor Croce provides an alternative energy analysis for single-fluid jet grouting, and he rightly draws attention to the need to deal with the energy at the nozzle rather than the pump. Development at Keller Ground Engineering during the last 10 years has, among other things, been directed at improving the hydraulic efficiency of the jet grout monitor. The diameters, particularly for clay soils, reported in the 1993 review to which Professor Croce refers are therefore somewhat conservative in the light of current experience.

My question relates to the energy analysis, which has been developed for the single-jet system. Clearly the method would have a much broader application if applied to the other methods. I should be interested in Professor Croce's comments on the applicability to the triple-jet system.

Authors' reply

The authors would like to thank the writers for their discussions, and for the opportunity to clarify some of the points raised in the paper.

The phenomena produced by jet grouting are very complex, and the degree of complexity increases with the number of fluids injected into the subsoil. The authors have thus chosen to tackle the single-fluid system as the first step of a broad research project aiming to provide rational design methods for all jet grouting systems.

The analysis of single-fluid jet grouting presented in the paper was based mainly on the experimental evidence acquired from a relatively small number of cases, where it was possible to perform accurate observations and measurements. Therefore the reported range of column diameters is restricted to the relevant energy values. For increasing energy, the diameter should correspondingly increase, as confirmed by the Japanese results mentioned by Dr Greenwood.

However, the relationship between jet energy and column diameter should be further investigated, and, by collecting a large number of field data, empirical design charts may eventually be produced. These charts should also account for the relevant soil properties: permeability for gravels, shear strength for finer soils. In this respect, Fig. 10 should not be considered as a design chart, although it shows the influence of jet energy and soil grading. General correlations may be attempted in the future by using the results of proper in-situ tests (e.g. borehole seepage tests for gravels, SPT for sands, CPT for clays).

Of course technological developments must be accounted for, and it seems logical that recent improvements in the hydraulic efficiency should result in larger column diameters, as reported by Dr Bell. However, this finding is consistent with the energy expression proposed by the authors, since higher hydraulic efficiency just results in larger values of the kinetic energy at the nozzles. It is also observed that jet grouting effectiveness has recently increased not only because of lower hydraulic losses in the system, but also because of new monitor config-urations obtained mainly by inclining the nozzles.

The authors believe that a thorough understanding of jet grouting mechanisms can be guided by theoretical modelling of grout–soil interaction. To this aim, it seems feasible to develop two different models, for soil permeation and for soil erosion respectively. In particular, for modelling soil permeation, turbu-lent flow regime and grout viscosity should be accounted for, and, for the analysis of soil erosion, drained and undrained shear strength should respectively be considered. In fine-grained soils, plasticity index may also play a role as suggested by Dr Greenwood, as far as it affects undrained shear strength.

With regard to double- and triple-fluid systems it is well known that when the disrupting jet (respectively grout or water) is surrounded by compressed air, the energy losses are substan-tially reduced. Therefore higher diameters are usually obtained, although the air effect fades out as the diameter gets larger, as

suggested by Dr Greenwood. It follows that the energy approach presented for single-fluid systems should also apply to double- and triple-fluid systems, as proposed by Dr Bell. In particular, the jet energy at the nozzles could still be calculated by equation (3), applied to the relevant grout or water jet, but different correlations are expected from experimental data. By comparing the correlations pertaining to different systems, for each soil type, it might be possible to evaluate their efficiency. Such evaluation could be aided by theoretical calculation of the energy dissipation that takes place in the ground.

Finally Dr Greenwood has pointed out some problems that may be encountered in the interpretation of experimental meas-urements by input/output balance. In fact very careful measure-ments are needed, and the proposed method is best suited for research. The method can be particularly useful, however, for investigating jet grouting effects on fine sands and silty sands, where direct observation of soilcrete structure is not feasible.

Closure

CHRIS RAISON, CHAIRMAN

There is very little I want to say now. I think the evidence, certainly to me, is that there has been an immense interest in this subject. During the organising of the symposium we had a huge response from potential authors who wished to contribute. Unfortunately because of time constraints and other difficulties those potential authors were not able to submit papers. But it does indicate the huge range of interest. This interest is not just from the UK, it is international. I hope that you all feel that the end product, the symposium that was published in December last year, was worth publishing. It certainly reflects and gives a good international feel. It certainly reflects the wide ranging interest in the various systems that are available for ground improvement.

I would like to thank the delegates who have attended the meeting. Again we have had over 160 people who have attended. That is very rewarding for the efforts of the organising committee. Really at this stage I would like to say thank you to my fellow committee members. There is Ken Watts, Barry Slocombe, Dr Ken Been, Dr Angus Skinner and I should not forget David Johnson. David has been out of the country for a little while and has flown back specifically for this meeting. Rather than let him get away without doing anything for the meeting I have asked him to give a vote of thanks to the authors. David if I could put upon you.

DAVID JOHNSON

As Chris has said I have got off quite lightly. My excuse is that I have been overseas. I agree that today's symposium has been a great success and in that regard there are various people that should be thanked. Chris has already thanked the members of the committee but we should particularly thank the keynote speakers Andrew Charles and Alan Bell.

I would like to briefly touch on a couple of items than Andrew Charles highlighted which I think are important. He highlighted the fact that there is a general lack of papers produced on this subject, technical papers. Really there should be more. Chris Raison has just mentioned that there was a very large number of papers that were submitted for this symposium and that it was impossible for them all to be accepted. I think that the authors that have been unsuccessful in getting papers published for this symposium really should be encouraged to publish them elsewhere. My personal view is that there should be, and I am wearing my contractors hat here, I think there should be more specialist contractors, academics, universities and organizations such as Building Research Establishment, that there really should be more collaborative projects to encourage more theoretical analysis of ground improvement in general.

We should of course thank various people who have presented today including contributors from the floor. Chris has already given thanks to members of the committee. We should not forget of course the staff from the Institution of Civil Engineers, Mary Henderson and her helpers who have been in the background organising the administration. Last but not least we should thank the microphone experts Rafael Monroe, Joshua Weng, Laura Pumfrey and Dominique Brightman, MSc students from Imperial College. We should express our appreciation in the usual way.

CHRIS RAISON

Just before I close the symposium we ought to also thank the panel members of the *Géotechnique* Advisory Panel who have put a lot of hard work themselves into looking at papers that were submitted and perhaps more importantly the referees that we used to review the papers in the first place. With that I will close the symposium and thank you all for attending.